Lecture Notes in Statistics

Edited by S. Fienberg, J. Gani, J. Kiefer, and K. Krickeberg

2

Mathematical Statistics and Probability Theory

Proceedings, Sixth International Conference, Wisła (Poland), 1978

Edited by W. Klonecki, A. Kozek, and J. Rosiński

Springer-Verlag
New York Heidelberg Berlin

Editors
Dr. Witold Klonecki, Dr. Andrzej Kozek, Dr. Jan Rosiński
Mathematical Institute of the Polish Academy of Science
Kopernika 18, 51-617 Wrocław, Poland

AMS Subject Classifications: 62-XX

Library of Congress Cataloging in Publication Data

Main entry under title:

Mathematical statistics and probability theory.

(Lecture notes in statistics; 2)
Includes bibliographies
1. Mathematical statistics—Congresses.
2. Probabilities—Congresses. I. Klonecki, Witold.
II. Kozek, A. III. Rosiński, Jan. IV. Series.
QA276.A1M3 519.5 80-13322

ISBN 0-387-**90493**-X Springer-Verlag New York Heidelberg Berlin
ISBN 3-540-**90493**-X Springer-Verlag Berlin Heidelberg New York

Printed in the United States of America.

9 8 7 6 5 4 3 2 1

Dedicated to
Professor Jerzy Neyman

Wielce Szanownemu i Kochanemu Panu
Profesorowi Jerzemu Neymanowi w dowód
poważania i wdzięczności

Witold Klonecki

FOREWORD

Since 1972 the Institute of Mathematics and the Committee of Mathematics of the Polish Academy of Sciences organize annually conferences on mathematical statistics in Wisła. The 1978 conference, supported also by the University of Wrocław, was held in Wisła from December 7 to December 13 and attended by around 100 participants from 11 countries. K. Urbanik, Rector of the University of Wrocław, was the honorary chairman of the conference.

Traditionally at these conferences there are presented results on mathematical statistics and related fields obtained in Poland during the year of the conference as well as results presented by invited scholars from other countries. In 1978 invitations to present talks were accepted by 20 eminent statisticians and probabilists. The topics of the invited lectures and contributed papers included theoretical statistics with a broad cover of the theory of linear models, inferences from stochastic processes, probability theory and applications to biology and medicine. In these notes there appear papers submitted by 30 participants of the conference.

During the conference, on December 9, there was held a special session of the Polish Mathematical Society on the occasion of electing Professor Jerzy Neyman the honorary member of the Polish Mathematical Society. At this session W. Orlicz, president of the Polish Mathematical Society, K.Krickeberg, president of the Bernoulli Society, R. Bartoszyński and K. Doksum gave talks on Neyman's contribution to statistics, his organizational achievements in the U.S. and his role as a founder of the IASPS, the forerunner of the Bernoulli Society(three of the talks appear in this volume).

We would like to thank all lecturers, including those at the session of the Polish Mathematical Society, all chairmen and participants for the contributions.

The organization of the conference was in very capable hands of Mrs. A. Huskowski and Mr. E. Mordziński.

W. Klonecki

A. Kozek

J. Rosiński

CONTENTS

SOME THOUGHTS ABOUT JERZY NEYMAN

by

Robert Bartoszyński

It is the second time within the last few years that I have the honour and privilege to have a talk about Professor Neyman and his contribution to statistics.

Let me start with few words of explanation of Neyman's biography. We heard from the speech of Professor Orlicz that Neyman is a grandson of an insurgent of 1863. Now, this information carries a very clear meaning to all Poles, but may perhaps be somewhat puzzling to non-Poles.

The point is that the Uprising of 1863 is a sort of holy event in the Polish history, and the knowledge that someone comes from a family whose members took part in it makes him automatically somehow dearer to the Poles.

Incidentally, I had been discussing with my colleagues whether or not such information about Neyman ought to be included in his biography. Not that anyone wanted to make Neyman less dear to us, of course; the question was: does Neyman really need that kind of "support"? His greatness comes from what he himself has done, and not from the merits of his family. Anyway, I am quite happy that the problem was resolved for me by Professor Orlicz, and that I could give these few words of explanation.

Now, Neyman's contributions to statistics are well known, and not likely to be underestimated by anyone who has any understanding of statistics. To put it most briefly, they consist of stating for the first time (together with E.S. Pearson) the principles of testing hypotheses, with the crucial concept of the power of the test; introducing the notion of confidence interval; and formulating the principles of optimization in sampling theory.

All this, as I said, is well known, and I repeat it merely because without mentioning these facts any talk about Neyman's contribution would not be complete.

What I wanted to present in some more detail today, are just two examples of statements of some problems connected closely with empirical domains. The aim is simply to illustrate the art - in which Neyman excels - of transforming the real-life problems into statistical ones.

The first of these problems concerns the so-called outliers [3]. An outlier is, roughly speaking, an element in the sample which is larger (say) than the remaining elements, to such a degree that one wonders if it is a genuine sample element, or perhaps results from an error of observation or error in recording the data.

To put it formally, let $y_1, y_2, \ldots, y_n$ be a sample from some underlying distribution F. We assume therefore that y_i's are independent and $F(t) = P(y_i \leq t)$, $i = 1, \ldots, n$. Assume further that F has a density f; we may then neglect the possibility of ties among elements of the sample. Let $x_1 < x_2 < \ldots < x_n$ be the sample $y_1, y_2, \ldots y_n$ ordered according to magnitude.

Suppose that x_n is an element which "appears too large"; we would like to devise a test for the hypothesis that it comes from the same distribution as the rest of the sample. It is intuitively clear that any test of such a hypothesis should be based on comparison of the "distance" from x_n to the "group" $x_1, \ldots, x_{n-1}$, with the "spread" of this group. The terms "distance" and "spread" were put in the quotation marks to stress the fact that the test statistics may be constructed in a number of ways, by taking as a distance between x_n and the rest of the sample such quantities as $x_n - \bar{x}$, $x_n - x_{n-1}$, etc., and as the measure of spread - quantities such as $x_{n-1} - x_1$, $x_{n-1} - x_2$, sample variance s^2, and so on.

An excellent review of test statistics used thus far may be found in Statistical Tables by Zieliński (1972). However, all these tests refer to the case where the underlying distribution F is normal.

Neyman and Scott suggest here a different approach to the problem. Let us take as the test statistic

$$t_n = \frac{x_n - x_{n-1}}{x_{n-1} - x_1},$$

and let us agree to say that the sample contains a (k,n)-outlier, if $t_n > k$, i.e. if $x_n - x_{n-1} > k(x_{n-1} - x_1)$. Naturally, any sample (here and in the sequel we tacitly assume that $n \geqslant 3$) will contain a (k,n)-outlier, but we are interested only in case of large k.

The probability $\pi(k,n;F)$ that a sample of size n will contain a (k,n)-outlier is easily seen to be equal

$$\pi(k,n;F) = \int_{-\infty}^{\infty}\int_{x}^{\infty} \left[F\left(\frac{y + kx}{k+1}\right) - F(x)\right]^{n-2} f(y)f(x)\,dy\,dx.$$

and, at least in principle, may be calculated for various k and n for a given F.

However, in practical situations, one seldom knows the distribution F. More often, it is known only that F belongs to some family $\mathcal{F}$ of distributions. For such cases, the quantity

$$\pi(k,n;\mathcal{F}) = \sup_{F\in\mathcal{F}} \pi(k,n;F)$$

is of obvious relevance. If, for the value k observed in the sample of size n, we have $\pi(k,n;\mathcal{F})$ small enough, we may have reasonable grounds to reject the element x_n from the sample.

Let us agree to say that the family $\mathcal{F}$ of distributions is (k,n)-outlier resistant if $\pi(k,n;\mathcal{F}) < 1$ and (k,n)-outlier prone if if $\pi(k,n;\mathcal{F}) = 1$. Moreover, a family $\mathcal{F}$ will be said to be completely outlier prone if $\pi(k,n;\mathcal{F}) = 1$ for all $k > 0$ and $n \geqslant 3$.

The results of Neyman and Scott may now be stated as follows.

Let F be any continuous distribution, and let $\mathcal{F}_1$ and $\mathcal{F}_2$ be the families obtained from F by translations and by changes of scale, i.e.

$$\mathcal{F}_1 = \{F(x-\Theta): -\infty < \Theta < \infty\}$$

and

$$\mathcal{F}_2 = \{F(x/\Theta): \Theta > 0\}.$$

We have then

Theorem 1. For any F, the families $\mathcal{F}_1$ and $\mathcal{F}_2$ are (k,n)-outlier resistant for $k > 0$ and $n \geqslant 3$.

It follows, in particular, that the family of all normal distributions is outlier resistant.

It is quite surprising that there exist families which are completely outlier prone. We have namely

Theorem 2. The family of all gamma distributions and the family of all lognormal distributions are completely outlier prone.

The practical consequence of this theorem appears rather shocking. Suppose that we take observations of some phenomenon about which we know only that it is governed by a gamma distribution, without any knowledge of its parameters. Then, no matter how "odd" the sample may appear, we cannot reject (on purely statistical grounds) any of its elements as outliers, since there exists a gamma distribution for which such a sample or even "worse", has probability of occurrence arbitrarily close to 1.

It is worth to mention here that, as shown by Green [1], a family of distributions is completely outlier prone if and only if it is (k, n)-outlier prone for some $k > 0$ and $n \geqslant 3$. An exhaustive classification of families of distributions with respect to their properties of outlier proneness and outlier resistance, may be found in Green [2].

Let us now turn to the second problem (which, incidentally, still awaits the solution).

Observing galaxies, one may distinguish a certain number of types, determined by shapes of galaxies (spiral, etc.). Let these types be $M_1, M_2, \ldots, M_N$. There is some ground to assume that they represent various stages of the evolution of galaxies and that the evolution proceeds always through a fixed sequence of types, $M_{i_1}, M_{i_2}, \ldots, M_{i_N}$. Thus, every galaxy is born (or possibly, emerges from some type which is not visible) in type M_{i_1}, after some time passes to M_{i_2}, and so on. The problem is that of determining the "true" permutation $i_1, \ldots, i_N$ of types of galaxies, i.e. that permutation which represents the order of their evolution.

Certainly, we cannot observe any act of transition, and the data on frequencies of various types of galaxies do not contain information about the permutation $i_1, \ldots, i_N$. However, there exist twin galaxies, and there are some premises for the hypothesis that both members of the twin pairs are born at the same time, and then develop independently of one another.

Let us assume that for each galaxy, the times $T_{i_1}, T_{i_2}, \ldots$ spent in particular types $M_{i_1}, M_{i_2}, \ldots$ are independent random variables with densities $f_{i_1}(t), f_{i_2}(t), \ldots$ Finally, let F be the distribution of the age of galaxies; the distribution F may be degenerate.

If $I = (i_1, \ldots, i_N)$ is the true permutation of types of galaxies, then the probability that a galaxy of age t will be of type M_j is

$$q_I(j,t) = P(T_{i_1} + \dots + T_{i_{r-1}} < t < T_{i_1} + \dots + T_{i_{r-1}} + T_{i_r}),$$

where $i_r = j$, so that

$$p_I(j) = \int_0^\infty q_I(j,t)\, dF(t)$$

is the probability that an observed galaxy will be of type M_j.

Similarly, the probability that one galaxy of a twin pair (say, "Eastern") will be of the type M_j, and the other of the type M_k, equals (by the assumed independence of evolution of members of the pair)

$$p_I(j,k) = \int_0^\infty q_I(j,t) q_I(k,t)\, dF(t).$$

Now, $\pi_I(j|k) = p_I(i,k)/p_I(k)$ is the conditional probability that Eastern galaxy will be of the type M_j, given that the other galaxy is of the type M_k, a quantity which is estimable through observations of frequencies of various combinations in twin galaxies.

The problem lies in constructing a reasonable estimator of the permutation I. More precisely, one can look for such estimators based on functions $H(x_1,\dots,x_N)$ of N variables, satisfying the following properties:

(a) H is symmetric in $x_1,\dots, x_N$.

(b) Let $H(k,I) := H(\pi_I(1|k), \pi_I(2|k),\dots, \pi_I(N|k))$. Then, for any choice of densities $f_i(t)$, distribution $F(t)$, permutation I, and $j,k = 1,\dots,N$, index j precedes index k in permutation I if and only if $H(j,I) < H(k,I)$.

If we knew such a function H, then ordering the values of H on empirical distributions of the type of Eastern galaxy, given the type of the other galaxy, would lead to a permutation, which can be reasonably taken as an estimator of the true permutation I.

It may be conjectured that the function H satisfying the above conditions is either $H(x_1,\dots,x_N) = \sum x_i^2$, or $H(x_1,\dots,x_N) = \sum x_i \log \frac{1}{x_i}$.

The intuitive justification here is quite obvious: the empirical distribution $f(1|k), f(2|k), \ldots, f(N|k)$, where $f(i|j)$ is the observed frequency of cases when the Eastern galaxy was of type M_i given that the other was of type M_j, should be more concentrated around the value k if k corresponds to one of the earlier types, and should be more spread out if k corresponds to some of the later types.

These two examples, as I said earlier, are aimed at illustrating the ingenuity and imaginativeness which go into transforming the real problem into a statistical one. I hope I have managed to some extent to show you certain features of Neyman's work in this respect.

At the end, I intended to say few words about Jerzy Neyman as a person. However, after some deliberation, I gave up. I could not do justice to this task, even if I were allowed to use my native tongue. I could only say that the unique atmosphere at the Statistical Laboratory of the University of California at Berkeley is due mostly to him - to his deep involvement in the problems of science and reality, his friendliness, and sense of humour. Jerzy Neyman was - and is - a moral authority for scientific community, a man who did not avoid to take risk and responsibility in difficult and controversial social and political issues.

References

[1] G r e e n, R. F., A note on outlier-prone families of distributions. Ann. of Stat., 2, 1293-1295 (1974).

[2] G r e e n, R. F., Outlier-prone and outlier-resistant distributions. Journal of American Statistical Association, 71,502-505 (1976).

[3] N e y m a n, J., and S c o t t, E. L., Outlier proneness of phenomena and of related distributions. In: J. S. Rustagi (ed.) Optimizing Methods in Statistics. New York, Academic Press, 413-430 (1971).

[4] Z i e l i ń s k i, R. Tablice Statystyczne (Statistical Tables), PWN, Warszawa 1972.

SOME REMARKS ON THE ACHIEVEMENTS OF PROFESSOR NEYMAN IN THE UNITED STATES

by

K. Doksum

University of California, Berkeley

I am greatly honored to be speaking to this distinguished audience. As Professor Neyman's colleague in Berkeley, I will talk about his achievements in the United States as well as some of his recent interests. In the early 1930's, Berkeley did not have a very good mathematics department, and the administration appointed a chairman to improve the situation. His name was Evans, and one of his greatest achievements was to bring Neyman to Berkeley in 1938. Today, the building that houses the Mathematics and Statistics Department in Berkeley is called Evans Hall.

The same year Neyman arrived in Berkeley, he founded the Statistical Laboratory. He still is the director of this laboratory and is involved in applications of statistics to astronomy, biology, health and meteorology. In 1955, Neyman started the Statistics department at Berkeley. Some of the well-known people he brought to this department at that time were E. L. Scott, L. Le Cam, E. L. Lehmann, D. Blackwell, M. Loeve, H. Scheffe and J. Hodges.

Neyman influenced the development of statistics in the United States not only by his fundamental research, but also by being a great teacher who attracted a lot of outstanding students. Some of the names are L. Le Cam, E. L. Lehmann, D. G. Chapman, C. L. Chiang and W. J. Bühler. We can draw a statistical tree where at the top, in the 0th generation, we have J. Neyman. In the 1st generation we have his students, in the 2nd generation the students of his students, and so on. A large proportion of the statisticians in the United States today can be found on this tree-not only on the United States, but also the rest of the world. For instance, 2/3 of the statistics Professors at Universities in Norway are on this tree. At this point, Neyman has had 39 students. One of these, Lehmann, has had 35 students; one of these 35, P. Bickel, has had 21 students, and so on.

Neyman argued successfully that statistics at the University should be taught by statisticians and not by professors in various applied disciplines that needed statistics. This made it possible to build a group of statisticians and later a department of statistics. Since Berkeley tends to serve as a model, this influenced the development of statistics throughout the United States.

In 1945, Neyman organized the Berkeley Symposium on Mathematical Statistics and Probability. After 1945, a Berkeley Symposium was held every year until 1970. The next is planned jointly with Stanford for 1980. The Berkeley Symposia differed substantially from most other scientific meetings in that they provided and extended period of contact between participants from various countries. They also promoted contacts between scholars with different interests and proved to be great successes.

I have known Neyman since 1963. In this period he has been interested in $C(\alpha)$ tests and their applications, astronomy and a statistical theory of clustering, cancer and problems of carcinogenesis, pollution and the synergistic combination of pollutants, and weather modification through the seeding of clouds. I would like to talk a little about one of his favorites, weather modification. In the laboratory it can be seen that the introduction of silver iodide produces condensation, and thus it is hoped that seeding clouds with silver iodide will produce rain. If this could be achieved, it would be very important because of all the areas of the earth that are dry, even though clouds do pass by. One problem with cloudseeding experiments was that they were not randomized. This introduced serious biases. For instance, one may imagine a pilot deciding not to seed a particular storm simply because it was big, and therefore dangerous to fly in. Neyman insisted (sometimes successfully) on randomized experiments, and he and the Statistical Laboratory participated in a cloudseeding experiment in the mountains in Northeast California near Lake Tahoe.

Another problem with cloudseeding experiments that Neyman found was that the cloudseeders and their proponents tended only to report the favorable part of the experiment. Neyman found that in some instances, when one looked at all the data, there was a substantial reduction of rain in the target area. This finding has gotten him involved in a certain amount of controversy. A recent very interesting discussion by Neyman can be found in the Journal of the American Statistical Association, 1979, p. 90.

The question I am most often asked when I am in Europe is: How is Professor Neyman? What I can say is that he is one of the most active members of our department. He teaches courses on models in applied fields, runs a seminar that meets at least once a week and administers research contracts. Every Wednesday he has a seminar and afterwards we are invited to the Faculty Club for drinks. After everyone is seated with their drinks, he calls for attention and declares: "It is time for a Polish toast: To all the ladies present and some of those absent".

ROLE OF JERZY NEYMAN IN THE SHAPING OF THE BERNOULLI SOCIETY

by

K. Krickeberg

Ladies and Gentelmen,

I have the honour of having been asked to say a few words about Jerzy Neyman's contribution to the founding of the Bernoulli Society. In fact, he contributed the main ideas to its very first origins.

Let me start, however, with a personal recollection in order to respond to R. Bartoszyński's call for a description of what he calls the "image" of Neyman. I met Jerzy Neyman for the first time in 1954 at the International Congress of Mathematicians in Amsterdam. I was just a graduate student, and I wanted to pursue further studies abroad, for example in Sweden or the United States. So I approached Neyman between two lectures, and he immediately agreed to listen to me. At this moment a very famous mathematician come up and said to Neyman that he had to talk to him, but Neyman replied, pointing to me who was completely unknown to him: "I am sorry, not now, because I have already an appointment with this gentleman".

His contribution to what became latter the Bernoulli Society is, of course, only a small part of his many activities in the domain of scientific planning and organisation, but like the founding of the Statistics Department and the Statistical Laboratory in Berkeley which was described before, it is a typical one which shows his clear vision and the way he used to transform his visions into reality.

The problem at hand in the early sixties was simply this:there was just no international society in the fields of mathematical statistics and probability. Of course, there were important national societies like the Institute of Mathematical Statistics in the United States or the Royal Statistical Society in the United Kingdom. There also existed the time honoured International Statistical Institute (ISI) which was originally the organization of the government statisticians. It is true in the mean-time it had already changed a lot, to a large extend again under Neyman's influence, and now counted also many mathematical statisticians among its members, but it was still a somewhat stiff and old-fashioned institution and, above all, membership was not open: there were only elected members in a limited number.

Now Neyman did something which is a good example for his diplomatic skills. He did not found a competing society, he did not start a fight with the ISI, but he gave the main impulses to the creation, in 1963, of a society which concentrated on a particular aspect, namely applications in the physical sciences. This was, of course, an especially important aspect to which he himself had contributed much as we have just heard, and which, unlike for example biometrics, had not yet been the object of an international association. This new society, the International Association for Statistics in the Physical Sciences, was affiliated with the ISI, and membership was open.

I think that Neyman already regarded this association as a nucleus of something larger to come. It was only 12 years later, when the time was ripe, that it was transformed, by the action of D.G. Kendall, into the present Bernoulli Society which considers itself responsible for the full area of mathematical statistics and probability. Again this was done not against but with the ISI: the Bernoulli Society is a section of the ISI. We may regard as symbolical the fact that it was formed during the ISI-session in Warsaw.

Finally, let me mention that at least two activities of the Bernoulli Society continue work of Neyman which was in the center of his interests. Firstly, the former IASPS is still there in the form of a standing committee of the Society. Secondly, the Bernoulli Society is determined to pay special attention to the development of our fields in countries and regions where this development has still been relatively weak. Let us recall that in the "tree" described by Doksum there were also the names of many people from the so-called "underdeveloped countries" - which I do not consider underdeveloped, but that is another story.

A MODEL FOR NONPARAMETRIC REGRESSION ANALYSIS OF COUNTING PROCESSES

by

Odd Aalen
University of Tromsø

1. Introduction

Often the focus of a statistical analysis is the occurrence of certain events. In clinical trials, for instance, patients may die or progress to other stages of the disease. Other examples may be found in biology, demography, and other fields.

If one observes events of several different types, then the observations may be described by a multivariate counting process $\underline{N}_t = (N_{1,t},\dots,N_{k,t})^T$ where $N_{j,t}$ is the number of events of type j that has occurred in the time interval $[0,t]$. One way of describing the probabilistic structure of such a counting process is by means of the intensity process $\underline{\Lambda}_t = (\Lambda_{1,t},\dots,\Lambda_{k,t})^T$ which may be defined by

$$\Lambda_{j,t} = \lim_{h\downarrow 0} \frac{1}{h} E(N_{j,t+h} - N_{j,t} | F_t),$$

where F_t is the collection of all events observable on the time interval $[0,t]$.

In Aalen (1978) we studied the multiplicative intensity model defined by letting $\Lambda_{j,t} = \gamma_{j,t} X_{j,t}$ where $\underline{\gamma}_t = (\gamma_{1,t},\dots,\gamma_{k,t})^T$ is an unknown function while $\underline{X}_t = (X_{1,t},\dots,X_{k,t})^T$ together with $\underline{N}_t$ is a stochastic process observable over some time interval.

In this paper we will introduce a matrix version of the multiplicative intensity model by writing

$$\underline{\Lambda}_t = \underline{Y}_t \, \underline{\alpha}_t,$$

where $\underline{\alpha}_t = (\alpha_{1,t},\ldots,\alpha_{r,t})^T$, $r<k$, is an unknown function while $\underline{Y}_t = (Y_{ij,t})$ is a $k \times r$ matrix of observable stochastic processes, $\underline{Y}_t$ being for each time a function of the past F_t.

This model is primarily intended for the study of regression in life testing. Consider, for example, patients in a clinical trial. One wants to study their mortality from a certain disease. The process $N_{i,t}$ has the value 0 as long as patient no.i is not observed to have died from the disease. If death from the disease occurs, and this is observed to take place, then the process jumps to 1 at that instant. As long as the patient is alive and under observation then $\underline{Y}_t$ is defined in the following way: $Y_{i,t}$ is equal to 1 while $Y_{ij,t}$, $j=2,\ldots,r$, indicate the values of explanatory variables (or covariates) that are considered to be of importance, like age, general health status etc.

When patient no. i has died or is not under observation, then $Y_{ij,t}$, $j=1,\ldots,r$, are all put equal to 0. The scheme defined here allows for censoring of quite general types (see [4]).

Such life-testing situations with covariate information have been the object of a number of recent studies, mostly parametric ones, but also a nonparametric (or rather semiparametric) one by Cox [9]. The method suggested in this paper is not meant as a competitor to the others, but as a supplementary approach giving more detailed information. Being a more genuinely nonparametric approach, our method allows one to assess possible changes in the influence of the covariates over time. (For a different approach to that problem, see [8]). Note also that the covariates are allowed to be stochastic processes depending on the previous development. (By the way, such a modification may also be introduced into Cox's model by using the theory of counting processes).

Our approach will be nonparametric in the sense that no assumption will be made about the functional form of $\underline{\alpha}$. However, apart from this, the set of acceptable $\underline{\alpha}$'s will naturally be restricted by the fact that each component of $\underline{\Lambda}_t$ must be nonnegative. Nevertheless, in estimating below the integrals $\int_0^t \alpha_{i,s} ds$, i = 1,...,r, we will perform an unrestricted linear estimation. This may result in the corresponding estimate of $\int_0^t \Lambda_{j,s} ds$ for some given values of $\underline{Y}$ being nondecreasing, i.e. implicitly $\Lambda_{j,t}$ may be estimated as being negative for some t. Similar situations are, however, well known from other parts of statistics. For instance, several variance estimators suggested in the context of variance component models have the property that they may assume negative values. How to improve our estimators in order to avoid problems of this kind is an open question which will not be taken up here. The practical importance of it, however, may not be great.

It should be mentioned that Cox [9] uses an exponential functional form for $\Lambda_{j,t}$ in order to have the positivity guaranteed. Our "linear model approach" has the advantage that it allows simple explicit computation of the estimators, and also, to some extent, assessment of their properties in the nonasymptotic case.

In Section 2 we will define our estimator and study its properties. Some testing problems will be considered in Section 3, while Section 4 will indicate extensions to other types of stochastic processes. Section 5 contains an application to a set of real data from a clinical trial. The data are taken from Prentice [16, 17]. The Appendices give some theoretical results that are useful in the rest of the paper.

2. Estimation

The approach will be closely analogous to that of Aalen [3], expcept for some simplifications. We will use the martingale based approach to counting processes, see Brémaud and Jacod [7] for a review. We will not go into details about this theory or its mathematical background, but just state briefly the needed assumptions and results. All processes are defined on the time interval [0, 1].

Formally a counting process $\underline{N}$ is defined as a vector of processes which are constant except for jumps of size +1. Two component processes are not allowed to jump at the same time. We also require $N_{i,o} = 0$ for all i.

The process $\underline{N}$ is supposed to be defined on a probability space (Ω, F, P). We further define $\{F_t\}$, $t \in [0, 1]$, as a right-continuous increasing sequence of sub-σ-algebras of F, and require N_t to be measurable with respect to F_t for all t. We assume $EN_{i,1} < \infty$ for all i. By introducing certain weak assumptions on the jump times of the N_i one ensures the existence of an intensity process $\underline{\Lambda}$ satisfying

(i) $$M_{i,t} = N_{i,t} - \int_0^t \Lambda_{i,s} ds, \quad i=1,\dots,k$$

are orthogonal square integrable martingales,

(ii) $$\langle M_{i,t}, M_{i,t} \rangle = \int_0^t \Lambda_{i,s} ds.$$

See e.g. Meyer [15] for the concepts applied here. We write $\underline{M}_t = (M_{1,t}, \dots, M_{k,t})^T$.

We will finally require $\underline{N}$ to have rightcontinuous sample functions while $\underline{\Lambda}$ together with $\underline{\alpha}$ and $\underline{Y}$ defined in the introduction are assumed to be predictable. We will assume that

$$\int_0^1 \alpha_{j,s} ds < \infty, \quad j=1,\dots,k.$$

We will consider estimation of the function $\underline{\beta} = (\beta_1, \ldots, \beta_r)^T$ given by

$$\beta_{j,t} = \int_0^t \alpha_{j,s} ds \qquad j=1,\ldots,r.$$

Now, it is fairly obvious that such estimation can only take place when rank $(\underline{Y}_t) = r$. If this is not the case, then given any realization of $\underline{Y}_t$ there will be an infinite number of $\underline{\alpha}_t$'s that give the same value of $\underline{\Lambda}_t$. Since the development of $\underline{N}$ is determined by $\underline{\Lambda}$ we clearly will have an identification problem.

Hence, we will exclude from consideration the values of t for which rank $(\underline{Y}_t) < r$. To do this we define:

$$J_t = \lim_{h \downarrow 0} 1 \left\{ \text{rank } (Y_{t-h}) = r \right\}$$

(The somewhat peculiar definition is needed in order to make J predictable). We now restrict ourselves to estimate

$$\underline{\beta}_t^* = \int_0^t J_s \underline{\alpha}_s ds.$$

We will suggest a class of unbiased estimators of $\underline{\beta}^*$. Let $\underline{Z}_t$ be a $r \times k$ matrix of predictable processes, such that, for each t, $\underline{Z}_t$ is a generalized inverse of $\underline{Y}_t$ (see e.g. [18], p.24). Since $\underline{Y}_t$ has full rank for the values of t we consider, $\underline{Z}_t$ will satisfy $\underline{Z}_t \underline{Y}_t = \underline{I}$ where $\underline{I}$ is the $r \times r$ identity matrix. Define

$$\hat{\underline{\beta}}_t = \int_0^t J_s \underline{Z}_s \, d\underline{N}_s.$$

We propose $\hat{\underline{\beta}}$ as an estimator of $\underline{\beta}^*$. The properties are given in the following theorem. See Appendix 1 for the extension of concepts of square integrable martingales and stochastic integrals to vector valued martingales. Let $\overline{\underline{N}}_t$ and $\overline{\underline{\Lambda}}_t$ be diagonal matrices with element (i,i) equal to $N_{i,t}$ and $\Lambda_{i,t}$ respectively.

Theorem 1. Assume

$$E\int_0^1 Z_{ij,s}^2 \, dN_{m,s} < \infty \quad \forall i,j,m.$$

Then $\hat{\underline{\beta}} - \underline{\beta}^*$ is a square integrable vector valued martingale with

$$\langle \hat{\underline{\beta}} - \underline{\beta}^*, \hat{\underline{\beta}} - \underline{\beta}^* \rangle_t = \int_0^t J_s \underline{Z}_s \underline{\bar{A}}_s \underline{Z}_s^T \, ds.$$

An unbiased estimator of the covariance matrix of $\hat{\underline{\beta}}_t - \underline{\beta}_t^*$ is given by

$$\left[\hat{\underline{\beta}} - \underline{\beta}^*, \hat{\underline{\beta}} - \underline{\beta}^*\right] = \int_0^t J_s \underline{Z}_s d\underline{\bar{N}}_s \underline{Z}_s^T .$$

P r o o f: It is well known that for any square integrable vector-valued martingale $\underline{M}$, $\langle \underline{M},\underline{M}\rangle - [\underline{M},\underline{M}]$ is a martingale. Hence the second part of the theorem follows immediately from the first. Put

$$\underline{M}_t = \underline{N}_t - \int_0^t \underline{Y}_s \underline{\alpha}_s ds.$$

The stochastic integral

$$\int_0^t J_s \underline{Z}_s \, d\underline{M}_s$$

is well defined if

$$E\int_0^1 Z_{ij,s}^2 \, \Lambda_{m,s} ds < \infty, \qquad \forall i,j,m .$$

By a result of Jacod ([11] section 2) this is implied by the condition of the theorem.

If the stochastic integral above can also be interpreted as a Riemann-Stieltjes integral, then we may compute

$$\int_0^t J_s \underline{Z}_s \, d\underline{M}_s = \hat{\underline{\beta}}_t - \underline{\beta}_t^*$$

and the theorem will follow from the properties of stochastic integrals. Such an interpretation is valid according to Doléans-Dade and Meyer([10], Prop. 3) if

$$E\int_0^1 |Z_{ij,s}|\, d|M_{m,s}| < \infty, \quad \forall i,j,m .$$

This holds if

$$E\left[\int_0^1 |Z_{ij,s}|\, dN_{m,s} + \int_0^1 |Z_{ij,s}|\, A_{m,s} ds\right] < \infty, \qquad \forall_{i,j,m} .$$

By Jacod ([11], Section 2) this is equivalent to

$$E\int_0^1 |Z_{ij,s}|\, dN_{m,s} < \infty$$

which, in view of the assumption $EN_{m,1} < \infty$, is implied by the condition of the theorem.

Remark 1. The martingale property implies that $\hat{\underline{\beta}}$ is an unbiased estimator of $\underline{\beta}^*$. By varying $\underline{Z}$ a large class of such estimators may be generated whenever $r < k_t$, the number of nonzero lines of $\underline{Y}_t$. It is natural to ask which choice of $\underline{Z}$ is to be preferred. One possibility is to invoke a formal least squares principle, and hence to minimize for each t the following expression which is to be regarded as a function of $\alpha_t dt$:

$$(d\underline{N}_t - \underline{Y}_t\, \underline{\alpha}_t dt)^T\, (d\underline{N}_t - \underline{Y}_t\, \underline{\alpha}_t dt) .$$

This gives

$$\underline{Z}_t = (\underline{Y}_t^T\ \underline{Y}_t)^{-1} \underline{Y}_t^T . \tag{1}$$

This choice of $\underline{Z}$ probably gives reasonable estimates. They will however not be optimal in general. This may be seen by considering the situation r=1. In that case (1) gives

$$\underline{Z}_t = \left(\sum_{i=1}^{k} Y_{i,t}^2\right)^{-1} \underline{Y}_t^T$$

which leads to

$$\hat{\beta}_{1,t} = \int_0^t \Big(\sum_{i=1}^{k} Y_{i,t}^2\Big)^{-1} d\Big(\sum_{i=1}^{k} Y_{i,t}N_{i,t}\Big) .$$

On the other hand, one may show by using the likelihood function (see e.g. [3]) that $(\underline{N},\underline{Y})$ collapses by a sufficiency reduction into $(\sum_{i=1}^{k} N_i, \sum_{i=1}^{k} Y_i)$. The estimator of $\hat{\beta}_1$ given above is not a function of this sufficient statistic unless $k_t = 1$ for all t.

The author has so far only performed preliminary investigations regarding the optimal choice of $\underline{Z}$. In the particular case r = k, then $\underline{Z}_t$ must be the unique inverse of $\underline{Y}_t$, whenever $J_t > 0$, and I believe that it leads in this case to a maximum likelihood estimator of $\underset{\sim}{\beta}$ in the sense of Kiefer and Wolfowitz (see Johansen, 1978). However, in the general case investigations seem to indicate that the maximum likelihood estimator may not exist, and that one can not find an estimator optimal (in any reasonable sense) for all $\underset{\sim}{\alpha}$. However, this latter conclusion is so far only a rough guess.

Remark 2. It is part of the definition of a counting process that only one of the component processes can jump at a time. Say that a jump occurs in N_i. Then $d\underline{\bar{N}}$ is a matrix of 0's except for element (i,i) which is equal to 1. Hence

$$\underline{Z}\ d\underline{\bar{N}} = \begin{pmatrix} 0\dots & Z_{i1}\dots & 0 \\ \cdot & & \cdot \\ \cdot & & \cdot \\ \cdot & & \cdot \\ 0\dots & Z_{ri}\dots & 0 \end{pmatrix}$$

and

$$\underline{Z}\ d\underline{\bar{N}}\ \underline{Z}^T = \begin{pmatrix} Z_{1i}^2 & Z_{1i}Z_{2i}\dots Z_{1i}Z_{ri} \\ Z_{2i}Z_{1i} & Z_{2i}^2\dots Z_{2i}Z_{ri} \\ \cdot & \cdot \\ \cdot & \cdot \\ Z_{ri}Z_{1i}\dots\dots & \dots\dots Z_{ri}^2 \end{pmatrix}$$

Asymptotic results may be formulated in various ways by exploiting the martingale property of $\hat{\underline{\beta}} - \overset{*}{\underline{\beta}}$. One may, for instance, use the approach of Aalen [3] or Aalen and Johansen [4]. Those results are formulated in terms of the variance process $\langle\cdot,\cdot\rangle$ of a martingale. In Appendix 2 we give weak convergence results, similar to those of Rebolledo [19], which depend instead on the quadratic variation process $[\cdot,\cdot]$. As we have seen, this process plays an important role in the estimation procedures.

3. Testing

3a. Testing of contrasts. Let $\underline{a}_t = (a_{1,t},\ldots,a_{r,t})$ be a bounded predictable function satisfying

$$\sum_{j=1}^{r} a_{j,t} = 0, \quad \forall t .$$

We want to test the hypothesis

$$H_1: \underline{a}_t \underline{\alpha}_t = 0, \quad \forall t .$$

A natural statistic for testing this is the following:

$$T_1 = \int_0^1 K_s \underline{a}_s \, d\hat{\underline{\beta}}_s,$$

where K is a predictable process denoting the weight given to different time intervals. Assume now that the assumption of Theorem 1 is in force. Then $\hat{\underline{\beta}} - \overset{*}{\underline{\beta}}$ is a square integrable vector martingale and hence, under H_1,

$$T_{1,t} = \int_0^t K_s \underline{a}_s \, d\underline{\beta}_s$$

will also be a square integrable martingale if

$$E \int_0^1 K_s^2 \, \underline{a}_s \, d\langle\hat{\underline{\beta}},\hat{\underline{\beta}}\rangle_s \, \underline{a}_s^T < \infty$$

(see the appendix for the theory of stochastic integrals of vector valued martingales). This allows the properties of T_1 under H_1 to be determined by martingale theory, in particular the asymptotic properties may be derived. See Theorem A3 of Appendix 2. For testing purposes one may then use asymptotic normality of T_1. An estimator of the variance under H_1 is given by

$$\left[T_1, T_1\right]_1 = \int_0^1 K_s^2 J_s \, \underline{a}_s \, \underline{Z}_s \, d\underline{N}_s \underline{Z}_s^T \, \underline{a}_s^T .$$

In order to get a reasonably powerful test one must have a limited set of alternatives in mind so that K can be chosen to make T_1 sensitive against those alternatives. If one wants to test H_1 against all possible alternatives, then a Kolmogorov-Smirnov test may be constructed in the manner indicated in Aalen ([1], Section 8).

3b. Testing the influence of explanatory variables. In order to check whether the Y_{ij}, i = 1,..., k, has any influence on the development of $\underline{N}$, one may test the hypothesis

$$H_2: \alpha_{j,t} = 0, \quad \forall t .$$

To discuss what kind of alternatives one may expect, we will consider the case of a clinical trial.

Consider an explanatory variable over the time interval where it influences survival. Usually one would expect that this influence is either positive (i.e. prolonging life) or negative over the whole time interval. Of course, it could be that there is a change occurring with time so that the explanatory variable has, say, a positive effect to begin with, and later a negative effect. However one would not expect this to be common. We will therefore consider testing H_2 against the alternatives

$A_1: \alpha_{j,t} < 0$ over some time interval and 0 otherwise.

or

A_2: $\alpha_{j,t} > 0$ over some time interval and 0 otherwise.

One may consider one-sided testing against A_1 or A_2 separately or a two-sided testing against A_1 and A_2 combined. It is natural to use a test statistic of the kind

$$\int_0^1 K_s \hat{\beta}_{j,s} ,$$

where K is some nonnegative predictable process analogous to that of the previous section. Asymptotic theory via construction of a martingale may be applied here too.

4. Some remarks on the extension to general stochastic processes

In this section we will briefly indicate a very general framework within which methods similar to those above may be applied. The ideas presented here have not been worked out in detail.

The basis for the extension is the famous Doob-Meyer decomposition of a stochastic process, which has received much attention recently. Intuitively, one considers the increment of the process over a small interval and decomposes it into the predictable and the unpredictable part. The former is the best guess based on the observation of the past, and the latter is simply the difference between the actual value and the best guess.

More formally, consider the time-continuous R^k-valued stochastic process $\underline{X}_t = (X_{1,t},\dots,X_{k,t})$ defined on some probability space (Ω,F,P), and assume that it is adapted to an increasing family of σ-algebras $\{F_t\}$. Under certain conditions a Doob-Meyer decomposition of the following form exists:

$$d\underline{X}_t = \underline{A}_t dt + d\underline{M}_t ,$$

where $\underline{A}_t$ is a predictable process (in the formal mathematical sense) while $\underline{M}_t$ is a martingale.

In the case of $\underline{X}_t$ being a multivariate counting process, $\underline{A}_t$ will be precisely the intensity process Λ_t. Another prominent example is the case where $\underline{M}_t$ is a vector of independent Wiener processes while $\underline{A}_t$ is uniquely determined by the value of $\underline{X}_t$. Such stochastic differential equations have been studied by many authors (see e.g. [5]).

Analogously to the counting process situation discussed in previous sections we suggest the model

$$\underline{A}_t = \underline{Y}_t \underline{\alpha}_t$$

with $\underline{Y}$ being a matrix of observable stochastic processes and $\underline{\alpha}$ being a vector of unknown functions. Since the methods we have developed for counting processes depend essentially on the martingale property of $\underline{M}$, they may in principle be extended to this general context.

Sometimes the model $\underline{A}_t = \underline{Y}_t \underline{\alpha}_t$ comes about in a natural way. Examples relating to counting processes are given by Aalen [3]. The Ornstein-Uhlenbeck process ([5], p. 134) is an example from the theory of stochastic differential equations. Often, however, the structure of the process will not be sufficiently known to specify $\underline{A}$ completely. Our linear model will then be an approximation based on the vague knowledge that one may have. This is analogous to the situation in usual linear model theory.

5. Application to a set of real data

We have applied the method of this paper to a set of data from the Veterans Administration Lung Cancer Study Group [16,17]. The regression variables included performance status (a measure of general medical condition on a scale 10, 20, ... , 90), disease duration prior to treatment, and age. By using the method of Cox [9], Prentice [17] found that performance status had a significant influence on survival together with the tumor cell type. We want to investigate how the influence of the performance status (which is measured at the entrance of the patient into the study) changes with time. For this purpose we use all the data except the group with cell type "standard, large" which seemed to have exceptionally long survival times. We assume that the intensity of death of patient no. i can be written

$$\Lambda_{i,t} = \alpha_{1,t} - \frac{1}{100} X_i \alpha_{2,t} ,$$

where X_i is the performance status. Hence $\alpha_{2,t}$ measures the influence of the performance status. We have then the model studied above with

$$\underline{Y}_t = \begin{pmatrix} I_1(t) & \cdots & -I_1(t)\frac{1}{100}X_1 \\ \vdots & & \vdots \\ I_k(t) & & -I_k(t)\frac{1}{100}X_k \end{pmatrix}$$

where $I_i(t) = 1$ if patient no. i is alive and under observation at time t. Otherwise $I_i(t) = 0$. Furthermore, k denotes the total number of patients considered.

In our model we ignore the influence of cell types and hence our results must be viewed as some kind of average over the different cell types. The data are given in Table 1.

The estimate $\hat{\beta}_{2,t}$ of the function

$$\beta_{2,t} = \int_0^t \alpha_{2,s} ds$$

is given in Figure 1. It indicates that $\alpha_{2,t}$ is large to begin with and then approaches 0 after about a 100 days. Hence the influence of the performance status on survival seems to fade after a while. We have not yet applied a significance test to check this.

In Figure 2, we give the estimates of

$$A_t^X = \beta_{1,t} - \frac{1}{100} X\beta_{2,t}$$

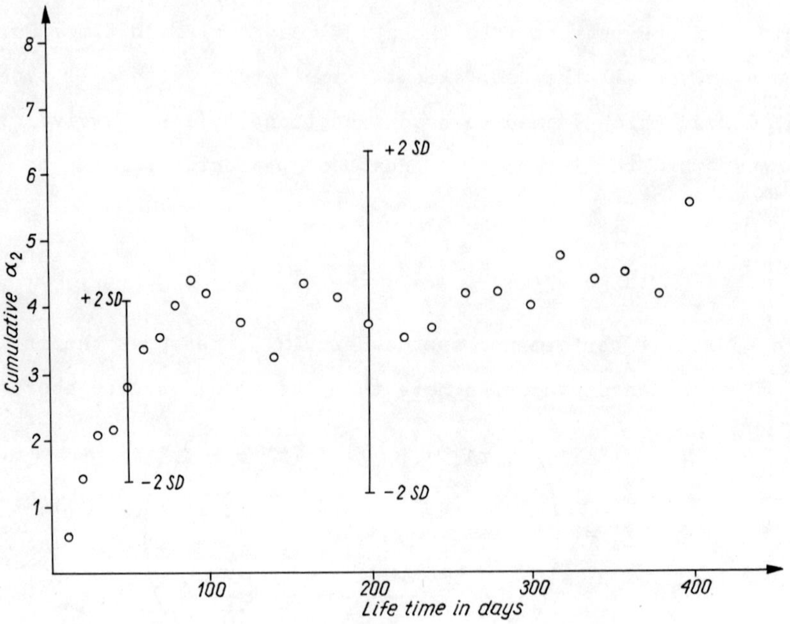

Fig. 1. Accumulated influence of performance status on survival(data from Prentice)

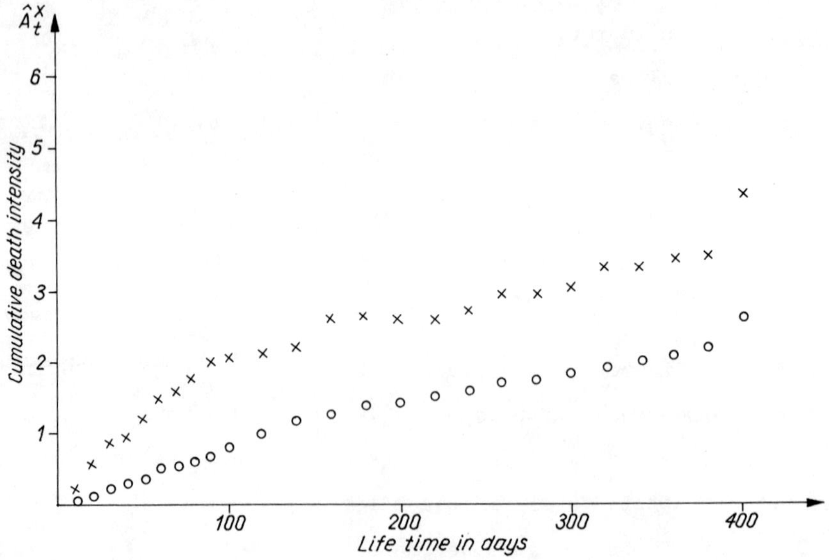

Fig. 2. Lung Cancer (data from Prentice), x - performance status 40, o - performance status 70

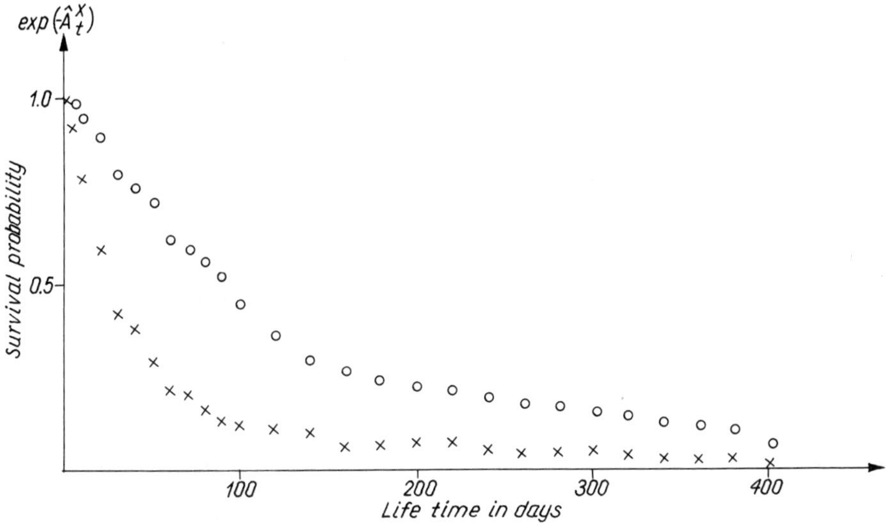

Fig. 3. Lung Cancer (data from Prentice), x - performance status 40, o - performance status 70

for the values X=40 and X=70. This figure also seems to indicate the fading influence of the performance status. In Figure 3 are given the estimates of the survival curves $\exp(-A_t^X)$ for X=40 and X=70.

This examples should indicate how the method suggested in this paper may give useful supplementary information to that acquired for instance by the method of Cox (1972). However, much more work has to be done in order to assess the real value of our approach. In order for it to work well, it may possibly be necessary that the number of observations compared to the number of regression variables is quite large.

T a b l e 1. Data from the Veterans Administration Lung Cancer Group (Prentice 1973). Survival of lung cancer patients in days (t). Performance status at entrance into study (x). (x = 10, 20, 30 - completely hospitalized, x = 40, 50, 60 - partial confinement to hospital, x = 70, 80, 90 - able to care for self)

t	x	t	x	t	x
72	60	7	50	24	60
411	70	63	50	99	70
228	60	392	40	8	80
126	60	10	40	99	85
118	70	8	20	61	70
10	20	92	70	25	70
82	40	35	40	95	70
110	80	117	80	80	50
314	50	132	80	51	30
100*	70	12	50	29	40
42	60	162	80	24	40
8	40	3	30	18	40
144	30	95	80	83*	99
25*	80	999	90	31	80
11	70	112	80	51	60
30	60	87*	80	90	60
384	60	231*	50	52	60
4	40	242	50	73	60
54	80	991	70	8	50
13	60	111	70	36	70
123*	40	1	20	48	10
97*	60	587	60	7	40
153	60	389	90	140	70
59	30	33	30	186	90
117	80	25	20	84	80
16	30	357	70	19	50
151	50	467	90	45	40
22	60	201	80	80	40
56	80	1	50	52	60
21	40	30	70	164	70
18	20	44	60	19	30
139	80	283	90	53	60
20	30	15	50	15	30
31.	75	25	30	43	60
52	70	103*	70	340	80
287	60	21	20	133	75
18	30	13	30	111	60
51	60	87	60	231	70
122	80	2	40	378	80
27	60	20	30	49	30
54	70	7	20		

*Censored survival.

References

[1] A a l e n, O., Nonparametric inference in connection with multiple decrement models. Scand. J. Statist., 3, 15-27 (1976).

[2] A a l e n, O., Weak convergence of stochastic integrals related to counting processes. Z. Wahrscheinlichkeitstheorie verw. Gebiete., 38, 261-277 (1977).

[3] A a l e n, O., Nonparametric inference for a family of counting processes. Ann. Statist., 6, 701-726 (1978).

[4] A a l e n, O. and J o h a n s e n, S., An empirical transition matrix for non-homogeneous markov chains based on censored observations. Scand. J. Statist., 5, 141-150 (1978).

[5] A r n o l d, L., Stochastic differential equations. Wiley, New York 1974.

[6] B i l l i n g s l e y, P., Convergence of probability measures. Wiley, New York 1968.

[7] B r é m a u d, P. et J a c o d, J., Processus ponctuels et martingales: Résultat récent sur la modélisation et le filtrage. Adv. Appl. Prob., 9, 362-416 (1977).

[8] B r o w n, C. C., On the use of indicator variables for studying the time-dependence of parameters in a response-time model. Biometrics., 31, 863-872 (1975).

[9] C o x, D. R., Regression models and life tables. J. Roy. Statist. Soc. Ser. B., 34, 187-220 (1972).

[10] D o l e á n s-D a d é, C. et M e y e r, P., Intégrales stochastiques par rapport aux martingales locales. Séminaire de probabilités IV. Lecture Notes in Mathematics. Springer-Verlag, Berlin 124, 77-107 (1972).

[11] J a c o d, J., Multivariate point processes: Predictable projection, Radon-Nikodym derivatives, representation of martingales. Z. Wahrscheinlichkeitstheorie und Verw. Gebiete., 31, 235-253 (1975).

[12] J o h a n s e n, S., The product limit estimator as maximum likelihood estimator. Scand. J. Statist., 5, 195-199 (1978).

[13] K u n i t a, H. and W a t a n a b e, S. On square integrable martingales. Nagoya Math. J., 30, 209-245 (1967).

[14] Mc L e i s h, D. L., Dependent central limit theorems and invariance principles. Ann. Prob., 2, 620-628 (1974).

[15] M e y e r, P., Un cours sur les integrales stochastiques. Séminaire de Probabilités X. Lecture Notes in Mathematics. Springer-Verlag, Berlin 511, 246-400 (1975).

[16] P r e n t i c e, R. L., Exponential survivals with censoring and explanatory variables. Biometrika, 60, 279-288 (1973).

[17] P r e n t i c e, R. L., Models for survival analysis. Paper presented at the regional meeting of WNAR, IMS, and ASA, Corvallis, Oregon, June 16, 1975.

[18] R a o, C. R., Linear statistical inference and its applications. Wiley, New York 1973.

[19] R e b o l l e d o, R., Central limit theorems for martingales. Technical report, Universite Paris VI (1978).

APPENDIX 1

The concepts used here may be found e.g. in Meyer [15]. Let $\underline{M}_t = (M_{1,t}, \ldots, M_{k,t})^T$ be a vector of square integrable martingales relative to the sequence of σ-algebras $\{F_t\}$ defined in Section 2. The covariance proces $\langle \underline{M}, \underline{M} \rangle_t$ is defined as the matrix containing $\langle M_i, M_j \rangle_t$ at the position (i,j). Let the quadratic covariation process $[\underline{M}, \underline{M}]_t$ be defined analogously. Let $\underline{H}_t = (H_{ij,t})$ be a $r \times k$ matrix of predictable processes such that $H_{ij,t}$ is a member of $L^2(M_{j,t})$ for each pair (i,j). The vector stochastic integral

$$\int_0^t \underline{H}_s d\underline{M}_s$$

is now defined as the obvious extension of the scalar stochastic integral. It enjoys the properties

$$\left\langle \int_0^t \underline{H}_s d\underline{M}_s, \int_0^t \underline{H}_s d\underline{M}_s \right\rangle = \int_0^t \underline{H}_s d\langle \underline{M}_s, \underline{M}_s \rangle \underline{H}_s^T$$

and

$$\left[\int_0^t \underline{H}_s d\underline{M}_s, \int_0^t \underline{H}_s d\underline{M}_s \right] = \int_0^t H_s d \left[\underline{M}_s, M_s \right] H_s^T.$$

APPENDIX 2

We will give some results on weak convergence of martingales which is useful for the statistical methods considered in this paper. The idea of the proof is to extend discrete time results of McLeish [14]. The same idea was used in [2] to derive results based on the variance process $\langle M,M \rangle$ of a square integrable martingale M. Here we will instead use the quadratic variation process $[M,M]$ which has been seen in this paper to play an important role in the estimation of variances. Further, by discretizing with respect to appropriate stopping times we get rid of some unnecessary conditions made in Aalen [2](see Requirement B), where the discretization were taken with respect to nonrandom times.

We will consider square integrable martingales M defined on the time interval $[0,1]$ and satisfying the following condition (define $\Delta M_t = M_t - M_{t-}$):

C. M has a finite number of discontinuities at totally inaccessible stopping times, and

$$E \sum_{t \in [0,1]} |\Delta M_t| < \infty \quad .$$

The concepts used here and below may be found for instance in Meyer [15].

In fact, Rebolledo [19] has given results similar to ours for a wider class of local martingales. The justification of our presentation is to show how continuous time results may be derived from the quite elementary results of McLeish [14]. Rebolledo uses considerably heavier theory, and his added generality, although nice from a mathematical point of view, may not be so important for statistical applications.

The martingales we consider will be elements of the space D of Billingsley [6] and weak convergence is to be understood in the usual sense. When we consider vector processes they are members of the Cartesian product D^m equipped with the corresponding product Skorohod topology. The martingales are defined on a probability space (Ω, F, P) relatively to an increasing rightcontinuous family of sub-σ-algebras of F.

<u>Theorem A1.</u> Let M_n be a sequence of square integrable martingales each satisfying condition C. Assume:

$$\text{(i)} \quad [M_n, M_n](t) \xrightarrow{p} t \qquad \forall t \in [0,1]$$

$$\text{(ii)} \quad \sup_t \Delta M_n(t) \xrightarrow{L_2} 0.$$

Then M_n converges weakly to the Wiener process.

<u>Theorem A2.</u> Let $\underline{L}_n = (M_{1,n}, \ldots, M_{k,n})$ be a sequence of processes such that for each n $M_{1,n}, \ldots, M_{k,n}$ are orthogonal square integrable martingales each satisfying condition C. Suppose that assumptions (i) and (ii) of Theorem A1 are satisfied for each sequence $M_{i,n}$, i=1,...,k. Then $\underline{L}_n$ converges weakly to a vector of independent Wiener processes.

Theorem A3. Let $\underline{L}_n$ be defined as in Theorem A2. Suppose that assumption (ii) of Theorem A1 is satisfied for each sequence $\{M_{i,n}\}$, i=1,...,k, and that in addition the following holds:

$$(i)' \quad \left[M_{i,n}, M_{i,n}\right](1) \xrightarrow{p} 1, \quad i=1,\dots,k.$$

Then $\underline{L}_n$ (1) converges in distribution to a vector of independent N(0,1) random variables.

We will prove Theorem A1 as a consequence of Theorem 3.2 of McLeish [14] .Theorem A2 then follows from a multivariate extension of McLeish's result. This extension can be formulated and proved in an analogous way to that given in the Appendix of Aalen [2] . We will not give the details of this. Finally, Theorem A3 can be proved in a similar manner, starting this time with Theorem 2,3 of McLeish [14].

P r o o f o f T h e o r e m A1: Our proof is partly based on ideas by Meyer [15] and Kunita and Watanabe [13]. Following Meyer [14] we decompose M_n into the compensated sum of jumps and the continuous part:

$$M_n(t) = M_n^c(t) + M_n^d(t)$$

where

$$M_n^d(t) = \sum_{s \leq t} \Delta M_n(s) - C_n(t).$$

Because of the total inaccessibility of jump times, C has continuous sample functions. Define the following stopping times for n=1,2,... and k=0,1,2...:

$$T_0^n = 0,$$

$$T_{k+1}^n = \inf\left\{t > T_k^n \mid \Delta M_n(t) \neq 0\right.$$

$$\left.\text{or } \left|M_n^c(t) - M_n^c(T_k^n)\right| \geq 2^{-n}\right\},$$

$$\text{or } \langle M_n^c, M_n^c\rangle(t) - \langle M_n^c, M_n^c\rangle(T_k^n) \geq 2^{-n},$$

$$\text{or } |C_n(t) - C_n(T_k^n)| \geq 2^{-n}\}.$$

Because of continuity we have:

$$|M_n^c(t) - M_n^c(T_k^n)| \leq 2^{-n},$$

$$\langle M_n^c, M_n^c\rangle(t) - \langle M_n^c, M_n^c\rangle(T_k^n) \leq 2^{-n},$$

and

$$|C_n(t) - C_n(T_k^n)| \leq 2^{-n}$$

whenever $T_k^n \leq t \leq T_{k+1}^n$. We also have $\lim T_k^n = \infty$ when $k \to \infty$ because of uniform continuity on each finite interval.

Put $S_k^n = T_k^n \wedge t$. Fix n and write

$$_kM = M_n(S_k^n),\ _kM^c = M_n^c(S_k^n),\ _kM^d = M_n^d(S_k^n),\ _k[M] = [M_n, M_n](S_k^n),$$

$$_k\langle M\rangle = \langle M_n, M_n\rangle(S_k^n)$$

and correspondingly for $_k\langle M^c\rangle$ and $_k\langle M^d\rangle$. Finally, put

$$_kC = C_n(S_k^n),\ \Delta_k M = \Delta M_n(S_k^n).$$

Following Kunita and Watanabe [13] we will now study the following quantity:

$$A_n = E\left\{\sum_{k=1}^{\infty}(_kM - {_{k-1}M})^2 - ([M,M](t) - [M,M](0))\right\}^2$$

$$= E\sum_{k=1}^{\infty}\left\{(_kM - {_{k-1}M})^2 - (_k[M] - {_{k-1}[M]})\right\}^2$$

$$= E\left[\sum_{k=1}^{\infty}\left\{({}_kM^c - {}_{k-1}M^c + {}_kM^d - {}_{k-1}M^d)^2 -\right.\right.$$

$$\left.\left. - ({}_k\langle M^c\rangle - {}_{k-1}\langle M^c\rangle + {}_k[M^d] - {}_{k-1}[M^d])\right\}^2\right] =$$

$$= E\left[\sum_{k=1}^{\infty}\left\{({}_kM^c - {}_{k-1}M^c)^2 + 2({}_kM^c - {}_{k-1}M^c)({}_kM^d - {}_{k-1}M^d) +\right.\right.$$

$$\left.\left. + ({}_kM^d - {}_{k-1}M^d)^2 - ({}_k\langle M^c\rangle - {}_{k-1}\langle M^c\rangle) - \Delta_k M^2\right\}^2\right].$$

Inserting ${}_kM^d - {}_{k-1}M^d = \Delta_k M + {}_kC - {}_{k-1}C$ we get:

$$A_n = E\left[\sum_{k=1}^{\infty}\left\{({}_kM^c - {}_{k-1}M^c)^2 + 2({}_kM^c - {}_{k-1}M^c)(\Delta_k M + {}_kC - {}_{k-1}C) +\right.\right.$$

$$\left.\left. + 2\Delta_k M({}_kC - {}_{k-1}C) + ({}_kC - {}_{k-1}C)^2 - ({}_k\langle M^c\rangle - {}_{k-1}\langle M^c\rangle)\right\}^2\right].$$

We use $\left(\sum_{i=1}^{k} x_i\right)^2 \leqslant k \sum_{i=1}^{k} x_i^2$, and get:

$$A_n \leqslant 6E\sum_{k=1}^{\infty}\left[({}_kM^c - {}_{k-1}M^c)^4 + 4({}_kM^c - {}_{k-1}M^c)^2\,\Delta_k M^2 +\right.$$

$$+ 4({}_kM^c - {}_{k-1}M^c)^2({}_kC - {}_{k-1}C)^2 + 4\Delta_k M^2({}_kC - {}_{k-1}C)^2 + ({}_kC - {}_{k-1}C)^4 +$$

$$\left. + ({}_k\langle M^c\rangle - {}_{k-1}\langle M^c\rangle)^2\right]$$

$$\leqslant \frac{6}{2^n} E\left[({}_kM^c - {}_{k-1}M^c)^2 + 4\Delta_k M^2 +\right.$$

$$\left. + 4({}_kM^c - {}_{k-1}M^c)^2 + 4\Delta_k M^2 + ({}_kC - {}_{k-1}C)^2 + ({}_k\langle M^c\rangle - {}_{k-1}\langle M^c\rangle)\right]$$

$$= \frac{6}{2^n}\left[6E(M_n^c(t)^2)+8E(M_n^d(t)^2)+ E\sum_{k=1}^{\infty}({}_kC-{}_{k-1}C)^2\right].$$

The two first expectations are finite.For the last one we have:

$$E\sum_{k=1}^{\infty}({}_kC-{}_{k-1}C)^2 \leqslant E\sum_{k=1}^{\infty}({}_kM^d-{}_{k-1}M^d)^2+E\sum_{k=1}^{\infty}\Delta M_k^2 + E\,2\sum_{k=1}^{\infty}|\Delta M_k({}_kC-{}_{k-1}C)|.$$

The two first expectations are finite. For the last one we have:

$$E\sum_{k=1}^{\infty}|\Delta M_k({}_kC-{}_{k-1}C)| \leqslant 2^{-n}E\sum_{k=1}^{\infty}|\Delta M_k| < \infty .$$

We have proved: $A_n < 2^{-n}B$ with B a finite positive constant. Hence, assuption (i) implies:

$$(1)\quad \sum_{k=1}^{\infty}\left[M_n(T_k^n\wedge t) - M_n(T_{k-1}{}^n\wedge t)\right]^2 \xrightarrow{P} t, \quad \forall t\in[0,1].$$

Consider now

$$D_n = E\left[\sup_k\left\{M_n(T_k^n)-M_n(T_{k-1}{}^n)\right\}^2\right].$$

$$D_n \leqslant 3E\left[\sup_t \Delta M^2(t) + \sup_k\left\{M_n^c(T_k^n)-M_n^c(T_{k-1}{}^n)\right\}^2 +\right.$$

$$\left. + \sup_k\left\{C_n(T_k^n) - C_n(T_{k-1}{}^n)\right\}^2\right] \leqslant \frac{6}{2^{-n}} + 3E\sup_t \Delta M_n^2(t)$$

$\to 0$ according to assumption (ii).

We now want to apply Theorem 3.2 of McLeish [14] with $X_{n,k} = M_n(T_k^n) - M_n(T_{k-1}{}^n)$. The martingale difference array assumption is fulfilled because of the T's being stopping times. Assumption (a) of McLeish has already been verified. It only remains to verify assumption (b). Define

$$k_n(t) = \sup\left\{j : T_j^n \leqslant t\right\}.$$

We have

$$\{k_n(t) \geq k\} = \{T_k^n \leq t\} \in F_{T_k^n} .$$

Hence $k_n(t)$ is a stopping time w.r.t. $\{F_{T_k^n}\}_{k=0}^{\infty}$. According to the remarks of McLeish ([14], p. 627), his Theorem 3.2 can then be used with this $k_n(t)$. We have to verify then that:

$$\sum_{k=1}^{k_n(t)} \left[M_n(T_k^n) - M_n(T_{k-1}^n)\right]^2 \xrightarrow{p} t .$$

The sum differs from that in (1) only by

$$\left[M_n(t) - M_n(T_{k_n(t)}^n)\right]^2$$

and we have to prove that this goes in probability to 0. This follows immediately from $D_n \rightarrow 0$.

ON SUPERPOSITIONS OF RANDOM MEASURES AND POINT PROCESSES

BY

Rimas Banys

Academy of Sciences of Lithuanian SSR, Vilnius

1. Limit Theorems for Superpositions of Integer-Valued Random Measures

Let $(\mathcal{X}, \mathcal{A})$ be a space, where $\mathcal{X} = \mathcal{X}_1 \times \mathcal{X}_2$, $\mathcal{A} = \mathcal{A}_1 \times \mathcal{A}_2$, $\mathcal{X}_2 = [0,\infty)$, while $\mathcal{A}_2$ is the σ-algebra of Borel subsets of $\mathcal{X}_2$.

By an integer-valued random measure we mean a system of random variables $Q(A) = Q(A,\omega)$ defined for every $A \in \mathcal{A}$ on a probability space $(\Omega, \mathcal{F}, \mathcal{P})$ and meeting the following conditions:

1. $Q(A) \in Z = \{\ldots,-1,0,1,\ldots\}$ for each $A \in \mathcal{A}$;
2. $P\{|Q(\mathcal{X}_1 \times A_2)| < \infty\} = 1$ for each bounded set, $A_2 \in \mathcal{A}_2$;
3. $P\{Q(\mathcal{X}_1 \times \{0\}) = 0\} = 1$;
4. $Q(\bigcup_m A_m) = \sum_m Q(A_m)$ a.s. for any disjoint sets $A_1, A_2, \ldots \in \mathcal{A}$.

Let $\mathcal{X}_1 = \{x_1, \ldots, x_N\}$. Putting

$$X_i(t) = Q(\{x_i\} \times [0,t)) \tag{1}$$

we establish a one-to-one correspondence between the integer-valued random measures Q and the multidimensional integer-valued random processes $X(\cdot) = (X_1(\cdot), \ldots, X_N(\cdot))$.

Integer-valued random measure Q is called a compound Poisson measure, if for all disjoint sets $A_p = \Gamma_p \times \Delta_p$, where $\Gamma_p \in \mathcal{A}_1$, while Δ_p is an interval, $1 \leq p \leq r$, the random variables $Q(A_p)$, $1 \leq p \leq r$, are independent and

$$E \exp\left\{ iuQ(A_p) \right\} = \exp\left\{ \sum_{s\neq 0} \lambda_s(A_p) \left(e^{ius}-1\right) \right\},$$

where the summation is over all nonzero integers. Moreover for any integer s the symbol $\lambda_s(\cdot)$ stands for a measure on $\mathcal{X}$ such that $\sum_{s\neq 0} |s| \lambda_s(A) < \infty$, for any $A \in \mathcal{X}$. If $\lambda_1(A) = \lambda(A)$ and if $\lambda_s(A)=0$ for $s \neq 1$ $A \in \mathcal{X}$, then Q is said to be Poisson.

The compound Poisson measure will be denote by $P_{\{\lambda_s\}}$.

The sequence $\{Q_n,\ n = 1,2,\ldots\}$ is said to converge to measure Q if for any disjoint sets $A_p = \Gamma_p \times \Delta_p$, $1 \leq p \leq r$, the distributions of $(Q_n(A_1),\ldots, Q_n(A_r))$ converge weakly to that of $(Q(A_1),\ldots,Q(A_r))$. We will write in this case $Q_n \to Q$.

Let

$$\left\{ Q_{n1},\ldots,Q_{nk_n},\quad n = 1,2,\ldots \right\} \tag{2}$$

be a triangular array of infinitesimal random measures such that for any Γ and Δ

$$\lim_{n\to\infty} \max_{1\leq m\leq k_n} P\left\{ Q_{nm}(\Gamma\times\Delta) \neq 0 \right\} = 0.$$

Now let

$$Q_n = \sum_{m=1}^{k_n} Q_{nm}.$$

Under the assumption that Q_{nm}, $1\leq m \leq k_n$, are independent, the following theorems can be established.

<u>Theorem 1.</u> If the random measures (2) are independent, then

$$Q_n \to P_{\{\lambda_s\}}$$

iff for any disjoint sets $A = \Gamma \times \Delta$, $A' = \Gamma' \times \Delta'$

$$\lim_{n\to\infty} \sum_{m=1}^{k_n} P\left\{Q_{nm}(A) \neq 0,\ Q_{nm}(A') \neq 0\right\} = 0 \tag{A}$$

and

$$\lim_{n\to\infty} \sum_{s\neq 0} \left| \sum_{m=1}^{k_n} P\left\{Q_{nm}(\Gamma \times \Delta) = s\right\} - \lambda_s(\Gamma\times\Delta) \right| = 0. \tag{B}$$

Theorem 2. If (A) holds, then the class of possible limiting measures of the sequences $\left\{Q_n,\ n=1,2,\ldots\right\}$ coincides with that of the compound Poisson measures.

Note that the theorems of Grigelionis [5] follow from Theorems 1 and 2.

Now we are going to prove a theorem which is analogous to a theorem on convergence of sums of weakly dependent random processes established in [4].

Given an arbitrary collection $\alpha = \left\{A_1,\ldots,A_r\right\}$ of disjoint sets $A_p = \Gamma_p \times \Delta_p$, $1 \leq p \leq r$, let $\mathcal{F}_m^{(n)}(\alpha)$ denote the σ-algebra generated by $(Q_{nk}(A_1),\ldots,Q_{nk}(A_r))$, $1 \leq k \leq m$.

Theorem 3. If the sequence (2) satisfies (A) and (B) and if there exists a subsequence of natural numbers $\varkappa_n$ such that for an arbitrary collection $\alpha = \left\{A_1,\ldots,A_r\right\}$ of disjoint sets $A_p = \Gamma_p \times \Delta_p$, $1 \leq p \leq r$, the following limits exist:

$$\lim_{n\to\infty} \sum_{m=\varkappa_n+1}^{k_n} E\left| P\left\{Q_{nm}(\Gamma_p\times\Delta_p) = s \,\middle|\, F_{m-\varkappa_n}^{(n)}(\alpha)\right\} - P\left\{Q_{nm}(\Gamma_p\times\Delta_p) = s\right\}\right| = 0,$$

$$1 \leq \quad \leq r$$

$$\lim_{n\to\infty} \sum_{m=\varkappa_n+1}^{k_n} P\left\{Q_{nm}(\Gamma_p\times\Delta_p) \neq 0,\ \bigcup_{l=m-\varkappa_n+1}^{m-1} (Q_{nl}(\Gamma_q\times\Delta_q) \neq 0)\right\} = 0,$$

$$1 \leq p,\ q \leq r$$

and

$$\lim_{n\to\infty} \max_{0\leq m\leq k_n-\varkappa_n} \sum_{k=m+1}^{m+\varkappa_n} P\left\{Q_{nk}\ (\Gamma_p \times \Delta_p) \neq 0\right\} = 0, \quad 1\leq p\leq \tau ,$$

then

$$Q_n \to P_{\{\lambda_s\}}.$$

To prove Theorem 3 we shall establish, in the same way as the main theorem on superpositions of random processes in [4], the following lemma.

Let

$$\vec{\xi}_{n1},\ldots, \vec{\xi}_{nk_n}, \quad n = 1,2,\ldots,$$

where $\vec{\xi}_{nm} = \left(\xi^{(1)}_{nm},\ldots, \xi^{(\tau)}_{nm}\right)$, $1 \leq m \leq k_n$ be a sequence of integer-valued infinitesimal random vectors so that

$$\lim_{n\to\infty} \max_{1\leq m\leq k_n} P\left\{\vec{\xi}_{nm} \neq \vec{0}\right\} = 0. \tag{3}$$

<u>Lemma.</u> Let $\vec{\xi}_{nm}$, $1\leq m\leq k_n$, $n = 1,2,\ldots$ satisfy (3) and let for $p\neq q$

(a) $$\lim_{n\to\infty} \sum_{m=1}^{k_n} P\left\{\xi^{(p)}_{nm} \neq 0,\ \ \xi^{(q)}_{nm} \neq 0\right\} = 0$$

and for $1\leq p\leq \tau$

(b) $$\lim_{n\to\infty} \sum_{s\neq 0} \left|\sum_{m=1}^{k_n} P\left\{\xi^{(p)}_{nm} = s\right\} - \lambda^{(p)}_s\right| = 0.$$

If there exists a subsequence of natural numbers $\varkappa_n$ such that for every nonzero s

(i) $$\lim_{n\to\infty} \sum_{m=\varkappa_n+1}^{k_n} E\left|P\left\{\xi^{(p)}_{nm} = s \,\middle|\, \mathcal{F}^{(n)}_{m-\varkappa_n}\right\} - P\left\{\xi^{(p)}_{nm} = s\right\}\right| = 0,$$

$1\leq p\leq \tau$

where $\mathcal{F}_l^{(n)}$ denotes the σ-algebra generated by $\vec{\xi}_{nm}$, $1 \leq m \leq l$,

(ii) $$\lim_{n\to\infty} \sum_{m=\varkappa_n+1}^{k_n} P\left\{\xi_{nm}^{(p)} \neq 0, \bigcup_{l=m-\varkappa_n+1}^{m-1} \left(\xi_{nl}^{(q)} \neq 0\right)\right\} = 0, \quad 1 \leq p, q \leq r$$

and

(iii) $$\lim_{n\to\infty} \max_{0 \leq m \leq k_n - \varkappa_n} \sum_{k=m+1}^{m+\varkappa_n} P\left\{\xi_{nk}^{(p)} \neq 0\right\} = 0,$$

then $\sum_{m=1}^{k_n} \vec{\xi}_{nm}$ converges in distribution to a random vector $\vec{\xi} = \left(\xi^{(1)}, \ldots, \xi^{(r)}\right)$ with independent components and such that

$$\mathcal{F}^{(p)}(u) = E\exp\left\{iu\,\xi^{(p)}\right\} = \exp\left\{\sum_{s\neq 0} \lambda_s^{(p)} \left(e^{ius} - 1\right)\right\}.$$

<u>P r o o f of the lemma:</u> Let

$$f_{nm}(\vec{u}) = E\exp\left\{i\langle\vec{u}, \vec{\xi}_{nm}\rangle\right\},$$

$$\mathcal{F}_m^{(n)}(\vec{u}) = E\exp\left\{i\langle\vec{u}, \sum_{k=1}^{m} \vec{\xi}_{nk}\rangle\right\},$$

and let

$$\mathcal{F}(\vec{u}) = \prod_{p=1}^{r} \mathcal{F}^{(p)}(\vec{u}),$$

where $\vec{u} = (u_1, \ldots, u_r) \in R^r$, while $\langle\vec{u}, \vec{v}\rangle = \sum_{p=1}^{r} u_i v_i$.

We need to show that

$$\lim_{n\to\infty} \mathcal{F}_{k_n}^{(n)}(\vec{u}) = \mathcal{F}(\vec{u}). \qquad (4)$$

Writting

$$\mathcal{F}_m^{(n)}(\vec{u}) = E\left(\exp\left\{i\langle\vec{u}, \sum_{k=1}^{m-1}\vec{\xi}_{nk}\rangle\right\}\cdot E_{\mathcal{F}_{m-1}^{(n)}}\, e^{i\langle\vec{u},\vec{\xi}_{nm}\rangle}\right), \tag{5}$$

we note that

$$E_{\mathcal{F}_{m-1}^{(n)}}\, e^{i\langle\vec{u},\ \vec{\xi}_{nm}\rangle} = f_{nm}\ (\vec{u}) + \sum_{\vec{s}} e^{i\langle\vec{u},\vec{s}\rangle} Q_{nm}(\vec{s}), \tag{6}$$

where

$$Q_{nm}(\vec{s}) = P\left\{\vec{\xi}_{nm} = \vec{s}\ \middle|\mathcal{F}_{m-1}^{(n)}\right\} - P\left\{\vec{\xi}_{nm} = \vec{s}\right\}.$$

Here $\vec{s}$ denotes a vector in R^{τ} with integer components.

Letting $\vec{e}_p = (\underbrace{0,\ldots,}_{p-1}1\underbrace{,\ldots,0}_{\tau-p})$ we may write

$$\sum_{\vec{s}} e^{i\langle\vec{u},\vec{s}\rangle} Q_{nm}(\vec{s}) = \sum_{\vec{s}\neq\vec{0}}\left(e^{i\langle\vec{u},\vec{s}\rangle}-1\right) Q_{nm}(\vec{s}) =$$

$$= \sum_{p=1}^{\tau}\sum_{s\neq 0}\left(e^{iu_p s} - 1\right) Q_{nm}(s\vec{e}_p) + \sum_{\vec{s}\neq\vec{0},\, s\vec{e}_p}\left(e^{i\langle\vec{u},\vec{s}\rangle}-1\right) Q_{nm}(\vec{s}) =$$

$$= \sum_{p=1}^{\tau}\sum_{s\neq 0}\left(e^{iu_p s}-1\right) Q_{nm}^{(p)}(s) + \mathcal{O}\left(\sum_{p\neq q}\left(P\left\{\xi_{nm}^{(p)} \neq 0,\ \xi_{nm}^{(q)} \neq 0\middle|\mathcal{F}_{m-1}^{(n)}\right\} + \right.\right.$$

$$\left.\left. + P\left\{\xi_{nm}^{(p)} \neq 0,\ \xi_{nm}^{(q)} \neq 0\right\}\right)\right). \tag{7}$$

Here

$$Q_{nm}^{(p)}(s) = P\left\{\xi_{nm}^{(p)} = s\middle|\mathcal{F}_{m-1}^{(n)}\right\} - P\left\{\xi_{nm}^{(p)} = s\right\}.$$

Hence

$$\mathcal{F}_m^{(n)}(\vec{u}) = f_{nm}(\vec{u})\,\mathcal{F}_{m-1}^{(n)}\ (\vec{u}) +$$

$$+ E\exp\left\{i\langle\vec{u}, \sum_{k=1}^{m-1}\vec{\xi}_{nk}\rangle\right\}\cdot\sum_{p=1}^{\tau}\sum_{s\neq 0}\left(e^{iu_p s}-1\right) Q_{nm}^{(p)}(s) +$$

$$+\sigma\Big(\sum_{p\neq q} P\Big\{\xi_{nm}^{(p)} \neq 0,\ \xi_{nm}^{(q)} \neq 0\Big\}\Big). \tag{8}$$

Making use of the above recurrent relation we obtain

$$\mathcal{F}_{k_n}^{(n)}(\vec{u}) = E\exp\Big\{i\langle\vec{u}, \sum_{m=1}^{k_n-1} \vec{\xi}_{nm}\rangle\Big\}\sum_{p=1}^{r}\sum_{s\neq 0}\Big(e^{iu_p s}-1\Big)Q_{nk_n}^{(p)}(s)+$$

$$+ f_{nk_n}(\vec{u})\, E\exp\Big\{i\langle\vec{u}, \sum_{m=1}^{k_n-2} \vec{\xi}_{nm}\rangle\Big\}\sum_{p=1}^{r}\sum_{s\neq 0}\Big(e^{iu_p s}-1\Big)Q_{nk_n-1}^{(p)}(s)+$$

$$+\ldots+\mathcal{F}_{\varkappa_n}^{(n)}(\vec{u}) \prod_{m=\varkappa_n+1}^{k_n} f_{nm}(\vec{u})+\sigma\Big(\sum_{p\neq q}\sum_{m=1}^{k_n} P\Big\{\xi_{nm}^{(p)}\neq 0,\ \xi_{nm}^{(q)}\neq 0\Big\}\Big). \tag{9}$$

The last summand of the right hand part of (9) tends to zero as $n\to\infty$ by virtue of (a). From Theorem 1 it follows that (a) and (b) imply the convergence of $\prod_{m=1}^{k_n} f_{nm}(\vec{u})$ to $\mathcal{F}(\vec{u})$. Condition (iii) implies that

$$\lim_{n\to\infty}\prod_{m=1}^{\varkappa_n} f_{nm}(\vec{u}) = 1 \quad\text{and}\quad \lim_{n\to\infty}\mathcal{F}_{\varkappa_n}^{(n)}(\vec{u}) = 1.$$

Thus

$$\lim_{n\to\infty}\mathcal{F}_{\varkappa_n}^{(n)}\prod_{m=\varkappa_n+1}^{k_n} f_{nm}(\vec{u}) = \mathcal{F}(\vec{u}). \tag{10}$$

Consequently, we need only to prove that the sum of the remaining summands in (9) tends to zero. For this purpose we evaluate the second summand of the right hand part of (8). Denoting it by J_{nm} we obtain

$$J_{nm} = E\exp\left\{i\langle\vec{u}, \sum_{k=1}^{m-\varkappa_n} \vec{\xi}_{nk}\rangle\right\}\left[E_{\mathcal{F}^{(n)}_{m-\varkappa_n}} \sum_{p=1}^{\tau} \sum_{s\neq 0}\left(e^{iu_p s} - 1\right) Q^{(p)}_{nm}(s) + \right.$$

$$\left. + E_{\mathcal{F}^{(n)}_{m-\varkappa_n}}\left(\exp\left\{i\langle\vec{u}, \sum_{l=m-\varkappa_n+1}^{m-1} \vec{\xi}_{nl}\rangle\right\} - 1\right)\sum_{p=1}^{\tau}\sum_{s\neq 0}\left(e^{iu_p s} - 1\right) Q^{(p)}_{nm}(s)\right].$$

Clearly for any $S > 0$

$$|J_{nm}| \leqslant C \sum_{p=1}^{\tau} P\left\{\xi^{(p)}_{nm} > S\right\} + \sum_{p=1}^{\tau} \sum_{\substack{|s|\leqslant S \\ s\neq 0}} E\left|P\left\{\xi^{(p)}_{nm} = s | \mathcal{F}^{(n)}_{m-\varkappa_n}\right\} - P\left\{\xi^{(p)}_{nm} = s\right\}\right| +$$

$$+ \sum_{p=1}^{\tau} \sum_{\substack{|s|\leqslant S \\ s\neq 0}} P\left\{\xi^{(p)}_{nm} = s, \sum_{l=m-\varkappa_n+1}^{m-1} \vec{\xi}_{nl} \neq \vec{0}\right\} +$$

$$+ P\left\{\sum_{l=m-\varkappa_n+1}^{m-1} \vec{\xi}_{nl} \neq \vec{0}\right\} \sum_{p=1}^{\tau} \sum_{\substack{|s|\leqslant S \\ s\neq 0}} P\left\{\xi^{(p)}_{nm} = s\right\}.$$

Hence by using (9)

$$\left|F^{(n)}_{k_n}(\vec{u}) - F(\vec{u})\right| \leqslant C \sum_{p=1}^{\tau} \sum_{m=1}^{k_n} P\left\{\xi^{(p)}_{nm} > S\right\} +$$

$$+ \sum_{p=1}^{\tau} \sum_{|s|\leqslant S} \sum_{m=\varkappa_n+1}^{k_n} E\left|P\left\{\xi^{(p)}_{nm} = s | \mathcal{F}^{(n)}_{m-\varkappa_n}\right\} - P\left\{\xi^{(p)}_{nm} = s\right\}\right| +$$

$$+ \sum_{p,q=1}^{\tau} \sum_{\substack{|s|\leqslant S \\ s\neq 0}} \sum_{m=\varkappa_n+1}^{k_n} P\left\{\xi^{(p)}_{nm} = s, \sum_{l=m-\varkappa_n+1}^{m-1} \xi^{(q)}_{nl} \neq 0\right\} +$$

$$+ \max_{\varkappa_n \leqslant m \leqslant k_n} P\left\{\sum_{l=m-\varkappa_n+1}^{m-1} \vec{\xi}_{nl} \neq \vec{0}\right\} \sum_{p=1}^{\tau} \sum_{\substack{|s|\leqslant S \\ s\neq 0}} \sum_{m=1}^{k_n} P\left\{\xi^{(p)}_{nm} = s\right\} +$$

$$+\left|\mathcal{F}^{(n)}_{\varkappa_n}(\vec{u})\prod_{m=\varkappa_n+1}^{k_n} f_{nm}(\vec{u}) - F(\vec{u})\right| + O\left(\sum_{p\neq q}\sum_{m=1}^{k_n} P\left\{\xi^{(p)}_{nm} \neq 0,\ \xi^{(q)}_{nm} \neq 0\right\}\right).$$

From (10) and the assumptions of the lemma it follows that the sum of the right hand part of this inequality tends to zero, which proves the lemma.

To prove Theorem 3 it sufficies to verify that $((Q_{nm}(A_1),\dots,\dots,Q_{nm}(A_r)), 1\leq m\leq k_n$, satisfy the assumptions of the lemma. We omit the details.

Theorem 3 can be generalized by using Bernstein's method, i.e. by eliminating some parts of the summands Q_{nm}, $1\leq m\leq k_n$, and by constructing a new sequence of enlarged summands [2], [3].

By using relation (1) one may easily obtain from known theorems concerning convergence of integer-valued random measures [2], [1] conditions insuring convergence of multidimensional integer-valued random processes.

2. On the Poisson Convergence of Superpositions of Point Processes

Let $\mathfrak{S}$ be a fixed locally compact second countable Hausdorff topological space with the σ-algebra $\mathcal{S}$ of Borel subsets and the ring $\mathcal{B}$ consisting of all bounded (i.e. relatively compact) sets in $\mathcal{S}$. We denote by $\mathcal{F}$ the class of all $\mathcal{S}$-measurable functions $f:\mathfrak{S}\to[0,\infty)$ and by $\mathcal{F}_c$ the subclass of all functions in $\mathcal{F}$ which are continuous and have a compact support. Moreover we denote $\mathcal{N}$ the class of all integer-valued Radon measures on $(\mathfrak{S},\mathcal{B})$. For any $\mu\in\mathcal{N}$ and $f\in\mathcal{F}$ put

$$\mu f = \int_{\mathfrak{S}} f(s)\,\mu(ds).$$

The class of all finite intersections of $\mathcal{N}$-sets of the form $\{\mu : u<\mu f<v\}$ with arbitrary $f\in\mathcal{F}_c$ and real u, v may serve as a base for a topology on $\mathcal{N}$ to be called vague one. Denote by $\mathcal{N}$

the σ-algebra of Borel subsets generated by this vague topology in $\mathcal{N}$. It is known [6] that $\mathcal{N}$ is also the σ-algebra generated by the mappings $\mu \longrightarrow \mu B$, $B \in \mathcal{B}$.

By a point process on $\mathfrak{S}$ we mean any measurable mapping of some fixed probability space $(\Omega, \mathcal{U}, P)$ into $(\mathcal{N}, \mathcal{N})$. The distribution of a point process ξ is by definition the probability measure $P\xi^{-1}$ on $(\mathcal{N}, \mathcal{N})$ which is given by

$$\left(P\xi^{-1}\right)M = P\left(\xi^{-1}M\right) = P\left\{\xi \in M\right\}, \quad M \in \mathcal{N}.$$

A point process ξ is called a Poisson process with intensity λ if for any disjoint $B_1, \ldots, B_n$, where $B_i \in \mathcal{B}$, $1 \leq i \leq n$, the random variables $\xi B_1, \ldots, \xi B_n$ are independent and

$$P\left\{\xi B = k\right\} = \frac{(\lambda B)^k}{k!} e^{-\lambda B},$$

where λ is a Radon measure on $(\mathfrak{S}, \mathcal{B})$.

Let

$$\left\{\xi_{nm}, \ 1 \leq m \leq k_n, \quad n = 1, 2, \ldots\right\} \tag{11}$$

be a triangular array of point processes such that for every $B \in \mathcal{B}$

$$\lim_{n \to \infty} \max_{1 \leq m \leq k_n} P\left\{\xi_{nm} B > 0\right\} = 0.$$

By convergence of point processes (denoted by $\xrightarrow{d}$) we mean convergence in distribution. Moreover let $\mathcal{F}_k^{(n)}$ denote the σ-algebra generated by $\xi_{n1}B, \ldots, \xi_{nk}B$, where $B \in \mathcal{B}$. Finally let

$$\xi_n = \sum_{m=1}^{k_n} \xi_{nm}.$$

<u>Theorem 4.</u> Suppose that ξ is a Poisson process with intensity λ and let the sequence of the point processes (11) be such that for every $B \in \mathcal{B}$

$$\lim_{n\to\infty} \sum_{m=1}^{k_n} P\{\xi_{nm} B>0\} = \lambda B,$$

and

$$\lim_{n\to\infty} \sum_{m=1}^{k_n} P\{\xi_{nm} B>1\} = 0.$$

If there exists a subsequence of natural numbers $\varkappa_n$ such that for any $B\in\mathcal{B}$

$$\lim_{n\to\infty} \sum_{m=\varkappa_n+1}^{k_n} E\left| P\{\xi_{nm} B>0|\mathcal{F}^{(n)}_{m-\varkappa_n}\} - P\{\xi_{nm} B>0\}\right| = 0,$$

$$\lim_{n\to\infty} \max_{0\le m\le k_n-\varkappa_n} \sum_{l=m+1}^{m+\varkappa_n} P\{\xi_{nl} B>0\} = 0,$$

and

$$\lim_{n\to\infty} \sum_{m=\varkappa_n+1}^{k_n} P\{\xi_{nm} B>0, \bigcup_{l=m-\varkappa_n+1}^{m-1} (\xi_{nl} B>0)\} = 0,$$

then

$$\xi_n \xrightarrow{d} \xi .$$

P r o o f: For a point process ξ, we may define the L-transform $L_\xi(f)$ by

$$L_\xi(f) = E\exp\{-\xi f\}, \quad f\in\mathcal{F}.$$

It is known [6] that $\xi_n \xrightarrow{d} \xi$ iff $L_{\xi_n}(f) \to L_\xi(f)$ for every $f\in\mathcal{F}_c$. Let

$$L^{(n)}_m(f) = L_{\sum_{i=1}^{m} \xi_{ni}}(f).$$

It is necessary to prove that $L_{k_n}^{(n)}(f) \rightarrow L_{\xi}(f)$, where $f \in \mathcal{F}_e$.
Because

$$L_m^{(n)}(f) = E \exp\left\{ - \sum_{k=1}^{m} \xi_{nk} f \right\} =$$

$$= E\left[\exp\left\{ - \sum_{k=1}^{m-1} \xi_{nk} f \right\} E_{\mathcal{F}_{m-1}^{(n)}} e^{-\xi_{nm} f}, \right.$$

Theorem 4 may be prove in the same way as Theorem 3 (compare[4]).

References

[1] B a n y s, R., Limit theorems for superpositions of multidimensional integer-valued random processes (in Russian), Liet.Matem. Rink., XIX, 1(1979).

[2] B a n y s, R., On convergence of superpositions integer-valued random measures (in Russian), XIX, 1(1979).

[3] B e r n s t e i n, S. N., Extension of limit theorem of probability to sums of dependent variables (in Russian). Uspechi Mat. Nauk, X, 65-114 (1944).

[4] B o r i s o v, I. S., Some limit theorems for sums of dependent random processes (in Russian), Sibirsk.Mat.Z., 1979 (to appear).

[5] G r i g e l i o n i s, B., Limit theorems for sums of random step processes (in Russian), Liet.Matem.Rink., X, 1, 29-49(1970).

[6] K a l l e n b e r g, O., Random Measures, Akademie-Verlag, Berlin, 1975.

APPLICATION AND OPTIMALITY OF THE CHI-SQUARE TEST OF FIT FOR TESTING ε-VALIDITY OF PARAMETRIC MODELS

by

Tadeusz Bednarski

Polish Academy of Sciences, Wrocław

1. Introduction

The chi-square test of fit provides a possible and widely used method to justify what is the underlying distribution. However real data frequently reveal at least small deviations from the ideal parametric model [6, 8, 9, 10]. Consequently for large samples the null hypothesis is rejected. It is clear that usual deviations from parametric models are small enough to preclude a large loss, even if in further inference we pretend that the model is correct. For the last few years a large number of procedures were devised to serve in such situations [1]. These are robust procedures desined to be resistant for small deviations from a given parametric model. The efficiency of robust estimates and tests is strongly connected with the size of possible contamination of the parametric model [1, 9, 10]. So it is important to have procedures for estimation or testing the size of this contamination.

The aim of this paper is to give a simple method of testing ε-validity of parametric models by use of the chi-square test of fit. In the traditional formulation of the problem of fit the null hypothesis states that observed variables have a distribution which comes from a fixed parametric family of probability measures $\{P_\Theta\}$,

$\theta \in \Theta \subset R^m$. In our formulation we allow a small discrepancy with the model. It is then convenient to consider a sequence of null and alternative hypotheses of the following form:

$$H_n : \mathcal{P}_{0n} = \bigcup_{\theta \in \Theta} \{p(\theta) + (p - p(\theta))/\sqrt{n} : p \in \mathcal{P}_{0,\theta}\}$$

$$K_n : \mathcal{P}_{1n} = \bigcup_{\theta \in \Theta} \{p(\theta) + (p - p(\theta))/\sqrt{n} : p \in \mathcal{P}_{1,\theta}\},$$

where $p(\theta) = [p_1(\theta), \ldots, p_k(\theta)]^T$, $p_i(\theta) > 0$, $\sum p_i(\theta) = 1$ and $\mathcal{P}_{i,\theta}$, i=0,1 are some subsets of the k-dimensional simplex, depending on θ. n stands for the number of i.i.d. observations. Asymptotic distributions of the chi-square statistic are well known in this situation [11]. We prove that this convergence is uniform and that the chi-square test is uniformly asymptotically minimax for a large class of testing problems, in the family of tests with convex acceptance regions. This is the subject of Section 2 and 3. In the last section we explain how the chi-square test can be applied in the problem of testing ε-validity of parametric models.

The following notation will be used in the sequel. A vector $p = (p_1, \ldots, p_k)^T$ will denote a column vector. For two vectors p and q we shall have $\sqrt{p} = (\sqrt{p_1}, \ldots, \sqrt{p_k})^T$ and $q/\sqrt{p} = (q_1/\sqrt{p_1}, \ldots, q_k/\sqrt{p_k})^T$. If A is a matrix and $x = Aq$ then $Aq/\sqrt{p}$ is defined as $x/\sqrt{p}$. For a vector q the matrix I_q is the diagonal one with entries q_i on the diagonal. S^k denotes the simplex in R^k, that is for $p \in S^k$ we have $p_i \geq 0$ for i=1, \ldots, k and $\sum p_i = 1$. Moreover, $\|\cdot\|$ is the usual norm in R^k.

2. The asymptotic optimality of the chi-square test of fit in the case of a single contaminated probability measure

Let $X_1, \ldots, X_n$ be a sample of size n from a fixed population, whose distribution function is $F(x)$. Let n_i be the number of observations in the sample falling into the i-th class interval for

$i=1, \ldots ,k$. Then the random vector $N = (n_1, \ldots ,n_k)^T$ has the multinomial distribution with parameters n and $p^T = (p_1, \ldots ,p_k)$, where p_i is the probability that X_j falls into the i-th class. We would like to verify the hypothesis that the distribution F belongs to a small neighbourhood of a fixed distribution F_0. That is, in terms of the multinomial variables, we want to test whether p belongs to a small neighbourhood of p_0. For this purpose we shall consider the following sequence of hypotheses:

$$H_n : \mathcal{P}_{0n} = \left\{ p \in S^k : \|(p - p_0)/\sqrt{p_0}\| \leq \varepsilon_0/\sqrt{n} \right\},$$

$$K_n : \mathcal{P}_{1n} = \left\{ p \in S^k : M/\sqrt{n} \geq \|(p - p_0)/\sqrt{p_0}\| \geq \varepsilon_1/\sqrt{n} \right\},$$

where $M > \varepsilon_1 > \varepsilon_0 > 0$ are constants, $p_0^T = (p_{01}, \ldots ,p_{0k})$ and $p_{0i} > 0$ for $i=1, \ldots ,k$.

Our aim is to find asymptotically optimal family of tests for this sequence of testing problems. A family of tests, we shall restrict to is the set of all zero-one tests with convex acceptance regions in R^k, that are functions of N. It has been proved that this family of tests is essentially complete for testing of fit in the case of every single null hypothesis [5] and that these tests are admissible for every n for H_n against K_n, see [12].

Let $\mathcal{F}$ denote the family of all test functions defined on R^k with convex acceptance regions.

Definition 2.1. We say that a sequence of tests $\{\varphi_n\} \subset \mathcal{F}$ is at the level α for testing problems H_n against K_n if

$$\limsup_n \left[\sup_{p \in \mathcal{P}_{0n}} E_p\, \varphi_n \right] \leq \alpha .$$

The vector p in $E_p\, \varphi_n$ denotes the parameter of the multinomial distribution for which the expectation is taken. Let $\{\varphi_n^*\}$ be a sequence of tests defined by

$$\varphi_n^*(N) = \begin{cases} 0, & \text{if } n \sum_{i=1}^{k} \left[(n_i/n - p_{0i})^2/p_{0i}\right] < c, \\ 1, & \text{if } n \sum_{i=1}^{k} \left[(n_i/n - p_{0i})^2/p_{0i}\right] \geqslant c, \end{cases}$$

where $p_0^T = (p_{01}, \dots, p_{0k})$ and $\lim_n \sup_{p \in \mathcal{P}_{0n}} E_p \varphi_n^* = \alpha$.

Theorem 2.1. Let $\{\varphi_n\} \subset \mathcal{F}$ be any sequence of tests at level α for testing problems H_n against K_n. Then we have

$$\liminf_n \left[\inf_{p \in \mathcal{P}_{1n}} E_p \varphi_n^* - \inf_{p \in \mathcal{P}_{1n}} E_p \varphi_n \right] \geqslant 0.$$

Let us denote by $\mathcal{F}_{\alpha,n}$ a family of tests $\varphi \in \mathcal{F}$ for which

$$\sup_{p \in \mathcal{P}_{0n}} E_p \varphi \leqslant \alpha .$$

Theorem 2.2. We have

$$\liminf_n \left[\inf_{p \in \mathcal{P}_{1n}} E_p \varphi_n^* - \sup_{\varphi \in \mathcal{F}_{\alpha,n}} \inf_{p \in \mathcal{P}_{1n}} E_p \varphi \right] \geqslant 0.$$

The first main step in proving the above theorems is the following lemma implied by Corollary 17.2 of [3]. Let a random vector $T_n(q)$ be defined by $T_n(q) = (N - nq)/\sqrt{nq}$, $q \in \mathcal{P} \subset S^k$ and let

$$\mathcal{P}_n(q) = \left\{ p \in S^k : \| (p-q)/\sqrt{q} \| \leqslant M/\sqrt{n} \right\}.$$

We assume that there is $\varepsilon > 0$ such that for every $p \in \mathcal{P}$, $p = (p_1, \dots, p_k)$, we have $p_i > \varepsilon$, $i=1, \dots, k$. Denote by $Q_{n,p,q}$ the distribution of T_n under $p \in \mathcal{P}_n(q)$. Let $\mathcal{C}$ be the class of all measurable convex subsets of R^k.

Lemma 3.1. We have

$$\lim_n \left\{ \sup_{q \in \mathcal{P}} \sup_{p \in \mathcal{P}_n(q)} \sup_{C \in \mathcal{C}} \left| Q_{n,p,q}(C) - \Phi_{n,p,q}(C) \right| \right\} = 0,$$

where $\Phi_{n,p,q}$ is the normal $N(\sqrt{n}(p-q)/\sqrt{q};\ I - \sqrt{q}\,\sqrt{q}^{\,T})$ measure.

The other part of the proof consists in finding a suitable solution for the testing problem obtained by taking asymptotic distributions of $T_n(p_0)$ under $p \in \mathcal{P}_{0n}$ and $p \in \mathcal{P}_{1n}$, [13].

3. Asymptotic optimality of the chi-square test of fit in the case of contaminated parametric model

The following assumptions are made about the parametric model.

1. The parameter space $\Theta \subset R^m$, $m < k$, is compact. The multinomial parameter $p(\Theta)$ is a one to one function of $\Theta \in \Theta$ and there is $\varepsilon > 0$ such that for all $\Theta \in \Theta$, $p_i(\Theta) > \varepsilon$ for $i=1, \ldots, k$.

2. The matrix $B(\Theta) = \left[(p_i(\Theta))^{-1/2}\, \partial p_i(\Theta)/\partial \Theta_i\right]$ has elements that are continuous functions of Θ.

3. The second partial derivatives of $\log\left[p_i(\Theta)\right]$ with respect to Θ_j exist and are continuous.

4. The matrix $B(\Theta)^T B(\Theta)$ is nonsingular for every $\Theta \in \Theta$

Let $M > 0$ be a constant and let $a(\Theta, p) = (p - p(\Theta))/\sqrt{p(\Theta)}$ for $p \in S^k$ and $\Theta \in \Theta$. Define

$$\mathcal{P}_n(\Theta, M) = \left\{ p \in S^k : \| a(\Theta, p) \| \leqslant M/\sqrt{n} \right\}.$$

5. The maximum likelihood estimator $\hat{\Theta}_n$ of Θ satisfies the condition: for every $M > 0$, $\varepsilon > 0$ and $\delta > 0$ there exists such n_0 that for all $n \geqslant n_0$

$$\sup_{\Theta \in \Theta} \ \sup_{p \in \mathcal{P}_n(\Theta, M)} P_p(\| \hat{\Theta}_n - \Theta \| > \varepsilon) < \delta.$$

The index p denotes the parameter of the multinomial distribution of the random vector N. This condition is satisfied in many situations for which the maximum likelihood estimator exists and it is asymptotically normal [2, 14]. The sequence of testing problems is defined now in the following way. Let

$$A(\Theta) = I - B(\Theta)\left[B(\Theta)^T B(\Theta)\right]^{-1} B(\Theta)^T$$

and let $L(\Theta)$ denote the linear space spanned by columns of the matrix $B(\Theta)$. Notice that $A(\Theta)$ is the projection on the kernel of $B(\Theta)$. For every n, n the number of i.i.d. observations, every $\Theta \in \Theta$ we define sets

$$\mathcal{P}_{0n}(\Theta) = \left\{p \in S^k : \| a(\Theta, p)\| \leq \varepsilon_0 / \sqrt{n}, \quad a(\Theta, p) \perp L(\Theta)\right\},$$

$$\mathcal{P}_{1n}(\Theta) = \left\{p \in S^k : M/\sqrt{n} \geq \| a(\Theta, p) \| \geq \varepsilon_1 / \sqrt{n}, \; a(\Theta, p) \perp L(\Theta)\right\},$$

where $\varepsilon_1 > \varepsilon_0$ and M are fixed constants and the symbol $\perp$ means "orthogonal to".

The sequence of hypotheses is then defined in the following way

$$H_n : \quad p \in \mathcal{P}_{0n} = \bigcup_{\Theta \in \Theta} \mathcal{P}_{0n}(\Theta),$$

$$K_n : \quad p \in \mathcal{P}_{1n} = \bigcup_{\Theta \in \Theta} \mathcal{P}_{1n}(\Theta).$$

In the sequel we assume that for n sufficiently large and every $p \in \mathcal{P}_{in}$, i=0,1, there exists $\Theta(p) \in \Theta$, a unique one, such that $a\left[\Theta(p), p\right] \perp L(\Theta(p))$. The contamination considered here may be understood as a cylinder around the curve $p(\Theta)$, $\Theta \in \Theta$.

The aim of this section is to find asymptotically optimal sequences of tests for the testing problems H_n versus K_n. Our considerations are restricted to tests $\varphi_n(T_n)$, where $T_n = \sqrt{n}(N/n - p(\hat{\Theta}_n))/\sqrt{p}(\hat{\Theta}_n)$ and $\hat{\Theta}_n$ is m.l.e. of Θ based on cell frequences. It is assumed that for every n, φ_n has a convex measurable acceptance region in the space of values of T_n, that is in R^k.

<u>Definition 3.1.</u> A sequence of tests $\{\varphi_n\}$ is at the level α for testing problems H_n against K_n if

$$\limsup_n \left[\sup_{p \in \mathcal{P}_{0n}} E_p \varphi_n\right] \leq \alpha.$$

Define a sequence of tests

$$\varphi_n^* = \begin{cases} 1, & \text{if} \quad \|T_n\|^2 > c(\alpha), \\ 0, & \text{if} \quad \|T_n\|^2 \leq c(\alpha), \end{cases}$$

where $c(\alpha)$ is the α critical value for the χ^2_{k-m-1} distribution with noncentrality parameter ε_0^2. As it can be seen $\{\varphi_n^*\}$ is at the level α for testing H_n and K_n.

Theorem 3.1. For every $\{\varphi_n\}$ at the level α for the testing problems $\mathcal{P}_{0n}$ against $\mathcal{P}_{1n}$ we have

$$\liminf_n \left[\inf_{p\in\mathcal{P}_{1n}} E_p\, \varphi_n^* - \inf_{p\in\mathcal{P}_{1n}} E_p\, \varphi_n \right] \geq 0.$$

Let us denote by $\mathcal{F}_{\alpha,n}$ the family of tests φ for which

$$\sup_{p\in\mathcal{P}_{0n}} E_p\, \varphi(T_n) \leq \alpha .$$

All tests in $\mathcal{F}_{\alpha,n}$ are as before assumed to have convex measurable acceptance regions in R^k.

Theorem 3.2. We have

$$\liminf_n \left[\inf_{p\in\mathcal{P}_{1n}} E_p\, \varphi_n^* - \sup_{\varphi\in\mathcal{F}_{\alpha,n}} \inf_{p\in\mathcal{P}_{1n}} E_p\, \varphi \right] \geq 0.$$

4. Applications

The theorems imply that the classical chi-square test of fit is asymptotically uniformly minimax for the testing problems $\mathcal{P}_{0n}$ against $\mathcal{P}_{1n}$ in the class of tests with convex measurable acceptance regions. Suppose that we have a sample of size n and we would like to verify the hypothesis $\mathcal{P}_{0n}$ with $\varepsilon/\sqrt{n} = .05$ against any alternative $\mathcal{P}_{1n}$. Then we take the critical value for the chi-square distribution with noncentrality parameter $(\sqrt{n}\times.05)^2$. The obtained test is approximately minimax for the class of tests and alternatives defined before. The method presented here is in particular applicable for testing ε-independence in contingency tables.

An alternative approach to this problem of testing contamination size is given in [4, 7], where estimates of Θ are obtained for fixed contamination sizes.

An extended version of this paper was submitted to the Math. Operationsforsch. Statist.

References

[1] A n d r e w s, D. F., B i c k e l, P. J., H a m p e l, F. R., H u b e r, P. J., R o g e r s, W. H., T u k e y, J. W., Robust Estimates of Location: Survey and Advances. Princeton, Princeton University Press, 1972.

[2] B a k a l a r c z y k, M., On asymptotic properties of maximum likelihood estimates for models with contamination. To be published (1979).

[3] B h a t t a c h a r y a, R. N., R a n g a R a o, R., Normal Approximation and Asymptotic Expansions. John Wiley and Sons 1976.

[4] B j ö r n s t a d, J. F., Inference theory in contingency tables. Statistical Research Report No. 2, University of Oslo 1975.

[5] B i r n b a u m, A., Characterizations of complete classes of tests of some multiparametric hypotheses with applications to likelihood ratio tests. Ann. Math. Statist., 26, 21-36 (1955).

[6] H a m p e l, F. R. Robust estimation: A condenced partial survey. Z. Wahrscheinlichkeitstheorie verw. Geb., 27,87-104(1973).

[7] H o d g e s, J. L., L e h m a n n, E., Testing the approximate

validity of statistical hypotheses. J. R. Statist. Soc., B, 2, 261-268 (1954).

[8] H u b e r, P. J., Robust estimation of a location parameter. Ann. Math. Statist., 35, 1753-1758 (1964).

[9] H u b e r, P. J., A robust version of the probability ratio test. Ann. Math. Statist., 36, 73-101 (1965).

[10] H u b e r, P. J., S t r a s s e n, V., The Neyman-Pearson lemma for capacities. Ann. Statist., 1, 251-263 (1973).

[11] K e n d a l l, M. G., S t u a r t, A., The Advanced Theory of Statistics. Vol. 2. Russian edition, Moscow 1973.

[12] L e d w i n a, T., On admissibility of tests for extended hypotheses of fit. Politechnika Wrocławska, Komunikat nr 110, Wrocław 1977.

[13] L e h m a n n, E. L., Testing Statistical Hypotheses. Wiley, New York 1959.

[14] P a r z e n, E., On uniform convergence of families of sequences of random variables. Univ. of California Publ. in Statist., 2, 23-54 (1954).

ON THE NOTION OF EFFICIENCY OF A BLOCK DESIGN

by

Tadeusz Caliński, Bronisław Ceranka and Stanisław Mejza
Academy of Agriculture, Poznań

A general definition of an orthogonal block design and, subsequently, of the efficiency of a block design are given. The efficiency of a block design is first considered for an individual estimable contrast of treatment parameters, then as a mean efficiency for all estimable contrasts. It appears that the two most common definitions of efficiency, one relating the precision of a block design to that of an equireplicate orthogonal block design with the same total number of plots, the other relating the precision to that of an orthogonal block design with the same numbers of treatment replications, are particular cases of the hither introduced general definition of efficiency.

AMS 1970 Subject Classification: 62K10

Key Words:
Balanced designs, Block designs, Efficiency, Orthogonal designs.

1. Introduction

There have been several suggestions about how to define the efficiency of a block design. The most common approach is to define the efficiency of a design as its precision relative to that of an orthogonal design. But there may be different ways of choosing an appropriate orthogonal design as the basis of comparison. There are two main distinct attitudes to that choice, equivalent only if the design under consideration is an equireplicate design. The first attitude is to choose for the comparison an equireplicate orthogonal block design with the same number of plots, the second is to choose an orthogonal block design with the same numbers of treatment replications, same as in the design under consideration.

In the present paper a generalized definition of an orthogonal block design is proposed and, subsequently, a generalized definition of the efficiency of a block design is suggested. It is then shown that the two rival attitudes to defining efficiency may be considered as particular cases of the present generalized approach.

2. Preliminaries

Let v treatments be applied to n experimental plots arranged in b blocks according to a block design with an incidence matrix $\underline{N} = [n_{ij}]$. The following common notation will be used: $\underline{N}\,\underline{1} = \underline{r} = [r_1, \dots, r_v]'$, $\underline{N}'\underline{1} = \underline{k} = [k_1, \dots, k_b]'$, $\underline{r}'\underline{1} = n = \underline{k}'\underline{1}$, where $\underline{1}$ is a conformable column vector of ones. It will be convenient to write $\underline{x}^{t\delta}$ for the diagonal matrix having as its diagonal elements the components of a vector $\underline{x}$, each raised to the power t (assuming t is such that x_i^t are all meaningful).

Let $\underline{\gamma} = [\gamma_1, \dots, \gamma_v]'$ denote the vector of treatment parameters.

Definition 1. A linear function of treatment parameters, $\underline{c}'\underline{\gamma}$, is called a contrast of treatment parameters, if $\underline{c}'\underline{1} = 0$.

Properties of a block design can be derived from patterns of the matrix $\underline{A} = \underline{r}^{\delta} - \underline{N}\,\underline{k}^{-\delta}\,\underline{N}'$, the rank of which will be denoted by $h (\leq v-1)$. Let $\underline{X}$ be an arbitrary (real) $v \times v$ positive definite symmetric matrix. Let λ_i be an eigenvalue and let $\underline{w}_i (\neq \underline{0})$ be a corresponding eigenvector of the matrix $\underline{A}$ with respect to $\underline{X}$, i.e. let $\underline{A}\,\underline{w}_i = \lambda_i \underline{X}\,\underline{w}_i$, $i = 1, 2,\dots, v$. The eigenvectors $\underline{w}_i$ can be chosen to be X-orthonormal in pairs, i.e. to satisfy the equalities $\underline{w}_i' \underline{X}\underline{w}_i = 1$ and $\underline{w}_i' \underline{X}\,\underline{w}_{i'} = 0$ for $i \neq i'$, where $i, i' = 1, 2,\dots, v$. Since $\underline{A}\,\underline{1} = \underline{0}$, the last (say) eigenvector $\underline{w}_v$ can be taken equal to $(\underline{1}'\underline{X}\,\underline{1})^{-1/2}\underline{1}$. Hence, $\underline{w}_i' \underline{X}\,\underline{1} = 0$ for $i = 1, 2,\dots, v-1$.

Definition 2. For any block design, contrasts of treatment parameters $\underline{c}_i'\boldsymbol{\gamma}$, $i = 1, 2,\dots, v-1$, are said to be X^{-1}-basic contrasts, if $\underline{c}_i = \underline{X}\,\underline{w}_i$, $i = 1, 2,\dots, v-1$, where $\underline{w}_1, \underline{w}_2,\dots, \underline{w}_{v-1}$ are any eigenvectors of the matrix $\underline{A}$ of the design with respect to the matrix $\underline{X}$ that are X-orthonormal in pairs and each X-orthogonal to $\underline{1}$.

It is easily seen that the vectors $\underline{c}_i$, $i = 1, 2,\dots, v-1$, that define X^{-1}-basic contrasts are X^{-1}-orthonormal in pairs. In fact, Definition 2 generalizes the notion of basic contrasts introduced by Pearce, Caliński and Marshall [8]. In the present paper $\underline{c}_i'\boldsymbol{\gamma}$, $i = 1, 2,\dots, v-1$, will always denote X^{-1}-basic contrasts.

It can be shown (see Caliński [2], Lemma 1) that in the intra-block analysis of a block experiment only such X^{-1}-basic contrasts are estimable which are defined by those pairwise X-orthonormal vectors $\underline{w}_1, \underline{w}_2,\dots, \underline{w}_h$ (say), which correspond to the non-zero eigenvalues of the matrix $\underline{A}$ with respect to the matrix $\underline{X}$, viz. $\lambda_1, \lambda_2,\dots, \lambda_h$. Also, it can be shown ([2], Theorem 2) that the variance of the best linear unbiased estimator (BLUE) of an X^{-1}-basic contrast $\widehat{\underline{c}_i'\boldsymbol{\gamma}}$, is of the form

$$\mathrm{var}(\widehat{\underline{c}_i'\boldsymbol{\gamma}}) = \sigma^2/\lambda_i, \quad i = 1, 2,\dots, h, \tag{2.1}$$

where σ^2 denotes the error variance in the intra-block analysis.

Furthermore, for the intra-block analysis it is known (see [2], Lemma 1 and Theorem 2) that if a contrast $\underline{c}'\boldsymbol{\gamma}$ is estimable in a block design, the vector $\underline{c}$ has a representation

$$\underline{c} = \sum_{i=1}^{h} l_i \underline{X}\, \underline{w}_i, \tag{2.2}$$

and the variance of the BLUE of $\underline{c}'\underline{\gamma}$ can then be expressed as

$$\mathrm{var}(\widehat{\underline{c}'\underline{\gamma}}) = \sigma^2 \sum_{i=1}^{h} l_i^2/\lambda_i, \tag{2.3}$$

where $l_1, \ldots, l_h$ are some appropriate coefficients.

Thus, from Definition 2 and from the formulae (2.2), (2.3) and (2.1), it follows that any estimable contrast $\underline{c}'\underline{\gamma}$ is a linear combination of some estimable X^{-1}-basic contrasts, i.e.

$$\underline{c}'\underline{\gamma} = \sum_{i=1}^{h} l_i \underline{c}_i' \underline{\gamma}, \tag{2.4}$$

and also that the variance of its BLUE is a linear combination of the variances of the BLUE's of the X^{-1}-basic contrasts.

3. A general definition of an orthogonal design

For further considerations in the paper it will be assumed that the matrix $\underline{X}$ is diagonal. With this in mind a general definition of an orthogonal design will now be introduced.

Definition 3. Let $\underline{X}$ be a (real) $v \times v$ positive definite diagonal matrix. A block design is said to be X^{-1}-orthogonal if its incidence matrix is of the form

$$\underline{N} = \underline{X}\,\underline{1}\,\underline{k}'/(\underline{1}'\underline{X}\,\underline{1}). \tag{3.1}$$

Note that if $\underline{N}$ is as in (3.1), then

$$\underline{c}'\underline{X}^{-1}\underline{N} = \underline{c}'\underline{1}\,\underline{k}'/(\underline{1}'\underline{X}\,\underline{1}) = 0$$

for any $\underline{c}$ such that $\underline{c}'\underline{1} = 0$. This means that any vector defining a contrast of treatment parameters is X^{-1}-orthogonal to the columns of the matrix $\underline{N}$ that has the form (3.1). This justifies the use of the adjective "$\underline{X}^{-1}$-orthogonal" in Definition 3.

Furthermore, note that for an X^{-1}-orthogonal block design the vector of treatment replications is

$$\underline{r} = \underline{N}\,\underline{1} = n\,\underline{X}\,\underline{1}/(\underline{1}'\underline{X}\,\underline{1}), \tag{3.2}$$

while

$$\underline{N}\,\underline{k}^{-\delta}\underline{N}' = n\,\underline{X}\,\underline{1}\,\underline{1}'\underline{X}/(\underline{1}'\underline{X}\,\underline{1})^2.$$

Hence, the matrix $\underline{A}$ of an X^{-1}-orthogonal block design is of the form

$$\underline{A} = \frac{n}{\underline{1}'\underline{X}\,\underline{1}}\left[\underline{X} - \underline{X}\,\underline{1}\,\underline{1}'\underline{X}/(\underline{1}'\underline{X}\,\underline{1})\right], \tag{3.3}$$

and so has only one non-zero eigenvalue with respect to $\underline{X}$. This eigenvalue is equal to $n/(\underline{1}'\underline{X}\,\underline{1})$ and has the multiplicity v-1.

Thus, referring (3.3) to Section 3 in [2], and particularly to Definition 1 and Corollary 4 there, the following lemma is obtained.

Lemma 1. An X^{-1}-orthogonal block design is an X^{-1}-balanced connected block design with λ, the unique non-zero eigenvalue of $\underline{A}$ with respect to $\underline{X}$, equal to $n/(\underline{1}'\underline{X}\,\underline{1})$.

Recall that a connected block design is such for which h = v-1.

A further result that follows from Lemma 1 and the formula (2.3), is Lemma 2.

Lemma 2. Let $\underline{c}_i'\underline{\tau}$, i = 1, 2,..., h, be estimable X^{-1}-basic contrasts for a block design. The variance of the BLUE obtainable from an X^{-1}-orthogonal block design for a contrast $\underline{c}'\underline{\tau}$ is of the form

$$\mathrm{var}\,(\widehat{\underline{c}'\underline{\tau}}) = \sigma^2\,\frac{\underline{1}'\underline{X}\,\underline{1}}{n}\sum_{i=1}^{h} l_i^2\,, \tag{3.4}$$

where $l_1,\ldots,l_h$ are as in the representation (2.2) of the contrast.

4. General definitions and results concerning efficiency

Now it is possible to introduce a general definition of the efficiency of a block design for a contrast.

Definition 4. The X^{-1}-efficiency factor of a block design for an estimable contrast of treatment parameters, $\underline{c}'\underline{\gamma}$, is defined as the ratio

$$E_{X^{-1}}(\underline{c}'\underline{\gamma}) = \frac{\mathrm{var}_0(\widehat{\underline{c}'\underline{\gamma}})}{\mathrm{var}_1(\widehat{\underline{c}'\underline{\gamma}})}, \tag{4.1}$$

where $\mathrm{var}_1(\widehat{\underline{c}'\underline{\gamma}})$ denotes the variance that the BLUE of $\underline{c}'\underline{\gamma}$ has in the design under consideration, while $\mathrm{var}_0(\widehat{\underline{c}'\underline{\gamma}})$ denotes the variance that the BLUE of $\underline{c}'\underline{\gamma}$ would have in an X^{-1}-orthogonal block design with the same number of plots, if it existed.

It should be noticed that if no X^{-1}-orthogonal design exists for the given $\underline{X}$ and n, the numerator in (4.1) is understood as a value obtained from (3.4) for these $\underline{X}$ and n, with $l_1,\ldots, l_h$ following from the representation (2.2) of $\underline{c}$ in the design under consideration.

In view of Definition 4 the main result of the present paper can now be established.

Theorem 1. The X^{-1}-efficiency factor of a block design for an estimable contrast of treatment parameters, $\underline{c}'\underline{\gamma}$, is of the form

$$E_{X^{-1}}(\underline{c}'\underline{\gamma}) = (\underline{1}'\underline{X}\,\underline{1}/n) \sum_{i=1}^{h} l_i^2 / \Big(\sum_{i=1}^{h} l_i^2/\lambda_i\Big), \tag{4.2}$$

where $l_i = \underline{c}'\underline{w}_i$, $i = 1, 2,\ldots, h$, and where $\underline{w}_1,\ldots, \underline{w}_h$ denote X-orthonormal eigenvectors of the matrix $\underline{A}$ of the design, with respect to the matrix $\underline{X}$, that correspond to the non-zero eigenvalues $\lambda_1,\ldots, \lambda_h$ of $\underline{A}$, with respect to $\underline{X}$.

P r o o f: The formula (4.2) results immediately from (4.1), (2.3) and (3.4), where $l_1,\ldots, l_h$ are directly obtainable from (2.2).

Corollary 1. The X^{-1}-efficiency factor of a block design for an estimable X^{-1}-basic contrast $\underline{c}_i'\underline{\tau}$ is of the form

$$E_{X^{-1}}(\underline{c}_i'\underline{\tau}) = \lambda_i \; \underline{1}'\underline{X}\;\underline{1}/n, \tag{4.3}$$

where λ_i is the non-zero eigenvalue of $\underline{A}$, with respect to $\underline{X}$, to which the eigenvector $\underline{w}_i$ appearing in $\underline{c}_i = \underline{X}\;\underline{w}_i$ corresponds ($i = 1, 2, \ldots, h$).

Apart from the efficiency of a block design for an individual estimable contrast, it may also be interesting to consider a mean efficiency of a block design, i.e. its mean efficiency with respect to all estimable contrasts. Since any estimable contrast of treatment parameters is a linear combination of estimable X^{-1}-basic contrasts, it seems justified to consider the overall efficiency of a design in terms of the efficiencies for those basic contrasts.

Definition 5. The mean X^{-1}-efficiency factor of a block design is defined as the harmonic mean of the X^{-1}-efficiency factors of the design for a complete set of estimable X^{-1}-basic contrasts, $\underline{c}_1'\underline{\tau}, \ldots, \ldots, \underline{c}_h'\underline{\tau}$, and will be denoted by $E_{X^{-1}}$; i.e.

$$E_{X^{-1}} = h/\left[\sum_{i=1}^{h} 1/E_{X^{-1}}(\underline{c}_i'\underline{\tau})\right]. \tag{4.4}$$

For abbreviation it will simply be called the X^{-1}-efficiency factor of a block design.

The following result is an immediate consequence of (4.4) and Corollary 1.

Theorem 2. The X^{-1}-efficiency factor of a block design is of the form

$$E_{X^{-1}} = (\underline{1}'\underline{X}\;\underline{1}/n)\; h/\sum_{i=1}^{h}\lambda_i^{-1}. \tag{4.5}$$

To see a justification for the use of the harmonic mean in Definition 5, note that (4.5) can also be written as

$$E_{X^{-1}} = \sum_{i=1}^{h} \operatorname{var}_0 (\widehat{\underline{c}_i' \underline{\gamma}}) / \sum_{i=1}^{h} \operatorname{var}_1 (\widehat{\underline{c}_i' \underline{\gamma}}),$$

where, similarly as in (4.1), $\operatorname{var}_1(\widehat{\underline{c}_i'\underline{\gamma}})$ denotes the variance of the BLUE of $\underline{c}_i'\underline{\gamma}$ in the design under consideration while $\operatorname{var}_0(\widehat{\underline{c}_i'\underline{\gamma}})$ denotes the variance that the BLUE would have in an X^{-1}-orthogonal block design with the same n, if such a design existed (i = 1, 2,..., h).

Here, again, the same remark as that made in connection with Definition 4 applies.

5. Some further results

A particular case of Theorem 2 is the following result applicable to any X^{-1}-balanced block design, i.e. a design for which the non-zero eigenvalues, $\lambda_1, \ldots, \lambda_h$, are all equal ([2], Theorem 3).

Corollary 2. The X^{-1}-efficiency factor of an X^{-1}-balanced block design is of the form

$$E_{X^{-1}} = (\underline{1}'\underline{X}\ \underline{1}/n)\lambda, \tag{5.1}$$

where $\lambda = h^{-1}\mathrm{tr}(\underline{A}\ \underline{X}^{-1})$ is the unique non-zero eigenvalue of the matrix $\underline{A}$ of the design with respect to $\underline{X}$.

P r o o f: The result (5.1) follows immediately from (4.5) if $\lambda_1 = \ldots = \lambda_h$, and it is known from the definition of λ_i's that

$$\sum_{i=1}^{h} \lambda_i = \mathrm{tr}(\underline{A}\underline{X}^{-1}),$$ which gives $\lambda = h^{-1}\mathrm{tr}(\underline{A}\ \underline{X}^{-1})$ in this case.

Another result, applicable to connected block designs only, is the following

Corollary 3. The X^{-1}-efficiency factor, $E_{X^{-1}}$, of a connected block design can equivalently be written in the form

$$E_{X^{-1}} = \bar{V}_0/\bar{V}_1,$$

where

$$\bar{V}_1 = \frac{1}{\binom{v}{2}} \sum_{1 \leq i < i' \leq v} x_i x_{i'} \operatorname{var}_1(\widehat{\gamma_i - \gamma_{i'}}),$$

and

$$\bar{V}_0 = \frac{1}{\binom{v}{2}} \sum_{1 \leq i < i' \leq v} x_i x_{i'} \operatorname{var}_0(\widehat{\gamma_i - \gamma_{i'}}),$$

with $x_1, \ldots, x_v$ being the diagonal (positive) elements of the matrix $\underline{X}$, and with $\operatorname{var}_1(\widehat{\gamma_i - \gamma_{i'}})$ and $\operatorname{var}_0(\widehat{\gamma_i - \gamma_{i'}})$ denoting the variances of the BLUE's of the elementary contrast $\gamma_i - \gamma_{i'}$ in the design under consideration and in an X^{-1}-orthogonal design with the same n, respectively, the latter being either real or notional.

P r o o f: From (2.2) and (2.3) and from the fact that for a connected block design $h = v-1$,

$$\operatorname{var}_1(\widehat{\gamma_i - \gamma_{i'}}) = \sigma^2 \sum_{\alpha=1}^{v-1} (w_{\alpha i} - w_{\alpha i'})^2/\lambda_\alpha,$$

where $w_{\alpha 1}, \ldots, w_{\alpha v}$ are the components of the eigenvector $\underline{w}_\alpha$ of $\underline{A}$ with respect to $\underline{X}$ corresponding to λ_α ($\alpha = 1, 2, \ldots, v-1$). Hence,

$$\sum_{1 \leq i < i' \leq v} x_i x_{i'} \operatorname{var}_1(\widehat{\gamma_i - \gamma_{i'}}) = \sigma^2 \sum_{1 \leq i < i' \leq v} x_i x_{i'} \sum_{\alpha=1}^{v-1} (w_{\alpha i} - w_{\alpha i'})^2/\lambda_\alpha$$

$$= \sigma^2 \sum_{\alpha=1}^{v-1} \lambda_\alpha^{-1} \left(\sum_{i=1}^{v} x_i\right) \left[\sum_{i=1}^{v} x_i w_{\alpha i}^2 - \left(\sum_{i=1}^{v} x_i w_{\alpha i}\right)^2 \Big/ \sum_{i=1}^{v} x_i \right]$$

$$= \sigma^2 \sum_{\alpha=1}^{v-1} \lambda_\alpha^{-1} (\underline{1}'\underline{X}\,\underline{1}) \left[\underline{w}_\alpha' \underline{X}\, \underline{w}_\alpha - \underline{w}_\alpha'\underline{X}\,\underline{1}/(\underline{1}'\underline{X}\,\underline{1}) \right]$$

$$= \sigma^2 (\underline{1}'\underline{X}\,\underline{1}) \sum_{\alpha=1}^{v-1} \lambda_\alpha^{-1} ,$$

and, similarly,

$$\sum_{1 \leq i < i' \leq v} x_i x_{i'} \operatorname{var}_0 (\widehat{\gamma_i - \gamma_{i'}}) = \sigma^2 (v-1)(\underline{1}'\underline{X}\,\underline{1})^2/n .$$

This gives

$$\bar{V}_0/\bar{V}_1 = (\underline{1}'\underline{X}\,\underline{1}/n) \left[(v-1)/ \sum_{i=1}^{v-1} \lambda_\alpha^{-1} \right],$$

which is exactly the same as $E_{X^{-1}}$ in (4.5) if $h = v-1$.

Corollary 3 gives a natural generalization of a result already known for $x_1 = \dots = x_v$ (as given in [9], Section 4.5), revealing in another way the generality of Definition 5.

Returning now to the general case of a block design, it will be useful to consider the range of values admitted by the factor $E_{X^{-1}}$. Obviously, $0 < E_{X^{-1}}$. Also, since the harmonic mean of a set of positive quantities cannot exceed their arithmetic mean, and both of the means attain the same value if and only if the quantities are all equal, the upper bound is given by the inequality

$$E_{X^{-1}} \leqslant \frac{\underline{1}'\underline{X}\,\underline{1}}{nh} \sum_{i=1}^{h} \lambda_i .$$

Furthemore, since the equality of λ_i's holds if and only if the design is X^{-1}-balanced, the following result is established.

<u>Lemma 3.</u> The X^{-1}-efficiency factor of any block design satisfies the inequality

$$0 < E_{X^{-1}} \leqslant \left[\underline{1}' \underline{X}\ \underline{1}/(nh) \right] tr(\underline{A}\ \underline{X}^{-1}), \qquad (5.2)$$

the equality on the right-hand side being held if and only if the design is X^{-1}-balanced.

An interesting interpretation of Lemma 3 is that among all block designs of the same n and the same mean $h^{-1}tr(\underline{A}\ \underline{X}^{-1})$ for a specified $\underline{X}$, the most X^{-1}efficient (i.e. of the highest X^{-1}-efficiency factor) is any X^{-1}-balanced block design. This is a generalization of Theorem 4.5.2 in [9].

Also, inequality (5.2) can, explicitly, be written as

$$0 < E_{X^{-1}} \leqslant \frac{1}{nh} \left(\sum_{i=1}^{v} x_i\right)\left(\sum_{i=1}^{v} \frac{r_i}{x_i} - \sum_{j=1}^{b} \frac{1}{k_j} \sum_{i=1}^{v} \frac{n_{ij}^2}{x_i}\right),$$

thus being evidently a generalization of Theorem 4.5.1 in [9].

Finally, it would be useful to know the upper bound for the quantity $\left[\underline{1}' \underline{X}\ \underline{1}/(nh) \right] tr(\underline{A}\underline{X}^{-1})$ itself, and the condition under which it is attained. These are provided by the following

<u>Theorem 3.</u> The X^{-1}-efficiency factor, $E_{X^{-1}}$, of a block design satisfies the inequality $0 < E_{X^{-1}} \leqslant 1$, with the equality

$$E_{X^{-1}} = 1 \qquad (5.3)$$

satisfied if the design is X^{-1}-orthogonal.

P r o o f: On account of Lemma 3, it is sufficient to prove that the inequality $[\underline{1}' \underline{X}\ \underline{1}/(nh)]\, tr(\underline{A}\ \underline{X}^{-1}) \leqslant 1$ holds for any X^{-1}-balanced block design, and that the equality (5.3) is attained if the design is X^{-1}-orthogonal. First consider an X^{-1}-balanced connected block design. It has

$$\underline{A} = \lambda \left[\underline{X} - \underline{X}\ \underline{1}\ \underline{1}'\underline{X}/(\underline{1}'\underline{X}\ \underline{1}) \right] \qquad (5.4)$$

with

$$\lambda = \frac{n - tr(\underline{N}\ \underline{k}^{-\delta}\underline{N}')}{\underline{1}'\underline{X}\ \underline{1} - \underline{1}'\underline{X}\ \underline{X}\ \underline{1}/(\underline{1}'\underline{X}\ \underline{1})}$$

(see [2], Corollary 4). The eigenvalue λ attains its maximum in the class of all X^{-1}-balanced connected block designs when $tr(\underline{N}\ \underline{k}^{-\delta}\underline{N}')$ is in its minimum. But $tr(\underline{N}\ \underline{k}^{-\delta}\underline{N}') = \sum_{i=1}^{v}\sum_{j=1}^{b} n_{ij}^2/k_j$, and $\sum_{j=1}^{b} n_{ij}^2/k_j \geq (\sum_{j=1}^{b} n_{ij})^2/\sum_{j=1}^{b} k_j$. This shows that the minimum is attained if $\sum_{j=1}^{b} n_{ij}^2/k_j = r_i^2/n$, i.e., in particular, if $\underline{N}\ \underline{k}^{-\delta}\underline{N}' = \underline{r}\ \underline{r}'/n$, the minimum then being equal to $\underline{r}'\underline{r}/n = \sum_{i=1}^{v} r_i^2/n$. Suppose that $\underline{N}\ \underline{k}^{-\delta}\underline{N}'$ is such, giving

$$\underline{A} = \underline{r}^{\delta} - \underline{r}\ \underline{r}'/n\ . \tag{5.5}$$

Comparing the off-diagonal elements of $\underline{A}$ in(5.5) with those of $\underline{A}$ in (5.4), it is found that $r_i/x_i = r_{i'}/x_{i'}$ for any $i \neq i'$ (= 1, 2, ..., v), i.e. $\underline{r} = a\ \underline{X}\ \underline{1}$, where a is a constant. But, since $\underline{1}'\underline{r} = n$, $a = n/(\underline{1}'\underline{X}\ \underline{1})$, thus giving

$$\underline{r}^{\delta} - \underline{r}\ \underline{r}'/n = (n/\underline{1}'\underline{X}\ \underline{1})\left[\underline{X} - \underline{X}\ \underline{1}\ \underline{1}'\underline{X}/(\underline{1}'\underline{X}\ \underline{1})\right],$$

or $\lambda = n/(\underline{1}'\underline{X}\ \underline{1})$. So it is shown that in the class of all X^{-1}-balanced connected block designs

$$\max_{\underline{N}} \frac{\underline{1}'\underline{X}\ \underline{1}}{n(v-1)}\ tr(\underline{A}\ \underline{X}^{-1}) = \max_{\underline{N}} \frac{\underline{1}'\underline{X}\ \underline{1}}{n}\ \lambda = 1. \tag{5.6}$$

Now consider any X^{-1}-balanced design, not necessarily connected. It has

$$\underline{A} = \lambda\left[\underline{X} - \sum_{i=h+1}^{v} \underline{X}\ \underline{w}_i\underline{w}_i'\underline{X}\right],$$

which can also be written

$$\underline{A} = \lambda\left[\underline{X} - \underline{X}\,\underline{1}\,\underline{1}'\underline{X}/(\underline{1}'\underline{X}\,\underline{1})\right] - \lambda \sum_{i=h+1}^{v-1} \underline{X}\,\underline{w}_i\underline{w}_i'\underline{X},$$

or $\underline{A} = \underline{A}_0 - \underline{A}_1$, with obvious definitions of $\underline{A}_0$ and $\underline{A}_1$. But for any $i \leqslant h$, the equality

$$\underline{w}_i'\,\underline{A}\,\underline{w}_i = \underline{w}_i'\,\underline{A}_0\,\underline{w}_i$$

holds, since $\underline{w}_i'\,\underline{A}_1\,\underline{w}_i = 0$ for these i's, thus showing that the unique non-zero eigenvalue of $\underline{A}$ with respect to $\underline{X}$ is the same as that of $\underline{A}_0$ with respect to $\underline{X}$. Hence the result (5.6) can be extended to the class of all X^{-1}-balanced block designs, for which it takes the form

$$\max_{\underline{N}} \frac{\underline{1}'\,\underline{X}\,\underline{1}}{nh}\,\mathrm{tr}(\underline{A}\,\underline{X}^{-1}) = \max_{\underline{N}} \frac{\underline{1}'\,\underline{X}\,\underline{1}}{n}\,\lambda = 1,$$

thus completing the proof.

6. Relation to the common definitions

Nothing has been said till now about the choice of the matrix $\underline{X}$. However, from the role played by the matrix, it seems natural to regard the diagonal elements of $\underline{X}$ as some weights assigned to the corresponding treatments. This attitude to x_i's will allow to reconcile the two main different approaches in defining efficiency of block designs.

If all treatments in an experiment are regarded as equally important, i.e. of equal weights, the identity matrix $\underline{I}$ is to be chosen for $\underline{X}$. Then an appropriate orthogonal block design will be an I-orthogonal design, viz. of the incidence matrix

$$\underline{N} = \underline{1}\,\underline{k}'/v. \tag{6.1}$$

A particular case of it is a randomized complete block design. Relative to (6.1), the I-efficiency factor of a block design for an estimable contrast $\underline{c}'\gamma = \sum_{i=1}^{h} l_i \; \underline{w}_i'\gamma$ is

$$E_I(\underline{c}'\underline{X}) = (v/n) \sum_{i=1}^{h} l_i^2/(\sum_{i=1}^{h} l_i^2/\lambda_i),$$

where $\lambda_1, \ldots, \lambda_h$ are the non-zero eigenvalues of the matrix $\underline{A}$ (with respect to I) of the design under consideration. Similarly, the (mean) I-efficiency factor of a block design is then

$$E_I = (v/n)h/ \sum_{i=1}^{h} \lambda_i^{-1} .$$

This way of defining efficiency of a block design, originated by Kempthorne [5], has been used by Kshirsagar [6], Atiqullah [1] and by Raghavarao [9].

On the contrary, if unequal weights are assigned to treatments, a diagonal matrix other than $\underline{I}$ is to be used for $\underline{X}$. Suppose that an experiment is designed in such a way that treatments are replicated proportionately to the chosen weights. Then $\underline{r}^{\delta}$ can be chosen for $\underline{X}$ and an appropriate orthogonal block design will be an $r^{-\delta}$-orthogonal design, viz. of the incidence matrix

$$\underline{N} = \underline{r}\,\underline{k}'/n. \tag{6.2}$$

Relative to (6.2), the $r^{-\delta}$-efficiency factor of a block design for an estimable contrast $\underline{c}'\underline{\gamma} = \sum_{i=1}^{h} l_i \; \underline{w}_i'\underline{r}^{\delta}\underline{\gamma}$ is

$$E_{r^{-\delta}}(\underline{c}'\underline{\gamma}) = \sum_{i=1}^{h} l_i^2/(\sum_{i=1}^{h} l_i^2/\lambda_i) , \tag{6.3}$$

where $\lambda_1, \ldots, \lambda_h$ are the non-zero eigenvalues of the matrix $\underline{A}$ of the design under consideration, with respect to $\underline{r}^{\delta}$. Similarly, the (mean) $r^{-\delta}$- efficiency factor of a block design is then

$$E_{r^{-\delta}} = h / \sum_{i=1}^{h} \lambda_i^{-1} .$$

This way of defining efficiency of a block design has been adopted by Pearce [7] and also used by Pearce et al. [8], Jarrett [4] and recently by Ceranka and Mejza [3].

It should become evident from the discussion above that the first approach is preferable in the case of equal weights, while the second in the case of weights being chosen proportionately to the treatment replications. In fact only in the latter case the $r^{-\delta}$-efficiency is fully justified. If the weights were some arbitrary positive numbers $x_1, \ldots, x_v$ not proportionate to $r_1, \ldots, r_v$, the diagonal matrix $\underline{X} = [x_i]$ would to be used instead of $\underline{r}^{\delta}$.

But there is one general advantage of the $r^{-\delta}$-efficiency over any other X^{-1}-efficiency. The $r^{-\delta}$-efficiency factor of a design for an individual contrast, as defined in (4.2), never exceeds 1, which may not be true for any other X^{-1}-efficiency.

References

[1] A t i q u l l a h, M., On a property of balanced designs. Biometrika, 48, 215-218 (1961).

[2] C a l i ń s k i, T., On the notion of balance in block designs. In: J. R. Barra et al., editors, Recent Developments in Statistics. North-Holland, Amsterdam, 365-374 (1977).

[3] C e r a n k a, B. and S. M e j z a, On the efficiency factor for a contrast of treatment parameters. Biom. J., 21, 99-102 (1979).

[4] J a r r e t t, R.G., Bounds for the efficiency factor of block designs. Biometrica, 64, 67-72 (1977).

[5] K e m p t h o r n e, O., The efficiency factor of an incomplete block design. Ann. Math. Statist., 27, 846-849 (1956).

[6] K s h i r s a g a r, A. M., A note on incomplete block designs. Ann. Math. Statist., 29, 907-910 (1958).

[7] P e a r c e, S. C., The efficiency of block designs in general. Biometrika, 57, 339-346 (1970).

[8] P e a r c e, S. C., T. C a l i ń s k i and T.F. de C. M a r s h a l l, The basic contrasts of an experimental design with special reference to the analysis of data. Biometrika, 61, 449-460 (1974).

[9] R a g h a v a r a o, D., Constructions and Combinatorial Problems in Design of Experiments. Wiley, New York (1971).

AN ASYMPTOTIC EXPANSION FOR DISTRIBUTIONS OF $C(\alpha)$ TEST STATISTICS

by

Dimitr M. Chibisov

Steklov Mathematical Institute, Moscow

1. Introduction

In the paper an asymptotic expansion (a.e.) for distribution functions (d.f's) of Neyman's $C(\alpha)$ test statistics to order $n^{-1/2}$ (with a remainder $O(n^{-1/2})$) is obtained under weaker conditions than previously known (Theorem 2.2). The proof is based on a special theorem giving an a.e. for the d.f. of a statistic admitting a stochastic expansion (Theorem 2.1).

The following notation will be used: let N be the set of natural numbers; R^p, $p \in N$, the space of row vectors like $\underline{x} = (x_1, \ldots, x_p)$; $\|\underline{x}\|^2 = \underline{x}\,\underline{x}'$, the prime denoting the transposition. Denote by B^p the Borel σ-field in R^p; $R = R^1$, $B = B^1$. We write τ for $n^{-1/2}$. The subscript i always runs over $\{1, \ldots, n\}$, the limits of summation from i=1 to i=n are omitted. For a set (event) A denote by 1_A its indicator-function, A^c the complement of A, and for $A \subset R^p$ let $\bar{A}$ be the closure of A. Let

$$S_r(x) = \left\{\underline{y} \in R^p : \|\underline{y} - \underline{x}\| \leqslant r\right\}, \quad S_r = S_r(\underline{0}).$$

Let $N(\mu, \sigma^2)$ (resp. $N(\underline{\mu}, \Sigma)$) denote the normal distribution on R (resp. R^p) with parameters μ, σ^2 (resp. $\underline{\mu}, \Sigma$). Denote by $\Phi(x)$ and $\phi(x)$, $x \in R$, the d.f. and the density of $N(0,1)$.

An a.e. for d.f.'s of C(α) test statistics to order τ was first obtained by the author, Chibisov [2]. Later a general a.e. was obtained by Chibisov [5], a summary of this result can be found in Chibisov [3].

This problem reduces (see [5]) to that of obtaining an a.e. for a random variable(r.v.)admitting a stochastic expansion, i.e. a r.v. Z_n of the following form. Let $(Y_{io}, \underline{Y}_i) = (Y_{io}, Y_{i1}, \ldots, Y_{ip})$ be n i.i.d. random vectors in R^{p+1},

$$S_n = \tau \sum Y_{io}, \quad \underline{T}_n = (T_{n1}, \ldots, T_{np}) = \tau \sum \underline{Y}_i; \tag{1.1}$$

let h_j, j=1,..., k, be polynomials of p+1 variables, and

$$Z_n = S_n + \sum_{j=1}^{k} \tau^j h_j(S_n, \underline{T}_n). \tag{1.2}$$

The result of Chibisov [5] is based on that of Chibisov [2] where an a.e. for the d.f. of Z_n to order τ^k is obtained under the following conditions:

(A) $$E\left|Y_{1l}\right|^{k+2} < \infty, \quad l = 0,1,\ldots,p;$$

(B) the distribution of $(S_n, \underline{T}_n)$ for n sufficiently large contains an absolutely continuous component.

These conditions were used to apply the result by Bikjalis [1] on an a.e. for $P\left\{(S_n, \underline{T}_n) \in A\right\}$ with a remainder term uniform in $A \in B^{p+1}$.

The condition (B) was reduced by Pfanzagl [17] to the Cramer's condition (C), i.e.

$$\lim_{a\to\infty} \sup \left[\left|f(s,\underline{t})\right|; \left\|s,\underline{t}\right\| \geq a\right] < 1, \tag{1.3}$$

$f(s,\underline{t})$ denoting the characteristic function (ch.f.) of $(Y_{10}, \underline{Y}_1)$.

Recently the author obtained a result [9,10] that weakens the condition (A). We present here its simple particular case applicable to C(α) test statistics (cf. (4.5), (4.6), (4.8) below). Let

$$Z_n = S_n + \tau h(\underline{T}_n) \qquad (1.4)$$

and for p = 2

$$h(\underline{T}_n) = c_1 T_{n1} T_{n2} + c_2 T_{n2}^2. \qquad (1.5)$$

Then the a.e. for the d.f. of Z_n to order τ given by Chibisov [4] with k=1 (and by Theorem 2.1 below) is valid under the following conditions:

(A′) $E Y_{1l} = 0$, $l=0,1,2$, $E|Y_{10}|^3 < \infty$, $E|Y_{1l}|^{r_l} < \infty$, $l=1,2$,

for some r_1, r_2 satisfying the inequalities:

$$2 \leqslant r_2 \leqslant 3, \quad 1/r_1 + 1/r_2 \leqslant 1;$$

(B′) The condition (C) (1.3).

In Theorem 2.1 of the present paper we consider Z_n of the form (1.4) with $h(\underline{x})$, $x \in R^p$, a continuous function increasing not faster than $C \|\underline{x}\|^2$ when $\|\underline{x}\|$ grows to infinity. An a. e. to order for the d. f. of Z_n is obtained under the following conditions:

(A″) $E Y_{1l} = 0$, $l = 0,1,\ldots, p$; $E|Y_{10}|^3 < \infty$

$E|Y_{1l}|^2 < \infty$, $l = 1,\ldots, p$.

(B″) The distribution of Y_{10} is non-lattice. (The exact formulation including a uniformity assertion is given in Section 2).

It is seen that the condition (A′) is weaker than (A″) but (B″) is weaker than (B′), i.e. these results don't cover each other.

The present result was obtained in 1974 and presented at the Meeting on Asymptotic Methods in Statistics in Oberwolfach (see [7]), and on the 3rd Conference on Mathematical Statistics in Wisla, 1975. It was not published then in a complete form because before the publication of the paper a new method was found that seemed to give a better result. However whereas the moment conditions in the result obtained (see [9,10]) are close to the best possible, the condition (C) on the joint distribution of $(Y_{01}, \underline{Y}_1)$ is not so. The present result is of interest as an example of the possibility of weakening this condition.

In Theorem 2.2 using Theorem 2.1 an a.e. to order τ for the d.f. of a $C(\alpha)$ test statistic is obtained under weaker conditions than in Chibisov [5]. The a.e. is obtained here only for the distribution under the null hypothesis. Though the distributions under alternatives can be treated in a similar way, in case of local alternatives (of order τ from H_0) that are of most interest better result can be obtained by the method of Eliseev [11].

2. Statement of the theorems

Let $(Y_{oi}, \underline{Y}_i) = (Y_{io}, Y_{i1}, \ldots, Y_{ip})$ be n i.i.d. random vectors in R^{p+1}, $p \in N$, with a distribution depending on a parameter $\theta \in \Theta$. Let a family of functions $h_\Theta : R^p \to R$, $\Theta \in \Theta$, be given. Define S_n, $\underline{T}_n$ by (1.1) and let

$$Z_{n,\Theta} = S_n + \tau h_\Theta(\underline{T}_n), \tag{2.1}$$

$$G_{n,\Theta}(x) = P_\Theta \left\{ Z_{n,\Theta} < x \right\}. \tag{2.2}$$

It will be assumed that $E_\theta Y_{1l} = 0$, $l=0,1,\ldots p$, $\Theta \in \Theta$. Put

$$\sigma_\theta^2 = E_\Theta Y_{10}^2, \quad \Sigma_\Theta = E_\theta (Y_{10}, \underline{Y}_1)'(Y_{10}, \underline{Y}_1), \quad \mu_{3\Theta} = E_\Theta Y_{10}^3. \tag{2.3}$$

Let $(S_\Theta, \underline{T}_\Theta)$ be a random vector in R^{p+1} distributed $N(0, \Sigma_\Theta)$.

Put

$$J_\Theta(x) = E\left[h_\Theta(\underline{T}_\Theta) \mid S_\Theta = x \right] \varphi(x/\sigma_\Theta)/\sigma_\Theta, \tag{2.4}$$

$$\Phi_{n,\theta}(x) = \Phi\left(\frac{x}{\sigma_\theta}\right) - \tau\left[\frac{\mu_{3\theta}}{6\sigma^3}\left(\frac{x^2}{\sigma_\theta^2} - 1\right)\varphi\left(\frac{x}{\sigma_\theta}\right) + J_\theta(x)\right]. \tag{2.5}$$

For a r.v. Y whose distribution depends on θ we shall say that $E_\theta Y$ converges uniformly in $\theta \in \Theta$ if

$$\lim_{a\to\infty} \sup_{\theta\in\Theta} E_\theta |Y| \, 1_{\{|Y|>a\}} = 0.$$

Theorem 2.1. Let the following conditions be satisfied:

(i) $E_\theta Y_{10} = 0$, $\theta \in \Theta$;

(ii) $E_\theta |Y_{10}|^3$ converges uniformly in $\theta \in \Theta$;

(iii) For $l=1,\dots,p$, $E_\theta Y_{1l} = 0$, $\theta\in\Theta$; EY_{1l}^2 converges uniformly in $\theta \in \Theta$;

(iv) $\inf\left[\lambda_\theta;\ \theta\in\Theta\right] > 0$ where λ_θ is the minimal characteristic root of Σ_θ;

(v) For any $0 < a_1 < a_2 < \infty$

$$\sup\left[|f_\theta(s)|;\ a_1 \leq |s| \leq a_2,\ \ \theta\in\Theta\right] < 1$$

with $f_\theta(s) = E_\theta \exp(isY_{10})$;

(vi) The family of functions $\{h_\theta,\ \theta\in\Theta\}$ is equicontinuous on any compact $K \subset R^p$;

(vii) There exists an $M > 0$ such that

$$|h_\theta(x)| \leq M(1 + \|x\|^2) \text{ for all } x \in R^p,\ \theta\in\Theta.$$

Then for $n \to \infty$

$$\sup\left[|G_{n,\theta}(x) - \Phi_{n,\theta}(x)|;\ x\in R,\ \theta\in\Theta\right] = o(\tau). \tag{2.6}$$

The conditions (ii), (iii), (iv) imply the following properties which will sometimes be used in the proof:

(ii') $\sup\left[\sigma_\theta^2;\ \theta\in\Theta\right] < \infty$.

(ii") $E_\theta Y_{10}^2$ converges uniformly in $\theta \in \Theta$.

(iii′) For $l=1,\dots,p$, $E_{\Theta}Y_{l1} = 0$, $\Theta\in\Theta$, and $\sup\left[E_{\Theta}Y_{l1}^2;\ \Theta\in\Theta\right]<\infty$.

(iv′) $\inf\left[\sigma_{\Theta}^2;\ \Theta\in\Theta\right]>0$.

Now we turn to $C(\alpha)$ tests. They were introduced by Neyman [16] where some statistical properties and a motivation for some restrictions (see condition (III) below) can be found; see also Chibisov [5] for some properties of $C(\alpha)$ tests obtained through a.e.'s and Pfanzagl and Wefelmeyer [18] for related general results. Here we consider only distributional questions concerning $C(\alpha)$ test statistics without discussing their statistical meaning.

Let $X_1,\dots,X_n$ be i.i.d. r.v.'s taking their values in a measurable space $(\mathcal{X},\mathcal{A})$ with a distribution depending on a parameter $\Theta\in\Theta$ where Θ is an open set in R^s, $s\in N$. Let a function $g:\mathcal{X}\times\Theta\to R$ and an estimate $\Theta_n:\mathcal{X}^n\to R^s$ be given (formally, in Theorem 2.2 Θ_n need not be an estimate, i.e. it can depend on Θ). A $C(\alpha)$ test statistic then is of the form $Z_n(\Theta_n)$, where

$$Z_n(\Theta) = \tau \sum g(X_i,\Theta). \tag{2.7}$$

The r.v. $Z_n(\Theta_n)$ is defined on the event $\{\Theta_n\in\Theta\}$; under the conditions of Theorem 2.2 $P_{\Theta}\{\Theta_n\notin\Theta\} = o(\tau)$.

State the conditions to be used in Theorem 2.2. Let K be a compact subset of Θ.

(I) The function g is $(\mathcal{A}\times\mathcal{B}^s)$-measurable and for each $x\in\mathcal{X}$ $g(x,\Theta)$ is twice continuously differentiable w.r.t. $\Theta_1,\dots,\Theta_s$ for all $\Theta\in\Theta$.

(II) There exist a function $\underline{q}=(q_1,\dots,q_s):\mathcal{X}\times\Theta\to R^s$ and random vectors $r_{n,\Theta}$ such that

$$n^{1/2}(\Theta_n-\Theta)=\tau\sum \underline{q}(x_i,\Theta)+r_{n,\Theta},\quad \Theta\in\Theta, \tag{2.8}$$

and for any $\varepsilon>0$

$$\sup\left[P_{\Theta}\{\|r_{n,\Theta}\|>\varepsilon\};\ \Theta\in K\right]=o(\tau). \tag{2.9}$$

Let

$$g_{\Theta_j}(x,\Theta) = \frac{\partial g(x,\Theta)}{\partial \Theta_j}, \quad \underline{g}^{(1)}(x,\Theta) = (g_{\Theta_1}(x,\Theta),\dots,g_{\Theta_s}(x,\Theta)), \tag{2.10}$$

$$g_{\Theta_j\Theta_k}(x,\Theta) = \frac{\partial^2 g(x,\Theta)}{\partial\Theta_j\,\partial\Theta_k}, \; j, k = 1,\dots,s, \tag{2.11}$$

$$\mu_{3\Theta} = E_\Theta g^3, \quad b_{jk,\Theta} = E_\Theta g_{\Theta_j\Theta_k}, \quad B_\Theta = (b_{jk,\Theta})_{j,k=1,\dots,s}, \tag{2.12}$$

where, say, $E_\Theta g^3 = E_\Theta g^3(X_1,\Theta)$ etc.

(III) $E_\Theta g = 0$, $E_\Theta g^2 = 1$, $E_\Theta \underline{g}^{(1)} = 0$, $E_\Theta \underline{q} = 0$ for all $\Theta \in \Theta$;

(IV) $E_\Theta |g|^3$, $E_\Theta \|\underline{g}^{(1)}\|^2$, $E_\Theta \|\underline{q}\|^2$, $E_\Theta |g_{\Theta_j\Theta_k}|^{3/2}$, j,k=1,..., s converge uniformly in $\Theta \in K$.

(V) There exist $\rho(\Theta)$ and $R(x,\Theta)$ such that $S_{\rho(\Theta)}(\Theta) \in \Theta$ for all $\Theta \in \Theta$, $\rho_K = \inf\,[\rho(\Theta);\, \Theta \in K] > 0$ and for $\Theta_0 \in \Theta$, j,k=1,...,s, $\Theta \in S_{\rho(\Theta_0)}(\Theta_0)$

$$|g_{\Theta_j\Theta_k}(x,\Theta) - g_{\Theta_j\Theta_k}(x,\Theta_0)| \le R(x,\Theta_0)\,\|\Theta - \Theta_0\| \tag{2.13}$$

and $E_\Theta R^{6/5}$ converges uniformly in $\Theta \in K$.

Denote by Σ_Θ the covariance matrix of the random vector $(g(X_1,\Theta), \underline{g}^{(1)}(X_1,\Theta), \underline{q}(X_1,\Theta))$ with respect to P_Θ. According to the partition of this vector into the subvectors of 1,s,s variables, partition Σ_Θ,

$$\Sigma_\Theta = \begin{pmatrix} \sigma^2 & \Sigma_{01} & \Sigma_{02} \\ \Sigma_{10} & \Sigma_{11} & \Sigma_{12} \\ \Sigma_{20} & \Sigma_{21} & \Sigma_{22} \end{pmatrix} \tag{2.14}$$

(we suppressed the subscript Θ at σ^2, Σ_{01}, etc.).

(VI) $\inf\,[\lambda_\Theta;\, \Theta \in K] > 0$ where λ_Θ is the minimal characteristic root of Σ_Θ.

(VII) For any $0 < a_1 < a_2 < \infty$

$$\sup \left[|f_\Theta(s)| ;\ a_1 \leqslant |s| \leqslant a_2,\ \Theta \in K \right] < 1,$$

where $f_\Theta(s) = E_\Theta \exp(isg)$.

Theorem 2.2. Let the conditions (I) to (VII) be satisfied with some compact $K \subset \Theta$. Then

$$\sup_{\Theta \in K} \sup_{x \in R} \left| P_\Theta \left\{ Z_n(\Theta_n) < x \right\} - \Phi_{n,\Theta}(x) \right| = O(\tau), \tag{2.15}$$

where

$$\Phi_{n,\Theta}(x) = \Phi(x) - \tau \left[\frac{\mu_{3\Theta}}{6} (x^2 - 1) \varphi(x) + J_{0,\Theta}(x) \right], \tag{2.16}$$

$$J_{0,\Theta}(x) = \left[tr \left(\Sigma_{12} + \frac{1}{2} B \Sigma_{22} \right) + \right. \tag{2.17}$$

$$\left. + \left(\Sigma_{01} \Sigma_{20} + \frac{1}{2} \Sigma_{02}\ B \Sigma_{20} \right) (x^2 - 1) \right] \varphi(x).$$

3. Proof of Theorem 2.1

Without loss of generality we assume that $\sigma_\Theta = 1$ for all $\Theta \in \Theta$. To simplify the notation we shall usually suppress the subscript Θ at P, h, G_n etc.

Let ξ be a r.v. independent of $\{(Y_{oi}, \underline{Y}_i)\}$ having a d,f. $V(x)$, a density $v(x)$ and a ch.f. $\omega(t)$ such that

$$v(x) = v(-x),\ \alpha^2 = E\xi^2 < \infty,\ \omega(t) = 0 \text{ for } |t| > 1. \tag{3.1}$$

One can obtain such distribution taking $\omega(t)$ proportional to the density of the 4 fold convolution of the uniform distribution on (-1/4, 1/4). Then

$$v(x) = 24(1-\cos(t/2))^2 / \pi t^4,\ \alpha^2 = 12.$$

However in what follows we use only the properties (3.1).

For $a > 0$, let

$$S_{n,a} = S_n + \tau\xi/a, \quad h_a(\underline{y},x) = h(y) + x/a, \tag{3.2}$$

$$Z_{n,a} = Z_n + 2\tau\xi/a = S_{n,a} + \tau h_a(\underline{T}_n, \xi). \tag{3.3}$$

Denote the d.f., the density and the ch.f. of $S_{n,a}$ by $F_{n,a}(x)$, $p_{n,a}(x)$ and $f_{n,a}(x)$ respectively; let

$$G_{n,a}(x) = P\{Z_{n,a} < x\}, \tag{3.4}$$

$$\eta_n = \sup_{x\in R} |G_n(x) - \Phi_n(x)|, \quad \eta_{n,a} = \sup_{x\in R} |G_{n,a} - \Phi_n(x)|. \tag{3.5}$$

Lemma 3.1. For any $\delta > 0$ there exists an $a = a(\delta)$ such that

$$\eta_n \leq 2\eta_{n,a} + \tau\delta.$$

The proofs of this and the other lemmas of this section will be given in sections 4 to 7.

We shall prove now that for any $a > 0$

$$\sup[\eta_{n,a};\ \Theta\in\Theta] = o(\tau). \tag{3.6}$$

By Lemma 3.1, this will imply the theorem.

It is easily seen that for any $x \in R$, $\gamma > 0$ and r.v.'s U,V the following inequality holds

$$|P(U+V<x) - [P(U+V<x,\ x-\gamma\leq U\leq x+\gamma) + P(U<x-\gamma)]| \leq P(|V|>\gamma). \tag{3.7}$$

Fix an arbitrary $\gamma > 0$ and let

$$\tilde{G}_{n,a}(x) = P\{Z_{n,a} < x,\ x-\gamma \leq S_{n,a} \leq x+\gamma\}. \tag{3.8}$$

Using (3.7) we have

$$|G_{n,a}(x) - [\tilde{G}_{n,a}(x) + F_{n,a}(x-\gamma)]| \leq P(|\tau h_a(S_n, \xi)| > \gamma). \tag{3.9}$$

It follows from the condition (ii) that as $x\to\infty$

$$\sup_{\Theta\in\Theta} P\{|T_{ln}|>x\}=o(x^{-2}), \qquad l=1,\ldots,p \tag{3.10}$$

(see Corollary 7.1). Using the condition (vii), (3.10) and the Chebyshev inequality we obtain

$$\sup_{\Theta\in\Theta} P(\tau|h_a(S_n,\xi)|>\gamma)\leq \sup_{\Theta\in\Theta} P(|h(S_n)|>\gamma n^{1/2}/2)+$$

$$+P(|\xi|>a\gamma n^{1/2}/2) = o(\tau). \tag{3.11}$$

The function $\tilde{G}_{n,a}(x)$ can be written as.

$$\tilde{G}_{n,a}(x)=\int_{x-\gamma}^{x+\gamma} P\{h_a(S_n,\xi)<n^{1/2}(x-u)|S_{n,a}=u\}p_{n,a}(u)du=$$

$$=\tau\int_{|z|\leq\gamma n^{1/2}} P\{h_a(S_n,\xi)<z|S_{n,a}=x-\tau z\}p_{n,a}(x-\tau z)dz. \tag{3.12}$$

Let

$$r_{n,a}(u,z)=[P\{h_a(T_n,\xi)<z|S_{n,a}=u\}-1_{(0,\infty)}(z)]p_{n,a}(u), \tag{3.13}$$

$$Q_{n,a}(x)=\int_{|z|\leq\gamma n^{1/2}} r_{n,a}(x-\tau z,z)\,dz. \tag{3.14}$$

Then (3.9), (3.11), (3.12) imply

$$\sup_{\Theta}|G_{n,a}(x)-[F_{n,a}(x)+\tau Q_{n,a}(x)]|=o(\tau). \tag{3.15}$$

Let

$$r_a(u,z)=[P\{h_a(T,\xi)<z|S=u\}-1_{(0,\infty)}(z)]\varphi(u), \tag{3.16}$$

$$\Psi_{n,a}(x)=\int_{|z|\leq\gamma n^{1/2}} r_a(x-\tau z,z)\,dz, \tag{3.17}$$

$$\Phi(x,\tau)=\Phi(x)-\tau(\mu_3/6)(x^2-1)\varphi(x). \tag{3.18}$$

We shall prove the following relations: for any $a>0$ as $n\to\infty$, uniformly in $x\in R$, $\Theta\in\Theta$

$$F_{n,a}(x) - \Phi(x,\tau) = o(\tau), \tag{3.19}$$

$$Q_{n,a}(x) - \Psi_{n,a}(x) \to 0, \tag{3.20}$$

$$\Psi_{n,a}(x) \to -J(x). \tag{3.21}$$

Then (3.15), (3.19) to (3.21) imply (3.6) and hence the theorem.

The relation (3.19) with F_n instead of $F_{n,a}$ is wellknown (see, e.g., Feller (1966), 16.4). It implies that $F_{n,a}(x) - \Phi_a(x,\tau) = o(\tau)$, where $\Phi_a(\cdot,\tau) = \Phi(\cdot,\tau) * V_{2\tau/a}(\cdot)$. Similarly to (4.5) (see below), $\Phi_a(x,\tau) - \Phi(x,\tau) = O(\tau^2)$ which implies (3.19).

Proof of (3.20). Let

$$\rho_{n,a} = \sup\left[\left|r_{n,a}(u,z) - r_a(u,z)\right|;\ \Theta\in\Theta,\ (u,z)\in R^2\right].$$

Lemma 3.2. Let the conditions (i), (ii″), (iii) to (vi) be satisfied. Then for any $a>0$ $\rho_{n,a}\to 0$ as $n\to\infty$.

Here and below in similar cases the assertion is that there exist versions of conditional probabilities having the stated property.

Let $b_n = \min\left(\rho_{n,a}^{-1/2},\ \gamma n^{1/2}\right)$. Split each of the functions $Q_{n,a}$ and $\Psi_{n,a}$ into three summands corresponding to integration in (3.14) and (3.17) over $-\gamma n^{1/2}\le z<-b_n$, $|z|\le b_n$ and $b_n<z\le\gamma n^{1/2}$ and denote them by superscripts (1), (2) and (3) respectively. Lemma 3.2 implies immediately that for all x, Θ

$$\left|Q_{n,a}^{(2)}(x) - \Psi_{n,a}^{(2)}(x)\right| \le 2\rho_{n,a}^{1/2} \to 0. \tag{3.22}$$

Now each of the functions $Q_{n,a}^{(j)}$, $\Psi_{n,a}^{(j)}$, $j=1,3$, will be estimated separately. Consider $\Psi_{n,a}^{(3)}(x)$ (see (3.16), (3.17)). Using the condition (vii) we have for $z>0$

$$P\left\{h_a(\underline{T},\xi)\geq z \mid S=u\right\}\leq P\left\{|\xi|\geq a\,z/2\right\}+$$

$$+\sum_{l=1}^{p} P\left\{T_l^2\geq (z-2M)/2Mp \mid S=u\right\}. \tag{3.23}$$

Since $b_n\to\infty$ one can find $0<C<1/2Mp$ and $n_1\in N$ such that $(z-2M)/2Mp > cz$ for $n\geq n_1$, $z\geq b_n$.

Then

$$\Psi^{(3)}_{n,a}(x)\leq \frac{1}{2\pi}\int_{b_n}^{\infty} P\left\{|\xi|>az/2\right\}dz+\sum_{l=1}^{p}\Psi_{nl}(x), \tag{3.24}$$

$$\Psi_{nl}(x)=\int_{b_n}^{\gamma n^{1/2}} P\left\{T_l^2>cz \,\middle|\, S=x-\tau z\right\}\varphi(x-\tau z)\,dz. \tag{3.25}$$

It follows from $b_n\to\infty$ and Chebyshev inequality that the integral in (3.24) tends to zero as $n\to\infty$.

Consider $\Psi_{n,l}$ for an arbitrary $l=1,\dots,p$. Note that for $|x|\geq\gamma$ and $|z|\leq\gamma n^{1/2}$ holds

$$\varphi(x-\tau z)\leq\varphi(|x|-\gamma). \tag{3.26}$$

By well known formulas,

$$E\left[T_l \mid S=u\right]=Au,\ \operatorname{var}\left[T_l \mid S=u\right]=B^2$$

with $A=\sigma_{ol}/\sigma^2$, $B^2=(1-\rho^2)\,\sigma_l^2$ where $\sigma=\sigma_\Theta$ (see (2.3)), $\sigma_l^2 = EY_{1l}^2$, $\sigma_{ol}=EY_{10}Y_{1l}$, $\rho=\sigma_{ol}/\sigma\sigma_l$.

By the conditions (ii′), (iii) and (iv), A and B are bounded uniformly in $\Theta\in\Theta$. We shall show that

$$\int_0^{\gamma n^{1/2}} P\left\{T_l^2>cz \,\middle|\, S=x-\tau z\right\}dz\leq 2c^{-1}\left(A^2x^2+B^2+A^2\gamma^2\right) \tag{3.27}$$

and for any $b>0$ uniformly in $|x|\leq b$, $\Theta\in\Theta$

$$\int_{b_n}^{\gamma n^{1/2}} P\left\{T_l^2>cz \mid S=x-\tau z\right\}dz\to 0 \quad \text{as} \quad n\to\infty. \tag{3.28}$$

Then (3.26), (3.27) will imply that for any $\varepsilon > 0$ there exists $b > 0$ such that $\sup\left[\Psi_{nl}(x);\ |x| > b\right] < \varepsilon$ and, by (3.28), $\sup\left[\Psi_{nl}(x);\ |x| \leq b\right] \to 0$ as $n \to \infty$, i.e. $\Psi_{nl}(x) \to 0$ uniformly in $x \in R$, $\Theta \in \Theta$. By (3.24), then

$$\Psi^{(3)}_{n,a}(x) \to 0 \tag{3.29}$$

as $n \to \infty$ uniformly in $x \in R$, $\Theta \in \Theta$.

For the proof of (3.27), (3.28), let X be a r.v. distributed $N(0,B^2)$. Then for $|z| \leq \gamma n^{1/2}$

$$P\left\{T_1^2 > cz \mid S = x - \tau z\right\} = P\left\{(Ax - A\tau z + X)^2 > cz\right\} \leq$$

$$\leq P\left\{(|Ax + X| + A\gamma)^2/c > z\right\}. \tag{3.30}$$

This implies that the left side of (3.27) is not greater than

$$E\left[(|Ax + X| + A\gamma)^2/c\right] \leq 2\ c^{-1}\ E\left[(Ax + X)^2 + A^2\gamma^2\right]$$

which gives (3.27). Moreover (3.28) follows from $b_n \to \infty$ and (3.30).

It is proved similarly that

$$\Psi^{-(1)}_{n,a}(x) \to 0 \tag{3.31}$$

as $n \to \infty$ uniformly in $x \in R$, $\Theta \in \Theta$.

Consider now $Q^{(3)}_{n,a}(x)$. In a similar way to (3.24) we obtain that we have to prove for any $l=1,\ldots,p$ the uniform in x, Θ convergence

$$\lim_{n\to\infty} \int_{b_n}^{\gamma n^{1/2}} P\left\{T_{nl}^2 > cz \mid S_{n,a} = x - \tau z\right\} p_{n,a}(x - \tau z)\,dz = 0. \tag{3.32}$$

Lemma 3.3. Let (Y_i, Z_i), $i=1,\ldots,n$, be i.i.d. random vectors, $S_n = \tau \sum Y_i$, $\Sigma_n = \sum Z_i$ and ξ, $S_{n,a}$ and $p_{n,a}(x)$ are defined as before. Assume that $\nu_r = E|Z_1|^r < \infty$ for some $0 < r \leq 2$. Then for any $v>0, \lambda>0$, $z > (\lambda^{r-1} n\ \nu_r)^{-r}$

$$\sup_{u\in R} P\left\{|\Sigma_n| > z \mid S_{n,a} = u\right\} p_{n,a}(u) \leq$$

$$\leq 2n\, P\left\{Z_1 > z/\lambda\right\}(n/(n-1))^{1/2} \sup_{u\in R} p_{n-1,a}(u) +$$

$$+\left(\frac{\lambda^{r-1} n \nu_r}{z^r}\right)^{\lambda} \frac{e^{\lambda}}{\pi}\left[\left(\frac{6\pi}{b(V) - 2hV^2\nu}\right)^{1/2}\left(1+\frac{\lambda}{n}\right)+\right.$$

$$\left. + 2an^{1/2}(\eta(V) + \lambda/n + h\nu)^n \int \omega(t)\,dt\right], \tag{3.33}$$

where $\nu = 0$ if $0 < r \leq 1$ and $\nu = E|Z_1|$ if $1 < r \leq 2$,

$$b(V) = \int_{|u|\leq V} u^2 d\, G_2(u), \quad G_2(u) = P\left\{Y_1 - Y_2 < u\right\}, \tag{3.34}$$

$$\eta(V) = \sup\left[|f(t)|;\ 1/V \leq |t| \leq a\right], \quad f(t) = E\exp(itY_1), \tag{3.35}$$

$$h = \lambda z^{-1} \log\left(z^r/\lambda^{r-1} n\, \nu_r\right). \tag{3.36}$$

Apply this lemma with $Y_i = Y_{i0}$, $Z_i = Y_{i1}$, $\Sigma_n = n^{1/2}\, T_{n1}$, $r = 2$, $x = (cnz)^{1/2}$ and an arbitrary $\lambda > 1$. Then $\sup\left[h;\ z \geq b_n,\ \Theta\in\Theta\right] \to 0$. By conditions (ii″) and (iv) one can choose $V > 0$ such that $\inf\left[b(V);\ \Theta\in\Theta\right] > 0$. By condition (v), there exists $n_1 \in N$ such that $\sup\left[\eta(V) + \lambda/n + h\nu;\ \Theta\in\Theta,\ n \geq n_1\right] \leq \eta_1 < 1$. Moreover, $\sup_u p_{n-1,a}(u) \to 1/2\pi$ (see (5.2) below). Therefore (3.33) implies that there exist $n_2 \in N$ and positive C_1, C_2, C_3 such that for all $\Theta \in \Theta$ and $n \geq n_2$

$$\sup_{u\in R} P\left\{|T_{n1}| > (cz)^{1/2} \mid S_{n,a} = u\right\} p_{n,a}(u) \leq$$

$$\leq C_1 n\, P\left\{Y_{11} > (cnz)^{1/2}/\lambda\right\} + C_2 z^{-\lambda} + C_3\, \eta_1^n\, n^{1/2}. \tag{3.37}$$

Apply (3.37) to estimate the integral in (3.32). Setting $y = cnz/\lambda^2$ we have

$$\int_{b_n}^{\gamma n^{1/2}} n\, P\left\{Y_{11} > (cnz)^{1/2}/\lambda\right\} dz \leq \lambda^2 C^{-1} \int_{nb_n c/\lambda^2}^{\infty} P\left\{Y_{11}^2 > y\right\} dy \tag{3.38}$$

and the right side of (3.38) tends to zero by the condition (iii). The convergence to zero of the integrals of $Z^{-\lambda}$, $\lambda>1$, and $\eta_1^n n^{1/2}$ is obvious. Thus we obtain that

$$Q_{n,a}^{(3)}(x) \to 0 \quad \text{as } n\to\infty \text{ uniformly in } x\in R,\ \theta\in\Theta. \tag{3.39}$$

It is proved similarly that $Q_{n,a}^{(1)}(x) \to 0$. Then (3.22), (3.29), (3.31), (3.39) imply (3.20).

Proof of (3.21). Note that

$$-\ J(x) = \int r_a(x,z)\,dz. \tag{3.40}$$

This follows from (2.4) and (3.16) by partial integration, see Feller [12], 5.6. To estimate $r_a(x,z)$ apply (3.23) and then (3.30) with $\tau z = \gamma = 0$. Then we get

$$\int_{|z|>\gamma n^{1/2}} r_a(x,z)\,dz \to 0 \quad \text{uniformly in } x,\ \Theta.$$

Thus (3.21) is equivalent to

$$\int_{|z|\le\gamma n^{1/2}} \left[r_a(x-\tau z,\ z) - r_a(x,z)\right]dz \to Q \tag{3.41}$$

uniformly in $x\in R$, $\Theta\in\Theta$. Writing r_a' for $(\partial/\partial x)r_a$ we have

$$\left| r_a(x-\tau z,z) - r_a(x,z)\right| \le \tau|z| \sup_x \left| r_a'(x,z)\right|. \tag{3.42}$$

Let for definiteness $z>0$. Then

$$r_a(x,z) = \int_{\{(\underline{u},y):h_a(\underline{u},y)\ge z\}} \varphi(x,\underline{u})\,\nu(y)\,du\,dy, \tag{3.43}$$

where $\varphi(x,\underline{u})$ is the density of $N(0,\Sigma)$. Write $\varphi(x,\underline{u}) = \varphi(x|\underline{u})\,\varphi(\underline{u})$ where the factors are the corresponding conditional and marginal densities. We have

$$\sup_x \left|(\partial/\partial x)\,\varphi(x|\underline{u})\right| = \sigma_0^{-2} \sup_x \left|x\,\varphi(x)\right| \tag{3.44}$$

where σ_0^2 is the conditional variance of S given $\underline{T}$, $\inf\left[\sigma_0^2;\Theta\in\Theta\right]>0$ by the condition (iv). As in the proof of Lemma 4.1 (see below) one can prove that (3.43) can be differentiated under the integral sign, and by (3.44) there exists a constant $C>0$ independent of $\Theta\in\Theta$ such that

$$|r_a'(x,z)|\leq C\int_{\{(\underline{u},y):h_a(u,y)\geq z\}}\varphi(\underline{u})\,\nu(y)\,d\underline{u}dy =$$

$$= C\,P\left\{h_a(\underline{T},\xi)\geq z\right\}. \tag{3.45}$$

Using (3.42), (3.45) and a similar estimate for $z<0$ we obtain on integrating by parts that the left side of (3.41) is not greater than

$$\frac{1}{2}\,C\tau E\left[h_a^2(\underline{T},\xi)\right]. \tag{3.46}$$

The expectation in (3.46) is bounded uniformly in Θ which proves (3.41) and hence (3.21) thus completing the proof of the theorem.

4. Proof of Theorem 2.2 will be given for $s = 1$; there are no difficulties in extending it to $s>1$. In case $s = 1$ we shall write $g^{(1)} = \partial g/\partial\Theta$, $g^{(2)} = \partial^2 g/\partial\Theta^2$, $b_\Theta^{(2)} = E_\Theta\, g^{(2)}$, instead of (2.10), (2.11) and (2.12). By the Taylor formula

$$Z_n(\Theta_n) = \tau\sum g(X_i,\Theta_n) = \tau\sum g(X_i,\Theta) + (\Theta_n-\Theta)\tau\sum g^{(1)}(X_i,\Theta) +$$

$$+\frac{1}{2}(\Theta_n-\Theta)^2\tau\sum g^{(2)}(X_i,\Theta) + \frac{1}{2}(\Theta_n-\Theta)^2\cdot\tau\sum\left[g^{(2)}(X_i,\Theta_{ni}^*)-g^{(2)}(X_i,\Theta)\right] \tag{4.1}$$

where $|\Theta_{ni}^* - \Theta|\leq|\Theta_n - \Theta|$. Let

$$Y_{io} = g(X_i,\Theta),\quad Y_{i,1} = g^{(1)}(X_i,\Theta),\quad Y_{i,2} = q(X_i,\Theta),$$

$$S_n = \tau\sum Y_{io},\quad T_{n,j} = \tau\sum Y_{i,j},\qquad j = 1,2, \tag{4.2}$$

$$S_n^{(2)} = \tau\sum\left[g^{(2)}(X_i,\Theta) - b_\Theta^{(2)}\right]. \tag{4.3}$$

For an arbitrary $\varepsilon > 0$ introduce events

$$E_{n,\Theta}=\left\{|S_n^{(2)}|\leq \varepsilon n^{1/2}\right\}\cap\left\{|r_{n,\Theta}|\leq \varepsilon\right\}\cap$$
$$\cap\left\{|\Theta_n-\Theta|\leq n^{-1/4}\right\}\cap\left\{\sum R(x_i,\Theta)\leq \varepsilon n^{5/4}\right\}.$$

The conditions (II), (IV), (V) and Corollary 7.1 imply that

$$\sup\left[P_\Theta(E_{n,\Theta}^c);\ \Theta\in K\right]=o(\tau). \tag{4.4}$$

By (4.1–4.3) we have on $E_{n,\Theta}$

$$Z_n^{**}\leq Z_n(\Theta_n)\leq Z_n^* \tag{4.5}$$

where

$$Z_n^*=S_n+\tau\Big[T_{n1}T_{n2}+\varepsilon|T_{n1}|+\frac{1}{2}T_{n2}^2(\varepsilon+b_\Theta^{(2)})+$$
$$+\left(\varepsilon|T_{n2}|+\frac{\varepsilon^2}{2}\right)\left(\varepsilon+|b_\Theta^{(2)}|\right)+\frac{1}{2}\left(|T_{n2}|+\varepsilon\right)^2\varepsilon\Big] \tag{4.6}$$

and Z_n^{**} has a similar form with ε replaced by $-\varepsilon$ at proper places. Let $D_x=\left\{Z_n(\Theta_n)<x\right\}$, $D_x^*=\left\{Z_n^*<x\right\}$. Then $D_x^*\cap E_{n,\Theta}\subset D_x\cap E_{n,\Theta}\subset D_x$ hence

$$P_\Theta\left\{Z_n(\Theta_n)<x\right\}\geq P_\Theta\left\{Z_n^*<x\right\}-P_\Theta(E_{n,\Theta}^c). \tag{4.7}$$

Now Z_n^* has the form of Z_n in (2.2) with p = 2, K as Θ and $h_{0,\Theta}(\underline{T}_n)+h_{\varepsilon,\Theta}(\underline{T}_n)$ as $h_\Theta(\underline{T}_n)$ where

$$h_{0,\Theta}(\underline{T}_n)=T_{n1}T_{n2}+\frac{1}{2}b_\Theta^{(2)}T_{n2}^2 \tag{4.8}$$

and $h_{\varepsilon,\Theta}$ contains all remaining terms in brackets in (4.6). Applying Theorem 2.1 to Z_n^* we obtain

$$\sup_{\Theta\in K}\left|P_\Theta\left\{Z_n^*<x\right\}-\Phi_{n,\Theta}(x,\varepsilon)\right|=o(\tau), \tag{4.9}$$

where $\Phi_{n,\Theta}(x,\varepsilon)$ is given by (2.5) with $\sigma=1$, $\mu_{3\Theta}$ given by (2.12) and $J_\Theta(x)=J_{0,\Theta}(x)+J_{\varepsilon,\Theta}(x)$ with

$$J_{\alpha,\Theta}(x) = E\left[h_{\alpha,\Theta}(\underline{T}) \mid S = x\right]\varphi(x), \quad \alpha = 0, \varepsilon,$$

$(S,\underline{T}) = (S, T_1, T_2)$ being a random vector distributed $N(\underline{0}, \Sigma_\Theta)$. By well known formulas for the moments of a conditional normal distribution, one establishes that $J_{0,\Theta}(x)$ (and its analogue in case $s > 1$) is given by (2.17) and $J_{\varepsilon,\Theta}(x) \to 0$ as $\varepsilon \to 0$ uniformly in $x \in R, \Theta \in K$. Since $\varepsilon > 0$ is arbitrary, (4.4), (4.7) and (4.9) imply that

$$\sup_{\Theta \in K} \sup_{x \in R} \left[P_\Theta\left\{Z_n(\Theta_n) < x\right\} - \Phi_{n,\Theta}(x)\right] \geq o(\tau), \tag{4.10}$$

where $\Phi_{n,\Theta}$ is given by (2.16). Now starting from the inequality $Z_n^{**} \leq Z_n(\Theta_n)$ we arrive in a similar way at an inequality which differs from (4.10) by reversing the inequality sign. Together with (4.10) this gives (2.15). Q.E.D.

5. Proof of Lemma 3.1

We first prove the following lemma. Let $\varphi(x,\underline{u})$ be the density of $N(\underline{0}, \Sigma_\Theta)$. Then $J_\Theta(x)$ (see (2.4)) can be written as

$$J_\Theta(x) = \int_{R^p} h_\Theta(\underline{u})\,\varphi(x,\underline{u})\,d\underline{u}. \tag{5.1}$$

Lemma 5.1. Under the conditions (i), (ii′), (iii′), (iv), (vii), the integral in (5.1) can be differentiated w.r.t. x under the integral sign and

$$\sup\left[\left|J_\Theta'(x)\right|;\ \Theta \in \Theta,\ x \in R\right] < \infty. \tag{5.2}$$

P r o o f. Using its explicite form, $\varphi(x,\underline{u})$ can be represented as

$$\varphi(x,\underline{u}) = \varphi_1(x)\,\varphi_2(\underline{u})\,\exp\left[x\,T(\underline{u})\right]$$

with φ_1 a density of a normal distribution and T a linear function. Then $J_\Theta(x) = \varphi_1(x)\left[c^+(x) - c^-(x)\right]$ where

$$c^{\pm}(x) = \int_{R^p} \exp\left[x\,T(\underline{u})\right] d\mu^{\pm}(\underline{u}), \quad d\mu^{\pm}(\underline{u}) = h_{\Theta}^{\pm}(\underline{u})\varphi_2(\underline{u})\,d\underline{u}.$$

Since

$$E_{\Theta}\left|h_{\Theta}(\underline{T})\right| = \int \varphi_1(x)\left[c^{+}(x) + c^{-}(x)\right]dx < \infty$$

the functions $c^{\pm}(x)$ are finite for almost all $x \in R$ and hence for all $x \in R$ because of convexity of a natural parametric space of an exponential family ([14] Ch.2, Lemma 7). The possibility of differentiation follows now from Theorem 9 of Ch.2 of the cited book. After differentiation of (5.1), the integrand multiplies by a linear function of $(x,\underline{u})$ whose coeffitients are elements of Σ_{Θ}^{-1}. They are bounded uniformly in $\Theta \in \Theta$ because the minors of Σ_{Θ} are bounded from above by (i), (ii′) and (iii′) and det Σ_{Θ} is bounded away from zero by (iv). Together with (vii) this proves (4.2).

Now we proceed to the proof of Lemma 3.1. Denote by V_b and v_b the d.f. and the density of $b\xi$ where ξ was introduced in Section 3, see (3.1). We shall use the following estimates: $E|\xi| \leqslant (E\xi^2)^{1/2} = \alpha$ and for $u > 0$

$$\int_u^{\infty} v_b(x)\,dx \leqslant \frac{bE|\xi|}{2u} \leqslant \frac{b\alpha}{2u}. \tag{5.3}$$

One has $G_{n,a} = G_n * V_{2\tau/a}$ (see (3.3)); let $\Phi_{n,a} = \Phi_n * V_{2\tau/a}$ and

$$\eta_n^a = \sup_x \left|G_{n,a}(x) - \Phi_{n,a}(x)\right|. \tag{5.4}$$

Let $\Psi(x) = (\mu_3/6)(x^2-1)\varphi(x) + J(x)$ so that $\Phi_n(x) = \Phi(x) + \tau\,\Psi(x)$ (see (2.5)). The conditions (ii), (iii′), (vii) and Lemma 5.1 imply that $\lim_{x \to \pm\infty} \Psi(x) = 0$ and

$$C_1 = \sup\left[\left|\Psi'(x)\right|;\ x \in R,\ \Theta \in \Theta\right] < \infty. \tag{5.5}$$

This implies, in turn, that

$$C_2 = \sup\left[|\Phi_n'(x)|;\ x \in R,\ n \in N,\ \Theta \in \Theta\right] < \infty.$$

By the smoothing lemma, Loeve (1960), 20.3b, using (5.3) we obtain

$$\eta_n^a \geq \frac{\eta_n}{2} - \frac{3\alpha C_2 \tau}{a}. \tag{5.6}$$

We shall show that for any $a > 0$

$$\sup\left[|\Phi_{n,a}(x) - \Phi_n(x)|;\ x \in R,\ \Theta \in \Theta\right] = O(\tau^2). \tag{5.7}$$

Then the assertion of Lemma 3.1 will follow from (5.6) and (5.7). We have

$$\Phi_{n,a}(x) = \int\left[\Phi(x-y) + \tau\Psi(x-y)\right] v_{2\tau/a}(y)\,dy, \tag{5.8}$$

$$\begin{aligned}\int \Phi(x-y)\, v_{2\tau/a}(y)\,dy &= \Phi(x) - \varphi(x)\int y\, v_{2\tau/a}(y)\,dy + \\ &+ \frac{1}{2}\int y^2 \varphi'(\tilde{x})\, v_{2\tau/a}(y)\,dy\end{aligned} \tag{5.9}$$

where $\tilde{x}$ lies between x and $x-y$. The first integral in the right side of (5.9) is 0, the second one is not greater than

$$\sup |\varphi'(x)|\alpha^2 (2\tau/a)^2 = O(\tau^2).$$

By (5.5),

$$\left|\int \psi(x-y)\, v_{2\tau/a}(y)\,dy - \psi(x)\right| \leq C_1 \int |y|\, v_{2\tau/a}(y)\,dy = O(\tau).$$

Putting these estimates into (5.8) we obtain (5.7). Q.E.D.

6. Proof of Lemma 3.2

We shall show that uniformly in u, z, Θ

$$\begin{aligned}&P\left\{h_a(\underline{T}_n, \xi) < z \mid S_{n,a} = u\right\} p_{n,a}(u) - \\ &- P\left\{h_a(\underline{T}, \xi) < Z \mid S = u\right\} \varphi(u) \to 0\end{aligned} \tag{6.1}$$

as $n \to \infty$. Letting $Z \to \infty$ in (6.1) we obtain that uniformly in u, Θ

$$p_{n,a}(u) \to \varphi(u). \tag{6.2}$$

Then Lemma 3.2 (see (3.13), (3.16)) will follow from (6.1), (6.2). (Of course (6.2) can be proved in a more direct way).

In connection with (6.1) we shall consider

$$F_{n,u}(A) = P\left\{(\underline{T}_n, \xi) \in A \,\middle|\, S_{n,a} = u\right\} p_{n,a}(u), \quad A \in B^{p+1}, \tag{6.3}$$

$$G_u(A) = P\left\{\underline{T} \in A \,\middle|\, S = u\right\} \varphi(u), \quad A \in B^p, \tag{6.4}$$

as measures on (R^{p+1}, B^{p+1}) and (R^p, B^p) depending on a parameter $u \in R$.

We establish first some preliminary results. Let F and G be measures on (R^k, B^k), $k \in N$. For $E \in R^k$ denote by E^ε its ε-neighbourhood, $E^\varepsilon = \left\{ \underline{x} : \inf\left[\|\underline{x} - \underline{y}\|;\ \underline{y} \in E\right] < \varepsilon\right\}$. The Levy-Prokhorov distance between F and G, $L(F,G)$, is defined as infimum of the set of positive numbers, ε, such that for any closed set $E \subset R^k$

$$F(E) < G(E^\varepsilon) + \varepsilon, \qquad G(E) < F(E^\varepsilon) + \varepsilon.$$

Let for any α from some set A, $F_{n,\alpha}$, $n \in N$, and F_α be measures on (R^k, B^k); put

$$f_{n,\alpha}(\underline{t}) = \int \exp(i\underline{t}\,\underline{x}')F_{n,\alpha}(d\underline{x}),$$

$$f_\alpha(\underline{t}) = \int \exp(i\underline{t}\underline{x}')F_\alpha(d\underline{x}).$$

A family $\{F_\alpha, \alpha \in A\}$ is called tight if $F_\alpha(S_b^c) \to 0$ as $b \to \infty$ uniformly in $\alpha \in A$.

<u>Lemma 6.1.</u> If $\{F_\alpha, \alpha \in A\}$ is tight, $\sup_\alpha F_\alpha(R^k) \leq C < \infty$, $f_{n\alpha}(\underline{t}) \to f_\alpha(\underline{t})$ for each $\underline{t} \in R^k$ uniformly in $\alpha \in A$ then $L(F_{n,\alpha}, F_\alpha) \to 0$ uniformly in $\alpha \in A$.

P r o o f: For any $\delta > 0$, distribution H and measures F, G on (R^k, B^k) the following inequality holds

$$L(F,G) \leq L(F * H,\ G * H) + \\ + \max\left[2\delta,\ (F(R^k) + G(R^k))\ H(S_\delta^c)\right] \qquad (6.5)$$

(its analogue for the Levy distance between distributions on R has been proved by Zolotarev ([19], p.226), this proof can be immediately extended to the present case).

Put $F = F_{n,\alpha}$, $G = F_\alpha$ in (6.5). For an arbitrary $\delta > 0$ take a distribution H such that $H(S_\delta^c) < \delta/C$ and its ch.f. vanishes outside some S_r, $0 < r < \infty$. Let $p_{n,\alpha,H}$ and $p_{\alpha,H}$ be the densities of $F_{n\alpha} * H$ and $F_\alpha * H$. We shall show first that

$$d(F_{n,\alpha} * H,\ F_\alpha * H) \equiv \frac{1}{2}\int \left| p_{n,\alpha,H}(x) - p_{\alpha,H}(x)\right| dx \to 0 \qquad (6.6)$$

as $n \to \infty$ uniformly in $\alpha \in A$. We have

$$\sup\left[\left|p_{n,\alpha,H}(x) - p_{\alpha,H}(x)\right|;\ x \in R^k, \alpha \in A\right] \leq \\ \leq \frac{1}{(2\pi)^k}\int_{S_r} \sup\left[\left|f_{n,\alpha}(t) - f_\alpha(t)\right|;\ \alpha \in A\right] dt \to 0 \qquad (6.7)$$

by the dominated convergence theorem. Split the integral in (6.6) into a sum of integrals, $I(S_b) + I(S_b^c)$, over S_b and S_b^c. We have $I(S_b^c) \leq F_{n,\alpha,H}(S_b^c) + F_{\alpha,H}(S_b^c)$. Take an arbitrary $\varepsilon > 0$ and find $d > 0$ such that $H(S_d^c) < \varepsilon$. Then $F_{\alpha,H}(S_b^c) \leq F_\alpha(S_{b-d}^c) + H(S_d^c)$ and the tightness of $\{F_\alpha\}$ means that there exists $b > 0$ such that $\sup_\alpha F_{\alpha,H}(S_b^c) < 2\varepsilon$. By (6.7), $I(S_b) \to 0$ uniformly in $\alpha \in A$; this implies that $F_{n,\alpha,H}(S_b) \to F_{\alpha,H}(S_b)$ uniformly in α. By assumption,

$$F_{n,\alpha}(R^k) = f_{n,\alpha}(0) \to f_\alpha(0) = F_\alpha(R^k).$$

Hence $\sup_\alpha F_{n,\alpha,H}(S_b^c) < 2\varepsilon$ for sufficiently large n. Since $\varepsilon > 0$ is arbitrary the relations obtained prove (6.6).

For any F and G, $L(F,G) \leqslant d(F,G)$, therefor it follows from (6.5), (6.6) and the choice of H that

$$\limsup_{n\to\infty} \sup_{\alpha} L(F_{n,\alpha}, F_{\alpha}) \leqslant 2\delta .$$

Since $\delta > 0$ is arbitrary this proves the lemma.

Lemma 6.2. Let $f(s,\underline{t})$ be the ch.f. of a random vector $(Y, \underline{Z})$, $Y \in R$, $Z \in R^k$, $k \geqslant 0$; $E|Y|^m < \infty$ for an integer $m \geqslant 0$. Then

$$f(s+\Delta,\underline{t}) = 1 + \sum_{j=1}^{m} f^{(j)}(s,\underline{t})\frac{\Delta^j}{j!} + \eta(\Delta,s,t)|\Delta|^m \tag{6.8}$$

where $f^{(j)} = \partial^j f/\partial s^j$,

$$\sup\left[|\eta(\Delta,s,\underline{t})|;\ (s,\underline{t}) \in R^{k+1}\right] \to 0 \quad \text{as } \Delta \to 0. \tag{6.9}$$

If the distribution of $(Y,\underline{Z})$ belongs to a family P such that $E|Y|^m$ converges uniformly in $P \in \mathcal{P}$ then the convergence in (6.9) is uniform in $P \in \mathcal{P}$. In case m=0 the sum in (6.8) is taken to be 0 and $0! = 1$.

The proof can be found in Chibisov (1973b), Lemma 7.

Lemma 6.3. If the conditions (i), (ii″), (iii), (iv), (v) are satisfied then

$$\sup\left[L(F_{n,u}, G_u \times V);\ u \in R,\ \theta \in \Theta\right] \to 0$$

as $n \to \infty$ (see (6.3), (6.4)).

P r o o f: Note that $G_u(R^p) = \varphi(u) \leqslant \varphi(0)$ and the family $\{G_u; u \in R, \theta \in \Theta\}$ is tight (for any $\varepsilon > 0$ there exists $b > 0$ such that $G_u(R^p) < \varepsilon$ for $|u| > b$ and the family of conditional distributions of $\underline{T}$ given $S=u$, $|u| \leqslant b$, is tight). This implies that the family $\{G_u \times V\}$ is bounded and tight. Denote by $f_{n,a}(s,\underline{t},v)$ the ch.f. of $(S_{n,a}, \underline{T}_n, \xi)$ We have

$$f_{n,a}(s,\underline{t},\nu) = f^n(\tau s,\tau\underline{t})\omega(\nu+\tau s/a),\ s,\nu\in R,\underline{t}\in R^p. \qquad (6.10)$$

Let

$$\psi_{n,a}(\underline{t},\nu;u) = \int \exp(i\underline{t}\underline{x}' + i\nu y)\ F_{n,u}(d\underline{x}, dy), \qquad (6.11)$$

$$\chi(\underline{t};u) = \int \exp(i\underline{t}\underline{x}')\ G_u(dx). \qquad (6.12)$$

By lemma 6.1, it is sufficient to show that for any $(\underline{t},\nu)$

$$\psi_{n,a}(\underline{t},\nu;u) \to \chi(\underline{t};u)\omega(\nu) \qquad (6.13)$$

as $n\to\infty$ uniformly in $u \in R$, $\Theta\in\Theta$.

Since

$$\int e^{ius}\,\psi_{n,a}(\underline{t},\nu;u)\,du = f_{n,a}(s,\underline{t},\nu)$$

and for any $\underline{t}$, ν, n

$$\int |f_{n,a}(s,\underline{t},\nu)|\,ds \leq \int |\omega(\nu+\tau s/a)|\,ds < \infty$$

we obtain

$$\psi_{n,a}(t,\nu;u) = \frac{1}{2\pi}\int e^{-ius}\,f_{n,a}(s,\underline{t},\nu)\,ds. \qquad (6.14)$$

In a similar way

$$\chi(\underline{t};\ u) = \frac{1}{2\pi}\int e^{-ius}\exp\left[-\tfrac{1}{2}(s,\underline{t})\textstyle\sum_{\Theta}(s,t)'\right]ds. \qquad (6.15)$$

The proof of (6.13) is now similar to the proof of the local central limit theorem. For fixed s, $\underline{t}$, ν

$$f_{n,a}\ (s,\underline{t},\nu) \to \omega(\nu)\ \exp\left[-\tfrac{1}{2}\ (s,\underline{t})\textstyle\sum_{\Theta}(s,\underline{t})'\right] \qquad (6.16)$$

uniformly in $\Theta\in\Theta$. Indeed, $\omega(\nu+\tau s/a)\to\omega(\nu)$ and the convergence for $f^n(\tau s, \tau\underline{t})$ is a standard fact; the uniformity in $\Theta\in\Theta$ can be shown by using in its proof the conditions (ii) and (iii) and Lemma

6.2 with k=0, $Y = sY_{10} + \underline{t}\,\underline{Y}_1'$, s=0, $h=\tau$, m=2. Moreover, for any $\underline{t}$ there exist $\delta = \delta(\underline{t}) > 0$ and $n_1 = n_1(\underline{t}) \in N$ such that for $|s| \leq \delta n^{1/2}$, $n \geq n_1$

$$\left| f_{n,a}(s,\underline{t},\nu) \right| \leq \exp\left[-\tfrac{1}{4}(s,\underline{t}) \sum_{\Theta} (s,t)' \right]. \tag{6.17}$$

As before one can show that it is possible to find δ and n_1 independent of $\Theta \in \Theta$.

Split the integral in (6.14) into a sum of integrals, $I_{n1}(t,\nu;u) + I_{n2}(t,\nu;u)$, over $|s| \leq \delta n^{1/2}$ and $|s| > \delta n^{1/2}$. Applying the dominated convergence theorem to the supremum w.r.t. u of the left side of (6.16) and using (6.17) we obtain

$$\sup \left[\left| I_{n1}(\underline{t},\nu;u)/2\pi - \chi(\underline{t};u)\,\omega(\nu) \right|;\ \Theta \in \Theta,\ u \in R \right] \to 0.$$

It remains to show that $I_{n2} \to 0$ uniformly in $\Theta \in \Theta$, $u \in R$. Due to the factor $\omega(\nu + \tau s/a)$ (see (6.10)), $f_{n,a}(s,\underline{t},\nu) = 0$ for

$$s \notin \left[(-1-\nu)an^{1/2},\ (1-\nu)an^{1/2} \right]$$

and anyway for $|s| > (1+|\nu|)a\, n^{1/2}$. By the condition (iv) there exist $n_1 \in N$ and $\eta_1 < 1$ such that

$$\sup \left[| f(s,\underline{0})|;\ n > n_1,\ \Theta \in \Theta,\ \delta \leq |s| \leq (1+|\nu|)a \right] \leq \eta_1;$$

the condition (ii′) and Lemma 6.2 imply that there exist $\delta_1 > 0$ and η, $0 < \eta < 1$, such that

$$\sup \left[|f(s,\underline{t})|;\ n > n_1,\ \Theta \in \Theta,\ \delta \leq |s| \leq (1+|\nu|)a,\ \|\underline{t}\| \leq \delta_1 \right] \leq \eta.$$

Thus for n such that $\tau \|\underline{t}\| \leq \delta_1$ and $n \geq n_1$

$$|f_{n,a}(s,\underline{t},\nu)| \leq \eta^n \,|\omega(\nu + \tau s/a)|$$

and

$$|I_{n2}| \leq \eta^n a n^{1/2} \int |\omega(s)|\, ds \to 0.$$

Q.E.D.

<u>Lemma 6.4.</u> Let $P_{n,\alpha}$, $n \in N$, $P_\alpha, \alpha \in A$ be measures on (R^{p+1}, B^{p+1}) such that $L(P_{n,\alpha}, P_\alpha) \to 0$ as $n \to \infty$ uniformly in $\alpha \in A$, P_α have uniformly in $\alpha \in A$ bounded densities w.r.t. Lebesgue measure and $\{P_\alpha, \alpha \in A\}$ is tight. Let a family of functions $h_\alpha : R^p \to R, \alpha \in A$, be equicontinuous on any compact from R^p. Let

$$D_\alpha(z) = \left\{ (\underline{y}, x) \in R^{p+1} : h_\alpha(y) + x < z \right\}.$$

Then

$$P_{n,\alpha}(D_\alpha(z)) \to P_\alpha(D_\alpha(z)) \tag{6.18}$$

as $n \to \infty$ uniformly in $z \in R, \alpha \in A$.

<u>Remark.</u> In fact in the proof the following consequence of the uniform boundedness of the densities will be used: for any $\varepsilon > 0$ there exists $\delta > 0$ such that

$$\lambda(E) < \delta \quad \text{implies} \quad \sup\left[P_\alpha(E);\ \alpha \in A \right] < \varepsilon,\ E \in B^{p+1}, \tag{6.19}$$

where λ is the Lebesgue measure on R^{p+1}.

P r o o f. Take an arbitrary $\varepsilon > 0$. Under the conditions of the Lemma one can find $n_1 = n_1(\varepsilon) \in N$ and $b = b(\varepsilon) > 0$ such that for any closed $E \subset R^{p+1}$, $n \geq n_1, \alpha \in A$

$$P_{n,\alpha}(E) < P_\alpha(E^\varepsilon) + \varepsilon, \quad P_\alpha(E) < P_{n,\alpha}(E^\varepsilon) + \varepsilon, \tag{6.20}$$

$$P_\alpha(K_b^c) < \varepsilon, \qquad P_{n,\alpha}(K_b^c) < \varepsilon$$

where $K_b = \left[-b, b\right]^{p+1}$. Immediately from the conditions of the lemma one can find n_1 and b_1 such that (6.20) and the first inequality in (6.21) hold with $\varepsilon/2$ instead of ε. Then $P_{n,\alpha}((K_b^c)^{\varepsilon/2}) < P_\alpha(K_b^c) + \varepsilon/2$ and putting $b = b_1 + \varepsilon/2$ we obtain (6.21).

Let $D_\alpha(z,b) = D_\alpha(z) \cap K_b$. Then we have by (6.21)

$$P_{n,\alpha}(D_\alpha(z)) < P_{n,\alpha}(D_\alpha(z,b)) + \varepsilon, \quad n > n_1. \tag{6.22}$$

We shall show that for any $\delta_1 > 0$ there exists $\varepsilon_1 > 0$ such that

$$\bar{D}_\alpha^{\varepsilon_1}(z,b) \subset D_\alpha(z + \delta_1, b + 1). \tag{6.23}$$

Since h_α are equicontinuous on $[-b, b]^p$, one can find ε_1, $0<\varepsilon_1<$ $< \delta_1/2$ such that

$$\sup_\alpha \left| h_\alpha(\underline{y}) - h_\alpha(\underline{y}_0) \right| < \delta_1/2 \text{ for } \underline{y}_0 \in [-b,b]^p, \| \underline{y}-\underline{y}_0 \| < \varepsilon_1. \tag{6.24}$$

If $(\underline{y},x) \in \bar{D}_\alpha^{\varepsilon_1}(z,b)$ then there exists a point $(\underline{y}_0, x_0) \in K_b$ such that $\| (\underline{y},x) - (\underline{y}_0, x_0) \| < \varepsilon_1$ and $h_\alpha(\underline{y}_0) + x_0 < z$. Then $\| \underline{y}-\underline{y}_0 \| < \varepsilon_1$, $|x - x_0| < \varepsilon_1$ and by (6.24)

$$h_\alpha(\underline{y}) + x < h_\alpha(\underline{y}_0) + \delta_1/2 + x_0 + \varepsilon_1 < z + \delta_1$$

i.e. $(\underline{y}, x) \in D_\alpha(z+\delta_1)$. Moreover $(\underline{y}, x) \in K_{b+\varepsilon_1}$ and taking $\varepsilon_1 < 1$ we obtain (6.23).

Note that

$$D_\alpha(z + \delta_1) \setminus D_\alpha(z) = \left\{ (\underline{y}, x) : z - h_\alpha(\underline{y}) \leqslant x < z - h_\alpha(\underline{y}) + \delta_1 \right\}$$

whence $\lambda\left[(D_\alpha(z + \delta_1) \setminus D_\alpha(z)) \cap K_{b+1} \right] \leqslant \delta_1 (2b+2)^p$. For $\varepsilon > 0$ find δ from (6.19) and put $\delta_1 = \delta(2b+2)^{-p}$. Then

$$P_\alpha\left[(D_\alpha(z + \delta_1) \setminus D_\alpha(z)) \cap K_{b+1} \right] < \varepsilon. \tag{6.25}$$

Take ε_1, $0 < \varepsilon_1 < \varepsilon$ such that (6.23) be fulfilled. Find $n_2 \geqslant n_1$ such that (6.20) holds for $n \geqslant n_2$ with ε_1 instead of ε. Then using (6.21), (6.22), (6.23) and (6.25) we obtain that for $n \geqslant n_2$

$$P_{n,\alpha}(D_\alpha(z)) \leqslant P_{n,\alpha}(\bar{D}_\alpha(z)) \leqslant P_{n,\alpha}(\bar{D}_\alpha(z,b)) + \varepsilon \leqslant$$

$$\leqslant P_\alpha(\bar{D}_\alpha^{\varepsilon_1}(z,b)) + 2\varepsilon \leqslant P_\alpha(D_\alpha(z+\delta_1, b+1)) + 2\varepsilon \leqslant$$

$$\leqslant P_\alpha(D_\alpha(z,b+1)) + 3\varepsilon \leqslant P_\alpha(D_\alpha(z)) + 4\varepsilon. \tag{6.26}$$

Applying the same arguments to $D_\alpha^c(z)$ we obtain a lower estimate for $P_{n,\alpha}(D_\alpha(z))$ similar to (6.26). Since $\varepsilon > 0$ is arbitrary, this proves the lemma.

Now Lemma 3.2 is proved by applying Lemmas 6.3 and 6.4 with $A = \Theta \times R$, $P_{n,\alpha} = F_{n,u}$, $P_\alpha = G_u \times V$, $\alpha = (0,u)$.

7. Proof of Lemma 3.3

First we shall prove a lemma which is a slight variation of Theorems 1 and 2 of Fuc and Nagaev [13] in case of i.i.d. summands.

Lemma 7.1. Let $Z_1, \dots, Z_n$ be i.i.d. r.v.'s., $\sum_n = \sum Z_i$, $\nu_r = E|Z_1|^r$ for some $0 < r \leqslant 2$ and $EZ_1 = 0$ if $1 < r \leqslant 2$. Then for any $x > 0$, $\lambda > 0$

$$P\left\{\sum\nolimits_n > x\right\} \leqslant n\, P\left\{Z_1 > x/\lambda\right\} + \left(\lambda^{r-1} en\, \nu_r\, x^{-r}\right)^{\lambda}. \tag{7.1}$$

P r o o f: Introduce the event $A_{n,y} = \left\{\max Z_i \leqslant y\right\}$. Then

$$P\left\{\sum\nolimits_n > x\right\} \leqslant P\left\{\sum\nolimits_n > x,\ A_{n,y}\right\} + P(A_{n,y}^c), \tag{7.2}$$

$$P\left\{A_{n,y}^c\right\} \leqslant n\, P\left\{Z_1 > y\right\},\quad P\left\{\sum\nolimits_n > x,\ A_{n,y}\right\} \leqslant$$

$$\leqslant e^{-hx} E\left[e^{h\sum_n} 1_{A_{n,y}}\right] = e^{-hx}\left[E\left(e^{h Z_1} 1_{\{Z_1 < y\}}\right)\right]^n \tag{7.3}$$

for any $h > 0$. Let $F(x) = P\left\{Z_1 < x\right\}$; $\mu = 0$ if $0 < r \leqslant 1$, $\mu = E Z_1$ if $1 < r \leqslant 2$ (the inequality (7.5) below with $\mu \neq 0$ will be used in the proof of Lemma 3.3). The inequalities contained in the proofs of the Theorems 1 and 2 of Fuc and Nagaev [13] and the inequality (in case $1 < r \leqslant 2$):

$$\int\limits_{|u| \leqslant y} u\, dF(y) = \mu - \int\limits_{|u|>y} u\, dF(u) \leqslant \mu + y^{-(r-1)} \nu_r \tag{7.4}$$

imply

$$E(e^{hZ_1} 1_{\{Z_1 < y\}}) \leq 1 + h\mu + e^{hy} \nu_r y^{-r}. \tag{7.5}$$

Let $y = x/\lambda$ and define h by (3.36). Then

$$e^{-hx} = \left(\lambda^{r-1} n\nu_r x^{-r}\right)^{\lambda}, \quad 1 + e^{hy} \nu_r y^{-r} = 1 + \lambda/n \leq e^{\lambda/n}. \tag{7.6}$$

Putting (7.6) into (7.5) (with $\mu = 0$) and using (7.2) and (7.3) we obtain (7.1). The argument is not valid if $h \leq 0$ but then the right side of (7.1) is greater than 1 and (7.1) holds trivially.

The following corollary concerns the case when the distribution of Z_i depends on Θ.

Corollary 7.1. If for some $0 < r \leq 2$ $E_\Theta |Z_1|^r$ converges uniformly in $\Theta \in \Theta$ and $E Z_1 = 0$ in case $1 < r \leq 2$ then for any sequence $\{x_n\}$ such that $n x_n^{-r} \to \infty$

$$\sup_{\Theta} P_\Theta \left\{ \sum_n > x_n \right\} = O(n x_n^{-r}). \tag{7.8}$$

P r o o f: Under the conditions of the lemma, $\sup \left[\nu_r; \Theta \in \Theta \right] < \infty$. Apply Lemma 7.1 with an arbitrary $\lambda > 1$. Then the second term in the right side of (7.1) is $O((n x_n^{-r})^{\lambda})$ uniformly in $\Theta \in \Theta$ and

$$\sup_{\Theta} P_\Theta \left\{ Z_n > x_n/\lambda \right\} \leq$$

$$\leq \sup_{\Theta} (\lambda/x_n)^r E_\Theta |Z_1|^r 1_{\{Z_1 > x_n/\lambda\}} = O(x_n^{-r}).$$

P r o o f o f L e m m a 3.3. Introduce $A_{n,y}$ as in the proof of Lemma 7.1. Similarly to (7.2), (7.3) we obtain

$$P\left\{ \sum_n > x \,\middle|\, S_{n,a} = u \right\} \leq n g_{n1}(y,u) + e^{-hx} g_{n2}(y,u), \tag{7.9}$$

$$g_{n1}(y,u) = P\left\{ Z_1 > y \,\middle|\, S_{n,a} = u \right\}, \tag{7.10}$$

$$g_{n2}(y,u) = E\left[\exp(h \textstyle\sum_n) 1_{A_{n,y}} \mid S_{n,a} = u\right]. \qquad (7.11)$$

It can be proved in a similar way to (6.4),(6.5) in Chibisov [2] that

$$\sup_{u \in R} g_{n1}(y,u) p_{n,a}(u) \leq P\{Z_1 > y\} (n/(n-1))^{1/2} \sup_{x \in R} p_{n-1,a}(x). \qquad (7.12)$$

Let $y = x/\lambda$. Then e^{-hx} is given by (7.6) and it remains to estimate $g_{n2}(x/\lambda, u)$. Let $\tilde{S}_{n,a} = n^{1/2} S_{n,a} = \sum Y_i + \xi/a$, $\tilde{p}_{n,a}$ be the density of $\tilde{S}_{n,a}$ and

$$\tilde{g}_{n2}(y,u) = E\left[\exp(h \textstyle\sum_n) 1_{A_{n,y}} \mid \tilde{S}_{n,a} = u\right].$$

Then $\tilde{p}_{n,a}(u\, n^{1/2})\, n^{1/2} = p_{n,a}(u)$ and

$$g_{n2}(y,u)\, p_{n,a}(u) = \tilde{g}_{n2}(y, un^{1/2})\, \tilde{p}_{n,a}(u\, n^{1/2})\, n^{1/2}. \qquad (7.13)$$

Let $G(u) = P\{Y_1 < u\}$ and introduce the following measures $(A \in B)$:

$$\pi_{n,y,a}(A) = \int_A \tilde{g}_{n2}(y,u)\, \tilde{p}_{n,a}(u)\,du =$$

$$= E\left[\exp(h \textstyle\sum_n)\, 1_{A_{n,y}}\, 1_A(\tilde{S}_{n,a})\right], \qquad (7.14)$$

$$\pi_y(A) = \int_A E\left[e^{h Z_1} 1_{\{Z_1 < y\}} \mid Y_1 = u\right] dG(u) = E\left[e^{h Z_1} 1_{\{Z_1 < y\}} 1_A(Y_1)\right]. \qquad (7.15)$$

<u>Lemma 7.2.</u> Let (U_1, V_1), (U_2, V_2) be independent random vectors, and

$$\pi(A) = E\left[U_1 U_2 1_A(V_1 + V_2)\right], \quad \pi_j(A) = E\left[U_j 1_A(V_j)\right], \quad j = 1,2.$$

Then $\pi = \pi_1 * \pi_2$

The proof is contained in the proof of Lemma 6.1 of Chibisov [2].

Lemma 7.2 extends in an obvious way to any finite number of random vectors. Its application to the set of vectors

$$\left\{\left(e^{hZ_i} 1_{\{Z_i < y\}}, Y_i\right), i=1,\dots, n, (1,\xi/a)\right\}$$

gives $\pi_{n,y,a} = \pi_y^{n*} * V_a$. It is seen from (7.14) that $\tilde{g}_{n,2}(y,\cdot)\tilde{p}_{n,a}(\cdot)$ is the density of $\pi_{n,y,a}$ with respect to the Lebesgue measure. Let $\nu_r(u) = E\left[|Z_1|^r \mid Y_1 = u\right]$; $\mu(u) = 0$ if $0 < r \leq 1$, $\mu(u) = E\left[Z_1 \mid Y_1 = u\right]$ if $1 < r \leq 2$. In a similar way to (7.5) we obtain

$$E\left[e^{hZ_1} 1_{\{Z_1 < y\}} \mid Y_1 = u\right] \leq 1 + h\,\mu(u) + e^{hy}\,\nu_r(u)\, y^{-r} =$$

$$= 1 + h\,\mu(u) + \lambda\nu_r(u)/n\,\mu_r.$$

Let

$$\bar{\pi}_{n,\lambda}(A) = \int_A (1 + h\mu(u) + \lambda\nu_r(u)/n\,\nu_r)\; d\,G(u). \qquad (7.16)$$

Then $\pi_y(A) \leq \bar{\pi}_{n,\lambda}(A)$ and $\pi_{n,y,a}(A) \leq \bar{\pi}_{n,\lambda,a}(A) = (\bar{\pi}_{n,\lambda}^{n*} * V_a)(A)$, $A \in B$. Therefore

$$\tilde{g}_{n,2}(y,u)\;\tilde{p}_{n,a}(u) \leq \bar{p}_{n,\lambda,a}(u), \quad u \in R, \qquad (7.17)$$

where $\bar{p}_{n,\lambda,a}$ is the density of $\bar{\pi}_{n,\lambda,a}$.

Let

$$r_{n,\lambda}(A) = \bar{\pi}_{n,\lambda}(A)/(1 + \lambda/n). \qquad (7.18)$$

Since $\int \nu_r(u)\,dG(u) = \nu_r$ and $\int \mu(u)\,dG(u) = 0$, it follows from (7.16) that $r_{n,\lambda}$ is a probability measure; denote by $\rho_{n,\lambda}(t)$ its ch.f. Then $\rho^n_{n,\lambda}(t)\,\omega_a(t)$ is the ch.f. of $r^{n*}_{n,\lambda} * V_a = (1+\lambda/n)^{-n}\bar{\pi}_{n,\lambda,a}$ Using that $(1+\lambda/n)^n \leq e^{\lambda}$ we obtain

$$\sup_u \bar{p}_{n,\lambda,a}(u) \leq \frac{e^{\lambda}}{2\pi} I_n, \quad I_n = \int |\rho_{n,\lambda}(t)|^n\,\omega_a(t)\,dt. \qquad (7.19)$$

In the sequel we shall suppress the subscript λ at $r_{n,\lambda}$ and $\rho_{n,\lambda}$. For a measure $r(\cdot)$ on B let $r^-(A) = r(-A)$, $A \in B$. Let $r_{n,2} = r_n * r_n^-$ and $\rho_{n,2}(t)$ be the ch.f. of $r_{n,2}$. Then $\rho_{n,2} = |\rho_n|^2$. Let for $v > 0$

$$b_n = b_n(v) = \int_{|u|\leqslant v} u^2 \, r_{n,2}(du). \tag{7.20}$$

By the truncation inequality ([15] 12.4.13′), for $|t| \leqslant 1/v$

$$\rho_{n,2}(t) \leqslant 1 - \frac{t^2}{3} \int_{|u|\leqslant 1/t} u^2 r_{n,2}(du) \leqslant 1 - \frac{b_n t^2}{3} \leqslant \exp\left(-\frac{b_n t^2}{3}\right). \tag{7.21}$$

Hence $|\rho_n(t)| = \rho_{n2}^{1/2}(t) \leqslant \exp(-b_n t^2/6)$ for $|t| \leqslant 1/v$ and splitting I_n into a sum of integrals, I_{n1} and I_{n2}, over $|t| \leqslant 1/v$ and $|t| > 1/v$ we obtain

$$I_{n1} \leqslant \int_{|t|\leqslant 1/v} \exp\left(-\frac{n\, b_n\, t^2}{6}\right) dt \leqslant \left(\frac{6\pi}{n\, b_n}\right)^{1/2}. \tag{7.22}$$

To estimate b_n from below, let $\nu(u) = E\left[\,|Z_1|\,\middle|\,Y_1 = u\right]$ if $1 < r \leqslant 2$ and $\nu(u) \equiv 0$ if $0 < r \leqslant 1$ and introduce the measure $H(A) = \int_A \nu(u)\, dG(u)$, $A \in B$; note that $H(R) = \nu$. Then we have from (7.16) that $\bar{\pi}_{n,\lambda}(A) \geqslant G(A) - h\,H(A)$, $A \in B$. Using (7.18) and (7.20) we obtain (see (3.34))

$$b_n \geqslant (1 + \lambda/n)^{-2} \int_{|u|\leqslant v} u^2 d\left[G_2 - h(G * H^- + G^- * H)\right] \geqslant$$
$$\geqslant (1 + \lambda/n)^{-2}(b(v) - 2\,h\,v^2 \nu). \tag{7.23}$$

Moreover it follows from (7.16) that $|\rho_{n,\lambda}(t)| \leqslant |f(t)| + \lambda/n + h\nu$ (see (3.35)). Therefore

$$\sup\left[|\rho_{n,\lambda}(t)|;\ 1/v \leqslant |t| \leqslant a\right] \leqslant \eta(v) + \lambda/n + h\nu$$

and since $\omega_a(t) = 0$ for $|t| > a$ we have

$$I_{n2} \leqslant (\eta(v) + \lambda/n + h\nu)^n\, a \int \omega(t)\, dt. \tag{7.24}$$

Now (3.33) follows from (7.9), (7.12), (7.7), (7.13), (7.17), (7.19), (7.22-24) and similar estimates for $-\sum_n$. Q.E.D.

References

[1] B i k j a l i s, A., Asymptotic expansions for the densities and distributions of sums of independent identically distributed random vectors. Litovsk. Mat. Sb., 8, 405-422 (1968) =Selected Transl. in Math. Statist. and Probability, 13, 213-234 (1973).

[2] C h i b i s o v, D. M., On the normal approximation for a certain class of statistics. Proc. 6th Berkeley Sympos.Math. Statist. and Prob., vol. 1, 153-174 (1972).

[3] - Asymptotic expansions for distributions of some test statistics for composite hypotheses. Teor. Ver. i Primen., 17, 3, 600-602 (1972).

[4] - An asymptotic expansion for the distribution of a statistic admitting an asymptotic expansion. Teor. Ver. i Primen., 17,4, 658-668 = Theor. Probability Appl., 17, 620-630 (1972).

[5] - Asymptotic expansions for Neyman's C(α) tests. Proc. 2nd Japan-USSR Sympos. on Prob. Theory (G. Maruyama and Yu.V.Prokhorov, eds.). Lecture Notes in Math., No. 330, Springer,Berlin, 16-45 (1973).

[6] - An asymptotic expansion for distributions of sums of a special form with an application to minimum contrast estimates. Teor. Ver. i Primen., 18, 4, 689-702 = Theor. Probability Appl., 18, 649-661 (1973).

[7] - Weakening the regularity conditions for some asymptotic expansions. Asymptotic Methods in Statistics,10.11-16.11.1974, Tagungsbericht N 44, Mathematisches Forschungsinstitut Oberwolfach, 6-7 (1974).

[9] -On an asymptotic expansion for the distribution of a statistic admitting a stochastic expansion. Teor. Ver. i Primen., 24, 1, 230-231 (1979).

[10] - Asymptotic expansion for the distribution of statistic admitting a stochastic expansion. Preprints in Statistics, 47, University of Cologne (1979).

[11] E l i s e e v, V. G., Asymptotic expansions under local alternatives. Teor. Ver. i Primen., 24, 1, 231-232 (1979).

[12] F e l l e r, W., An Introduction to Probability Theory and Its Applications. Vol. II. Wiley, New York 1966.

[13] F u c, D. H. and N a g a e v, S. V., Probability inequalities for sums of independent random variables. Teor. Ver. i Primen., 10, 4, 660-675 (1971).

[14] L e h m a n n, E. L. Testing Statistical Hypotheses. Wiley, New York 1959.

[15] L o è v e, M., Probability Theory. Princeton, van Nostrand 1960.

[16] N e y m a n, J., Optimal asymptotic tests of composite statistical hypotheses. Probability and Statistics (The Harald Cramér Volume). Uppsala, Almquist and Wiksells, 213-234 (1959).

[17] P f a n z a g l, J., Asymptotically optimum estimation and test procedures. Proc. Prague Sympos. on Asymptotic Statistics 3-6 September 1973, Prague, vol. I, 201-272 (1974).

[18] P f a n z a g l, J. and W e f e l m e y e r, W., An asymptotically complete class of tests. Z. Wahrscheinlichkeitstheorie and Verw. Gebiete, 45, 49-72 (1978).

[19] Z o l o t a r e v, V. M., Estimates for differences of distributions in Levy metric, Trudy of Steklov Math. Inst., 112, 224-231 (1971).

PROPERTIES OF REALIZATIONS OF RANDOM FIELDS

by

Z. Ciesielski

Polish Academy of Sciences, Sopot

Summary

To each integer $\mu \geq 0$ there corresponds dyadic partition $T_\mu = (j\,2^{-\mu},\ j = 0, \ldots, 2^\mu)$ of $I = \langle 0, 1\rangle$. Let $T^d_\mu = T^{(1)}_\mu \times \ldots \times T^{(d)}_\mu$ with $T^{(i)}_\mu = T_\mu$, and for a given multi-index $\alpha = (\alpha_1, \ldots, \alpha_d)$ let

$$T^d_\mu(\alpha) = \left\{(t_1, \ldots, t_d) \in T^d_\mu:\ t_i \leq 1 - \alpha_i 2^{-\mu},\ i = 1, \ldots, d\right\}.$$

The progressive difference of $f: T^d_\mu \to R$ of order α and corresponding to the increment $2^{-\mu}$ is denoted by $\Delta^\alpha_{d,\mu} f(t)$, $t \in T^d_\mu(\alpha)$. The main result states: If the random field $\left\{X(t): t \in T^d_\mu,\ \mu \geq 0\right\}$ satisfies the inequality $(0 < \gamma \leq m\beta,\ m \leq 2^\mu,\ C > 0)$

$$E\left|\Delta^{me_i}_{d,\mu} X(t)\right|^\beta \leq C\,2^{-\mu(d+\gamma)}, \quad t \in T^d_\mu(me_i), \quad i = 1, \ldots, d,$$

with $e_i = (\delta_{i,j},\ j = 1, \ldots, d)$, then it has continuous extension to I^d such that for $\alpha\beta < \gamma$

$$P\left\{\omega_m(X(\cdot);\ h) = O(h^\alpha)\right\} = 1,$$

where

$$\omega_m(f;\ h) = \sup\left\{\left|\Delta^m_u f(t)\right| :\ t,\ t+mu \in I^d,\ |u| \leq h\right\}.$$

This solves the author's problem formulated in [2].

1. Introduction

The aim of this note is to extend the result on stochastic processes obtained by the author in [2] to several time variables. The idea of the proof is essentially the same as in [2] although the generalization is not straight forward. The proof depends on some new results on spline aproximation in several variables of which an outline is given below. It is clear from the content of this paper that the result presented here has extensions in various directions e.g. to Besov spaces and to the anisotropic fields. We confined ourselves to the Lipschitz case in the L_∞ norm.

2. Extrapolation and approximation

For given closed domain $\Omega \subset R^d$ and $h \in R^d$ let

$$\Omega(h) = \left\{ t \in \Omega : t + \lambda h \in \Omega, \lambda \in I \right\}.$$

The space of uniformly continuous bounded functions on Ω is denoted by $C(\Omega)$,

$$C^{\beta}(\Omega) = \left\{ f : D^{\alpha} f \in C(\Omega), \alpha_i \leq \beta_i, \quad i=1,\dots,d \right\}$$

$$C^m(\Omega) = C^{\beta}(\Omega) \quad \text{for} \quad \beta = (m,\dots,m).$$

Here and later on m denotes a non-negative integer, and

$$\| f \| (\Omega) = \sup \left\{ | f(f) | : t \in \Omega \right\}.$$

The basic unit vectors in R^d are denoted by $e_1, \dots, e_d$. The modulus of smootheness of order m in the direction $u \in S^{d-1}$ is defined by the formula

$$\omega_0^{(u)} (f; \delta)_{\Omega} = \| f \| (\Omega), \tag{2.1}$$

$$\omega_m^{(u)} (f; \delta) = \sup_{0 < h \leq \delta} \| \Delta_u^m f \| (\Omega(mhu)),$$

where $0 < \delta < \rho/m$, $\rho = \operatorname{diam}\Omega$, and

$$\Delta_u^m f(t) = \sum_{j=0}^{m} (-1)^{m+j} \binom{m}{j} f(t+ju).$$

Moreover we define

$$\omega_m(f;\delta)_\Omega = \sup_{|u|=1} \omega_m^{(u)}(f;\delta)_\Omega$$

and use the convention

$$\omega_m^{(i)}(f;\delta)_\Omega = \omega_m^{(e_i)}(f;\delta)_\Omega$$

In what follows the letter Ω will be quite often supressed in the case $\Omega = I^d$, and it will be kept otherwise.

Let now for integer $\mu \geq 0$

$$T_\mu = (j\, 2^{-\mu},\ j = 0,\dots, 2^{\mu}) ,$$

$$T_\mu^d = T_\mu^{(1)} \times \dots \times T_\mu^{(d)},\quad T_\mu^{(i)} = T_\mu.$$

For multi-index α let

$$T_\mu^d(\alpha) = \left\{ t \in T_\mu^d : t + \alpha 2^{-\mu} \in T_\mu^d \right\}.$$

The progressive difference operator of order α on $\left\{ T_\mu^d,\ \mu \geq 0 \right\}$ is defined inductively as follows

$$\Delta_{d,\mu}^{\alpha} : C(T_\mu^d) \longrightarrow C(T_\mu^d(\alpha)),$$

$$\Delta_{d,\mu}^{0} f = f,$$

$$\Delta_d^{\alpha + e_j} f = \Delta_{2^{-\mu} e_j}^{1} \Delta_{d,\mu}^{\alpha} f.$$

In each of the spaces $C(T_\mu^d)$ and $C(T_\mu^d(\alpha))$ we use the maximum norm $\|\ \|$. Now, for given integers $\mu \geq 0$, $2^{\mu} \geq m \geq 0$, an extension operator (extrapolation) $L_\mu : C(T_\mu) \longrightarrow C(I)$ is defined as in [6]

$$L_\mu f = \begin{cases} Q_j f(t) \text{ for } j-1 \leq 2^\mu t < j, \quad j = 1, \ldots, 2^\mu - m, \\ P_{2^\mu - m} f(t) \text{ for } 2^\mu - m \leq 2^\mu t \leq 2^\mu; \end{cases}$$

where $P_i f$ is the algebraic polynomial of degree $\leq m$ interpolating f at the points $j\,2^{-\mu}$, $j = i, \ldots, i+m$, and $Q_i f$ is the unique algebraic polynomial of degree $\leq 2m+1$ such that

$$D^k(Q_i f)_{t=i2^{-\mu}} = D^k(P_i f)_{t=i2^{-\mu}}, \quad 0 \leq k \leq m,$$

$$D^k(Q_i f)_{t=(i-1)2^{-\mu}} = D^k(P_{i-1} f)_{t=(i-1)2^{-\mu}}, \quad 0 \leq k \leq m.$$

The extension (extrapolation) operator $L^d_\mu : C(T^d_\mu) \longrightarrow C(I^d)$ is now defined by the formula

$$L^d_\mu = L_1^{\varepsilon^1} \circ \ldots \circ L_d^{\varepsilon^d}, \quad \varepsilon^i = \sum_{j=i+1}^{d} e_j,$$

where for $\varepsilon \in \{0, 1\}^d$ with $\varepsilon_j = 0$

$$L_j^\varepsilon : C(T_\varepsilon) \longrightarrow C(T_{\varepsilon+e_j}),$$

$$T_\varepsilon = T_{\varepsilon_1} \times \ldots \times T_{\varepsilon_d}, \quad \varepsilon = (\varepsilon_1, \ldots, \varepsilon_d),$$

$T_{\varepsilon_i} = T_\mu$ or I as $\varepsilon_i = 0$ or 1 respectively, and for $f \in C(T_\varepsilon)$, $t \in T_{\varepsilon+e_j}$,

$$L_j^\varepsilon f(t_1, \ldots, t_d) = (Lf(t_1, \ldots, t_{j-1}, t_{j+1}, \ldots, t_d))(t_j).$$

<u>Lemma 2.2.</u> (J. Ryll [6]). Let L^d_μ be defined as above and let $f \in C(T^d_\mu)$. Then,

(i) $L^d_\mu f \in C^m(I^d)$,

(ii) $(L^d_\mu f)\big|_{T^d} = f$,

(iii) $\| D^\alpha (L^d f)\| \leq C(m,d)\, 2^{|\alpha|\mu} \| \Delta^\alpha_{d,\mu} f \|$

for $\alpha_j \leq m$, $j = 1, \ldots, d$.

We are now interested in the approximation problem i.e. find a good estimate for $\left\| f - L_\mu^d (f \big|_{T^d}) \right\|$, $f \in C(I^d)$. For this purpose let

$$K_\mu^d f = L_\mu^d (f \big|_{T_\mu^d}),$$

$$\varepsilon(i) = \sum_{j \neq i} e_j$$

$$K_\mu^{(i)} f = L_i^{\varepsilon(i)} (f \big|_{T_{\varepsilon(i)}}).$$

Clearly,

$$K_\mu^{(i)} : C(I^d) \longrightarrow C^m(I^d), \quad i = 1, \ldots, d.$$

and

$$K_\mu^d = K_\mu^{(1)} \ldots K_\mu^{(d)} .$$

It follows by Lemma 2.2, d = 1, that the operators $K_\mu^{(i)}$: $C(I^d) \longrightarrow C(I^d)$, $d \geqslant 1$, are bounded uniformly in μ.

Using the identity (E = identity operator)

$$E - K_\mu^d = (E - K_\mu^{(1)}) + \ldots + (K_\mu^{(1)} \ldots K_\mu^{(d-1)} - K_\mu^{(1)} \ldots K_\mu^{(d)})$$

we obtain for $f \in C(I^d)$

$$\left\| f - K_\mu^d f \right\| \leqslant C(d,m) \sum_{j=1}^{d} \left\| f - K_\mu^{(j)} f \right\|.$$

Now, the one-dimensional Theorem 2.1 of [2] implies for $2^\mu \geqslant m$

$$\left\| f - K_\mu^{(j)} f \right\| \leqslant C(m)\, \omega_m^{(j)}(f, 2^{-\mu}),$$

and therefore we have

<u>Proposition 2.3.</u> For $f \in C(I^d)$ and $2^\mu \geqslant m$

$$\left\| f - K_\mu^d f \right\| \leqslant C(d,m) \sum_{j=1}^{d} \omega_m^{(j)} (f; 2^{-\mu}).$$

Lemma 2.4. For $m \geq 0$, $d \geq 1$, there is a constant $C(d,m)$ such that for $f \in C(I^d)$

$$\left\| K^d_{\mu+1} f - K^d_{\mu} f \right\| \leq C(d,m) \sum_{j=1}^{d} \left\| \Delta^{me_j}_{d,\mu} f \right\|.$$

P r o o f: Since $(K^d_{\mu+1} f)\Big|_{T^d_{\mu+1}} = f\Big|_{T^d_{\mu+1}}$

and $T^d_{\mu+1}$ is finer than T^d_{μ}, and K^d_{μ} is determined by $f\Big|_{T^d_{\mu}}$ it follows that $K^d_{\mu} = K^d_{\mu} K^d_{\mu+1}$. Thus, for $g = K^d_{\mu+1} f$ by Proposition 2.3 and by Lemma 2.2 we have

$$\begin{aligned} \left\| K^d_{\mu+1} f - K^d_{\mu} f \right\| &= \left\| g - K^d_{\mu} g \right\| \\ &\leq C(d,m) \sum_{j=1}^{d} \omega^{(j)}_{m} (g;\ 2^{-\mu}) \\ &\leq C(d,m)\ 2^{-m\mu} \sum_{j=1}^{d} \left\| D^{me_j} K^d_{\mu+1} f \right\| \\ &\leq C(d,m) \sum_{j=1}^{d} \left\| \Delta^{me_j}_{d,\mu} f \right\|. \end{aligned}$$

Corollary 2.3. If f is defined on $\bigcup_{\mu \geq 0} T^d$ and $(2^{\mu_0} \geq m)$

$$\sum_{\mu \geq \mu_0} \sum_{j=1}^{d} \left\| \Delta^{me_j}_{d,\mu} f \right\| < \infty,$$

then the series

$$\sum_{\mu \geq \mu_0} (L^d_{\mu+1} f - L^d_{\mu} f)$$

converges uniformly on I^d.

3. Besov spaces on cubes

Let $\Omega \subset R^d$ be a closed domain, let $\underline{r} = (r_1, \dots, r_d)$, $r_i > 0$, $1 \leqslant q \leqslant \infty$ be given. We are going to consider the anisotropic Besov space $B_q^{\underline{r}}(\Omega) \equiv B_{\infty,q}^{\underline{r}}(\Omega)$ corresponding to the $L_\infty(\Omega)$ norm. In the case of $\Omega = I^d$ we simply write $B_q^{\underline{r}}$ for $B_q^{\underline{r}}(\Omega)$. For a given integer vector $\underline{m} = (m_1, \dots, m_d)$, $m_i \geqslant 0$, $C^{\underline{m}}(\Omega)$ is defined as in the first section. One way of defining the space $B_q^{re_j}$ is by interpolation and it is identified with $\left[C(I^d), C^{me_j}(I^d)\right]_{\frac{r}{m}, q}$, $0 < r < m$, with the norm for finite q

$$\| f \|_{r,q}^{(j)'} = \| f \| + \left(\int_0^\infty \left(\frac{K(f, t^m; C, C^{e_j m})}{t^r} \right)^q \frac{dt}{t} \right)^{1/q}$$

and for $q = \infty$

$$\| f \|_{r,\infty}^{(i)'} = \| f \| + \sup_{t > 0} t^{-r} K(f, t^m; C, C^m),$$

where K is the Peetre functional.

With the help of the Johnen's paper [4] we can show that

$$K(f, t^m; C(I^d), C^{me_j}(I^d)) \sim \omega_m^{(j)}(f; t).$$

The proof of this equivalence is based on the construction of extension operators $T^{(j)}: C(I^d) \to C(I^{(j)})$, with $I^{(j)} = I^d \cup \{t: t_i \in I \text{ for } i \neq j, -t_j \in I\}$, and such that $(mh \leqslant 1)$

$$\begin{aligned} &\| T^{(j)} f \| (I^{(j)}) \leqslant C(m) \| f \|, \\ &\omega_m^{(j)}(T^{(j)} f; h)_{I^{(j)}} \leqslant C(m)\, \omega_m^{(j)}(f; h). \end{aligned} \tag{3.1}$$

This leads to an equivalent norm in $B_q^{re_j}$

$$\| f \|^{(j)}_{r,q} = \| f \| + \left(\int_0^\infty \left(\frac{\omega_m^{(j)}(f;\, t)}{t^r} \right)^q \frac{dt}{t} \right)^{1/q}.$$

The anisotropic Besov space is now defined as

$$B_q^{\underline{r}} = B_q^{r_1 e_1} \cap \ldots \cap B_q^{r_d e_d}$$

with the norm

$$\| f \|_{\underline{r},q} = \sum_{j=1}^{d} \| f \|^{(j)}_{r_j,q}.$$

The isotropic Besov space $B_q^r\,(I^d)$ over I^d can be defined as the real interpolation space $\left[C(I^d),\ C^m(I^d)\right]_{\frac{r}{m},q}$, $0 < r < m$, $1 \leqslant q \leqslant \infty$, in which there is equivalent norm

$$\| f \|_{r,q} = \| f \| + \left(\int_0^\infty \left(\frac{\omega_m(f;\, t)}{t^r} \right)^q \frac{dt}{t} \right)^{1/q}.$$

Notice, the space B_∞^r is simply the $Lip(r,\, I^d)$ i.e. the space of f such that $\omega_m(f;\, h) = O(h^r)$ for some integer $m > r$.

Lemma 3.2. The space $B_q^{(r,\ldots,r)}(I^d)$, $r > 0$, $1 \leqslant q \leqslant \infty$, has $\| f \|_{r,q}$ as one of its equivalent norms.

Outline of the proof. Adapting from [4] the method of extending functions in smooth way we can show that there is linear bounded $T\colon C^{\overline{m}}\,(I^d) \rightarrow C^{\overline{m}}(R^d)$ such that for $f \in C(I^d)$ $(Tf)\big|_{I^d} = f$ and for $j = 1,\ldots,d$

$$\omega_m^{(j)}(Tf;\, h) \leqslant C(d,m)\,(h^m \| f \| + \omega_m^{(j)}(f;\, h))$$

for $0 < h \leqslant d^{1/2}/m$. This implies

$$\| Tf \|_{r_j,q}\,(R^d) \leqslant C(d,m,\underline{r})\, \| f \|^{(j)}_{r_j,q},\quad j = 1,\ldots,d.$$

where $m > r_j > 0$.

In particular this gives continuity of the extension operator

$$T:\ B_q^{(r,\dots,r)}(I^d) \longrightarrow B^{(r,\dots,r)}(R^d).$$

It now follows by a result of Solonnikov (see [5], p. 222) that $B_q^{(r,\dots,r)}(R^d) = B_q^r(R^d)$. Consequently

$$\|f\|_{r,q} \leqslant \|Tf\|_{r,q}\ (R^d) \leqslant C(d,m,r) \sum_{j=1}^{d} \|Tf\|_{r,q}^{(j)}\ (R^d)$$

$$\leqslant C(d,m,r) \sum_{j=1}^{d} \|f\|_{r,q}^{(j)},$$

and the opposite inequality is trivial.

4. Spline Bases in B_q^r.

We consider the spline orthonormal system $\{f_j^{(m)},\ j \geqslant -m\}$ defined as in [1] and [3]. For given integer $\mu \geqslant 0$ let

$$P_\mu^{(m)} f = \sum_{j=-m}^{2^\mu} (f,\ f_j^{(m)})\ f_j^{(m)}.$$

The operators $P_\mu^{(m)} : C^m(I) \longrightarrow C^m(I)$ are simultaneous and therefore their d-fold products $P_{d,\mu}^{(m)} = P_\mu^{(m)} \dots P_\mu^{(m)}$ are simultaneous operators in $C^m(I^d)$, and their norms are bounded uniformly in μ in each $C^k(I^d)$, $0 \leqslant k \leqslant m$. Thus we have

<u>Proposition 4.1.</u> For each $m \geqslant 0$ we have in $C^m(I^d)$ the following Schauder decomposition

$$E = \sum_{\mu=\mu_0}^{\infty} Q_\mu, \quad 2^{\mu_0} \geqslant m, \tag{4.2}$$

where $Q_{\mu_0} = P_{d,\mu_0}^{(m)}$, $Q_\mu = P_{d,\mu}^{(m)} - P_{d,\mu-1}^{(m)}$ for $\mu > \mu_0$.

Corollary 4.3. The decomposition (4.2) is a Schauder decomposition in $B_q^r(I^d)$, $0<r<m$, $1 \leqslant q < \infty$.

Theorem 4.4. Let $0<r<m$, $1 \leqslant q \leqslant \infty$. Then in $B_q^r(I^d)$ we have an equivalent norm

$$^1\|f\|_{r,q} = \Big(\sum_{\mu=\mu_0}^{\infty} (2^{r\mu}\|Q_\mu\|)^q\Big)^{1/q}, \quad q<\infty.$$

$$^1\|f\|_{r,q} = \sup_{\mu \geqslant \mu_0} (2^{r\mu}\|Q_\mu\|), \quad q=\infty.$$

Outline of the proof. Using elementary inequalities one shows $^1\|f\|_{r,q}$ is equivalent to

$$^2\|f\|_{r,q} = \|f\| + \Big(\sum_{\mu=0}^{\infty} (2^{\mu r}\|f - P_{d,\mu}^{(m)} f\|^q)\Big)^{1/q}.$$

Since $\|f - P_{d,\mu}^{(m)} f\|$ is equivalent to the best approximation it follows that

$$\|f - P_{d,\mu}^{(m)} f\| \sim \sum_{i=1}^{d} \|f - P_\mu^{(m,i)} f\|, \tag{4.5}$$

where $P_\mu^{(m,i)} = E_1 \dots E_{i-1} P_\mu^{(m)} E_i \dots E_d$, and E_i is the identity operator acting in the i-th variable. Again elementary computation gives

$$^2\|f\|_{r,q} \sim \sum_{i=1}^{d} {}^2\|f\|_{r,q}^{(i)} \sim \sum_{i=1}^{d} {}^1\|f\|_{r,q}^{(i)}$$

with

$$^2\|f\|_{r,q}^{(i)} = \|f\| + \Big(\sum_{\mu=0}^{\infty} (2^{\mu r}\|f - P_\mu^{(m,i)} f\|)^q\Big)^{1/q},$$

$$^1\|f\|_{r,q}^{(i)} = \Big(\sum_{\mu=\mu_0}^{\infty} (2^{r\mu}\|Q_\mu^{(1)}\|)^q\Big)^{1/q},$$

$$Q_{\mu_0}^{(i)} = P_{\mu_0}^{(m,i)}, \quad Q_{\mu}^{(i)} = P_{\mu}^{(m,i)} - P_{\mu-1}^{(m,i)}, \quad \mu > \mu_0.$$

Following the proofs in one-dimensional case (see [1]) we can prove the inequalities

$$\| f - P_{\mu}^{(m,i)} \| \leq C(m)\, \omega_m^{(i)}(f, 2^{-\mu}), \quad \mu \geq \mu_0 ,$$
$$\| \Delta_{he_i}^{m} Q_{\mu}^{(i)} f \| \leq C(m)(h\, 2^{-\mu})^m \| Q_{\mu}^{(i)} f \|, \quad 0 < mh \leq 1. \tag{4.6}$$

More or less standard argument in combination with Proposition 4.1 and inequalities (4.6) gives

$$\| f \|_{r,q}^{(i)} \sim {}^{1}\| f \|_{r,q}^{(i)}, \quad i = 1, \ldots, d. \tag{4.7}$$

To complete the proof we use (4.7) and Lemma 3.2.

<u>Theorem 4.8.</u> Let q, $1 \leq q \leq \infty$, $d \geq 1$ and $r > 0$ be given, and let $f \in C(I^d)$. Then $f \in B_q^r(I^d)$ if and only if for an integer $m > r$

$$^{3}\| f \|_{r,q} = \| f \| + \sum_{i=1}^{d} \Big(\sum_{2^{\mu} \geq m} (2^{\mu r} \| \Delta_{d,\mu}^{me_i} f \|)^q \Big)^{1/q} < \infty, \tag{4.9}$$

and $^{3}\| f \|_{r,q}$ is equivalent norm in B_q^r.

P r o o f: It is clear that $f \in B_q^r$ implies (4.9). Let us now assume that (4.9) holds. For the operators K_{μ}^{d} according to Proposition 2.3 we have

$$\| f - K_{\mu}^{d} f \| \leq C \sum_{i=1}^{d} \omega_m^{(i)}(f; 2^{-\mu}),$$

and by Lemma 2.2, (iii),

$$\| D^{me_i} K_{\mu}^{d} f \| \leq C\, 2^{m\mu} \| \Delta_{d,\mu}^{me_i} f \|$$

Using the uniform boundedness of $\{ P_{d,\mu}^{(m)}, \mu \geq 0 \}$, (4.5) and (4.6) we get

$$\left\| f - P^{(m)}_{d,\mu} f \right\| \leqslant \left\| f - K^d_\mu f \right\| +$$

$$+ \left\| K^d_\mu f - P^{(m)}_{d,\mu} K^d_\mu f \right\| + \left\| P^{(m)}_{d,\mu} (f - K^d_\mu f) \right\|$$

$$\leqslant C \Big(\left\| f - K^d_\mu f \right\| + \sum_{i=1}^{d} \omega^{(i)}_m (K^d_\mu f ; 2^{-\mu}) \Big)$$

$$\leqslant C \Big(\sum_{\nu=\mu}^{\infty} \left\| (K^d_{\nu+1} - K^d_\nu) f \right\| + \sum_{i=1}^{d} \left\| \Delta^{me_i}_{d,\mu} f \right\| \Big),$$

whence by Lemma 2.4.

$$\left\| f - P^{(m)}_{d,\mu} f \right\| \leqslant C \sum_{i=1}^{d} \sum_{\nu=\mu}^{\infty} \left\| \Delta^{me_i}_{d,\mu} f \right\|. \tag{4.10}$$

This and (4.9) imply that ${}^2\|f\|_{r,q} < \infty$, and consequently that ${}^1\|f\|_{r,q} < \infty$, and by Theorem 4.4 that $f \in B^r_q$.

5. The main result

We now consider a real valued random field (r.f.) $\left\{X(t); t \in \bigcup_{\mu \geqslant 0} T^d_\mu\right\}$.

Theorem 5.1. Let for given m and $C > 0$, $0 < \gamma \leqslant m\beta$,

$$E \left| \Delta^{me_i}_{d,\mu} X(t) \right|^\beta \leqslant C\, 2^{-\mu(d+\gamma)}, \tag{5.2}$$

hold for $t \in T^d_\mu(me_i)$, $i = 1, \ldots, d$, $2^\mu \geqslant m$. Then the r.f. X has continuous extension to I^d and

$$P\left\{\omega_m(X(\cdot); h) = O(h^\alpha)\right\} = 1 \quad \text{for} \quad 0 < \alpha\beta < \gamma. \tag{5.3}$$

P r o o f: We are going to show at first that the r.f. $\{X(t), t \in T^d_\mu, \mu \geqslant 0\}$ has continuous extension to I^d. It follows by (5.2) that

$$E \left\| \Delta_{d,\mu}^{me_i} X \right\|^{\beta} \leqslant C\, 2^{-\mu\gamma}, \quad i = 1,\dots, d$$

and this implies the convergence of

$$\sum_{\mu} \left\| \Delta_{d,\mu}^{me_i} X \right\|, \quad i = 1,\dots, d,$$

with probability 1. Thus by Corollary 2.3 the r.f. $(2^{\mu_0} \geqslant m)$

$$X(t) = L^{d}_{\mu_0} X(t) + \sum_{\mu=\mu_0}^{\infty} (L^{d}_{\mu+1} - L^{d}_{\mu})\, X(t)$$

is the continuous extension to I^d of the given r.f. Therefore by Proposition 4.1. we have in $C(I^d)$

$$X(\cdot) = \sum_{\mu=\mu_0}^{\infty} Q_{\mu} X(\cdot)$$

with probability 1. Applying now (4.10) we find in the same way as in [2] that

$$E \left\| Q_{\mu} X \right\|^{\beta} \leqslant C\, 2^{-\mu\gamma},$$

whence we infer

$$P\left\{ \left\| Q_{\mu} X \right\| \geqslant 2^{-\mu\alpha} \right\} \leqslant C\, 2^{\mu(\alpha\beta-\gamma)},$$

and therefore with probability 1

$$\left\| Q_{\mu} X \right\| = O(2^{-\alpha\mu}),$$

i.e. by Theorem 4.4. $X(\cdot) \in B^{\alpha}_{\infty}$ and this completes the proof.

In the gaussian case as a consequence we obtain

Theorem 5.4. Let $\left\{ X(t);\ t \in T^d, \mu \geqslant 0 \right\}$ be a gaussian r.f. such that

$$E \left| \Delta_{d,\mu}^{me_i} X(t) \right|^2 \leqslant C\, 2^{-\mu\lambda}, \quad 2^{\mu} \geqslant m,$$

holds for $t \in T^d_{\mu}(me_i)$, $i = 1,\dots, d$, with some constants $C > 0$, $0 < \lambda < 2m$. Then X has continuous extension to I^d such that

$$P\left\{\omega_m(X(\cdot);\ h) = O(h^{\alpha})\right\} = 1 \quad \text{for} \quad 0 < 2\alpha < \lambda .$$

References

[1] C i e s i e l s k i, Z., Constructive function theory and spline systems. Studia Math., 53, 277-302 (1975).

[2] - Approximation by splines and its application to Lipschitz classes and to stochastic processes. Teoria približenii funkcii, izd. "Nauka", Moskwa 1977, pp.397-404. (Approximation Theory of Function. Proc. of the Conference in Kaluga 1975.Ed., by "Nauka" Moscow 1977).

[3] - and D o m s t a, J., Construction of orthonormal basis in $C^m(I^d)$ and $W_p^m(I^d)$. Studia Math., 41, 211-224 (1972).

[4] J o h n e n, H., Inequalities connected with the moduli of smoothness. Matematicki Vestnik, 19, 290-303 (1973).

[5] N i k o l s k i i, S. M., Approximation of functions of several variables and embedding theorems. Ed. by "Nauka", Moscow 1977 (in Russian).

[6] R y l l, J., Interpolating bases for spaces of differentiable functions. Studia Math., 63, 125-144 (1978).

MONOTONE DEPENDENCE FUNCTION:

BACKGROUND, NEW RESULTS AND APPLICATIONS

by

Jan Ćwik, Teresa Kowalczyk, Adam Kowalski, Elżbieta Pleszczyńska, Wiesław Szczesny and Teresa Wierzbowska

Polish Academy of Sciences, Warsaw

1. Introduction

The monotone dependence function $\mu(X,Y)$ is a function- valued measure of dependence of a random variable X on a random variable Y; more precisely, μ is a parameter assigning to any (X,Y) with nondegenerate marginal distributions and finite expectation of X a continuous function from (0,1) into [-1,1] such that $\mu(p; X,Y)$ is for any $p \in (0,1)$ a suitably normalized expected value of X under the condition that Y exceeds its p-th quantile.

The monotone dependence function was defined in case of continuous marginals in [4] and its generalization for discrete marginals was given in [2]. Let us recall that definition (for the sake of simplicity in the continuous case):

$$\mu(p;\ X,Y) = \begin{cases} \mu^+\ (p;\ X,Y), \text{ if } \mu^+(p;\ X,Y) \geqslant 0, \\ -\mu^+\ (p;\ -X,Y), \text{ otherwise,} \end{cases}$$

where

$$\mu^+(p;\ X,Y) = (E(X|Y > y_p) - EX)/(E(X|X > x_p) - EX),$$

x_p and y_p being any p-th quantiles of X and Y, respectively.

In view of the following properties, $\mu(X,Y)$ is a convenient and efficiently computable descriptive statistic:

$$\forall p \in (0,1) \qquad -1 \leq \mu(p;\ X,Y) \leq 1; \tag{1.1}$$

$$\mu(p;\ X,Y) \equiv \begin{cases} 1(-1), \text{ iff there exists a nondecreasing} \\ \text{(nonincreasing) function } f\colon R \to R \\ \text{such that } X = f(Y) \text{ a.e.,} \\ 0, \quad \text{iff } E(X|Y) = EX \text{ a.e.;} \end{cases} \tag{1.2}$$

for any real a and b, $a \neq 0$

$$\mu(p;\ aX + b,\ f(Y)) = \begin{cases} \operatorname{sgn}(a)\mu(p;\ X,Y) \text{ if } f \text{ is } Y - \text{a.e.} \\ \qquad\qquad \text{increasing,} \\ -\operatorname{sgn}(a)\ \mu(1-p;\ X,Y) \text{ if } f \text{ is } Y - \text{a.e. decreasing,} \end{cases} \tag{1.3}$$

if $\Phi_X, \Phi_{X'}, \Phi_Y, \Phi_{Y'}$ are distribution functions of X, X', Y, Y' and $\Phi_X = \Phi_{X'}$, $\Phi_Y = \Phi_{Y'}$, then

$$\begin{aligned} &\forall p \in (0,1) \quad \mu(p;\ X,Y) \geq \mu(p;\ X',\ Y') \text{ iff} \\ &\forall p \in (0,1) \quad E(X|Y > y_p) \geq E(X'|Y > y_p) \end{aligned} \tag{1.4}$$

and the equality $\mu(X,Y) = \mu(X',Y')$ holds iff $E(X|Y)$ and $E(X'|Y')$ have the same distribution.

These properties constitute a suitable background for some further extensions; in particular the multivariate case was studied in [6]. For several distributions the analytic formulae of the monotone dependence function are already known; the case of the multivariate lognormal distribution is of a particular interest [9].

In Sec. 2 of this paper we are concerned with the connections between the regression functions and the monotone dependence functions; a full treatment of these problems is given in [3]. Apart from those new theoretical results, an example of application, namely in the field of chronological ordering, is discussed in Sec. 3. The content of Sec. 3 is a short outline of [5] and [8].

2. Shape of the monotone dependence function

For any random variable ξ let Φ_ξ be the distribution function of ξ and let $a_\xi = \inf\{X: \Phi_\xi(x) > 0\}$, $b_\xi = \sup\{x: \Phi_\xi(x) < 1\}$ (in particular, a_ξ may be equal to $-\infty$ and b_ξ to $+\infty$). Let B denote the set of random variables (X,Y) for which the marginal distribution functions are continuous and EX exists. For any $(X,Y) \in B$ there exist increasing functions f^+ and f^- from R into R such that random variables $Y^+ = f^+(Y)$ and $Y^- = f^-(Y)$ satisfy the equalities $\Phi_{Y^+} = \Phi_X$ and $\Phi_{Y^-} = \Phi_{-X}$. For any $(X,Y) \in B$ let r^+ (r^-) be a real function defined on (a_X, b_X) ((a_{-X}, b_{-X})) such that $r^+(Y^+)$ ($r^-(Y^-)$) is a representant of the conditional expectation of X given Y^+ (Y^-). Finally, for any $s \in R$ it will be convenient to denote $\Phi_X(s)$ by p_s^+ and $\Phi_{-X}(s)$ by p_s^-. We shall write shortly $\mu(p)$ whenever it is clear which random variables are considered.

Theorem 1. Let $(X,Y) \in B$ and $s \in (a_X, b_X)$. If

(i) $\mu(p_s^+) > 0$ and if there exists r^+ such that

(ii) r^+ restricted to (a_X, s) is continuous, nondecreasing and convex (concave), and

(iii) the line $L^+(t) = \mu(p_s^+)t + (1-\mu(p_s^+))$ EX has exactly one intersection with r^+ at $t_0 \in (a_X, s) \cap (a_X, EX)$ and $L^+(t) > r^+(t)$ (<) for $t \in (t_0, s]$ then μ is increasing (decreasing) on $(0, p_s^+)$. Moreover, μ is zero on $(0, p_s^+)$ iff there exists r^+ constant on (a_X, s) and under (i) μ is positive and constant on $(0, p_s^+)$ iff there exists r^+ equal to L^+ on (a_X, s).

P r o o f. It suffices to prove the theorem under the assumption that $EX = EY^+ = 0$. Note that if r^+ is nondecreasing on (a_X, s), then

$$\int_{a_X}^{t} r^+(y)\, d\Phi_{Y^+} < 0 \qquad \text{for} \quad t \in (a_X, s)$$

since

$$\int_{a_X}^{b_X} r^+(y)\,d\Phi_{Y^+} = EX = 0.$$

It follows from the definition of μ^+ that for any $p \in (0, p_s^+)$

$$0 < \int_{a_X}^{t_p} r^+(t)\,d\Phi_{Y^+} \Big/ \int_{a_X}^{t_p} t\,d\Phi_{Y^+} \tag{2.1}$$

and

$$\frac{d}{d_p}\mu^+(p) \geqslant 0 \ (\leqslant 0) \quad \text{iff} \quad r^+(t_p) \leqslant \mu(p)t_p \tag{2.2}$$

with the equality on the left-hand side of (2.2) holding iff both expressions on the right-hand side are equal.

To show the first part of the theorem two cases will be considered in turn.

1° $s \leqslant 0$. By (iii)

$$r^+(s) < \mu(p_s^+)s \quad (>). \tag{2.3}$$

Let s_o be the least point in $[a_X, s)$ such that μ is increasing (decreasing) on $(p_{s_o}^+, p_s^+)$. In view of (i) and (2.3) the continuity of r^+ and μ implies the existence of s_o, and its equality to a_X will be shown now. Assume that $a_X < s_o < s$.

Then by (2.2) there exists $s_1 \in (a_X, s_o]$ such that

$$\mu(p_{s_1}^+)s_1 < \mu(p_s^+) \ (>) \text{ and } r^+(s_1) \geqslant \mu(p_{s_1}^+)\cdot(s_1).\ (\leqslant). \tag{2.4}$$

The line $y = \mu(p_{s_1}^+)\cdot t$ satisfies

$$\int_{a_X}^{s_1} r^+(t)\,d\Phi_{Y^+} = \int_{a_X}^{s_1} \mu(p_{s_1}^+)t\,d\Phi_{Y^+}.$$

Obviously, the line intersects r^+ at least once; but as $\mu(p_{s_1}^+) < \mu(p_s^+)$ and r^+ is convex (concave), there is exactly one intersection at a point $t_1 < s_1$ and $r^+(t) < \mu(p_{s_1}^+)t$ (>) for $t \in (t_1, s_1]$ which contradicts (2.4).

2° $s > 0$. By (iii)

$$r^+(0) < 0 \quad (>).$$

Two cases are possible:

a) $r^+(s) \leqslant 0$ $\quad (r^+(s) \geqslant s)$,

b) $r^+(s) > 0$ $\quad (r^+(s) < s)$.

The proof under the assumption a) is obvious, so assume b). Let $t_1 \in (0,s)$ be such that $r^+(t_1) = 0$ $(r^+(t_1) = t_1)$. If follows from a) that $\mu(p)$ is increasing (decreasing) on the interval $(0, p_{t_1})$. By similar considerations as in the case 1° $\mu(p)$ is increasing (decreasing) on the interval (p_{t_1}, p_s). Hence the proof of the first part of the theorem is completed. The second part follows immediately from (2.2) and

$$\int_{a_X}^{t_p} r^+(t)\, d\,\Phi_{Y^+} = \int_{a_X}^{t_p} \mu(p)t\, d\,\Phi_{Y^+} \quad \text{for } p \in (0, p_s^+).$$

<u>Corollary.</u> Let $(X,Y) \in B$. If there exists r^+ which is continuous, nonlinear, nondecreasing and convex (concave), then $\mu(X,Y)$ is positive and increasing (decreasing). Moreover, linearly increasing r^+ exists iff $\mu(X,Y)$ is constant and positive while constant r^+ exists iff $\mu(X,Y)$ is identically equal to zero.

P r o o f: It can be verified that for any $s \in (a_X, b_X)$ the assumptions of Th. 1 are satisfied. An independent proof will be published in [3].

It follows from Theorem 1 applied to $(-X, Y^-)$ that for $s \in (a_{-X}, b_{-X})$ such that

(i′) $\mu(p_s^-) < 0$

and for which there exists r^- such that

(ii′) r^- restricted to $(a_{-X}, s]$ is continuous, nonincreasing and convex (concave), (iii') the line $L^{-}(t) = \mu(p_s^{-})t + (1 + \mu(p_s^{-}))$ EX has exactly one intersection with r^{-} at $t_o \in (a_{-X}, s] \cap (a_{-X}, -EX)$ and $L^{-}(t) > r^{-}(t)$ (<) for $(t_o, s]$, then μ is increasing (decreasing) on $(0, p_s^{-})$. Moreover, μ is zero on $(0, p_s^{-})$ iff there exists r^{-} constant on (a_{-X}, s) and under (i') μ is negative and constant on $(0, p_s^{-})$ iff there exists r^{-} equal to L^{-} on (a_{-X}, s).

The dual versions for (s, b_X) can be easily deduced from Theorem 1 and its corollaries.

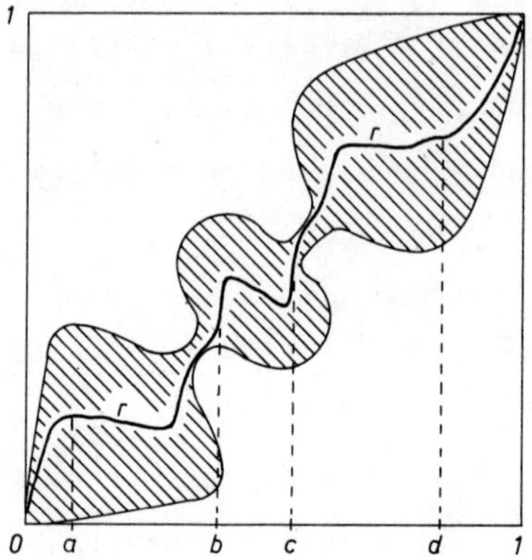

Fig. 1. Regression function r and the support of some uniform bivariate distribution P

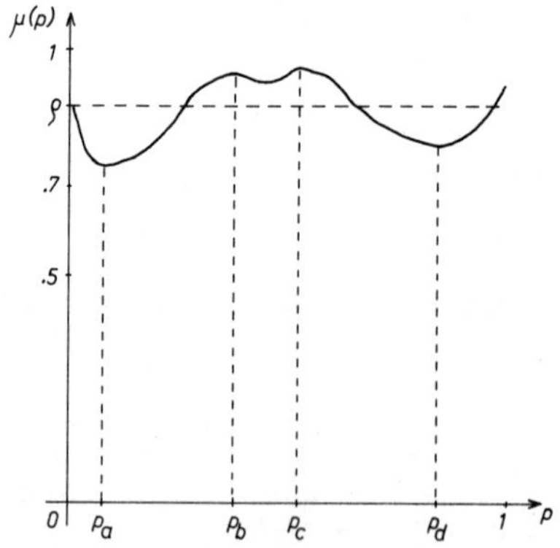

Fig. 2. Monotone dependence function $\mu(p)$ and correlation coefficient ρ of distribution P (cf. Fig. 1)

Example. Suppose that the distribution of (X,Y) is uniform over the area indicated on Fig. 1. Intuitively, dependence of X on Y is positive and it is changing from stronger to weaker and from weaker to stronger three times in turn. The regression function r, presented on Fig. 1, coincides with r^+ since the marginal distributions are equal. It is easy to verify that the assumptions of suitable versions of Theorem 1 are fulfilled on $(0,a)$ and $(d,1)$. Accordingly, the function $\mu(X,Y)$, given on Fig. 2, is decreasing on $(0,p_a)$ and increasing on $(p_d,1)$, where p_a and p_d denote the values of marginal distribution function at a and d, respectively. The behaviour of the function $\mu(X,Y)$ in the interval (p_a,p_d) reflects the fluctuations of the correspondence of X on Y in the underlying distribution.

3. Chronological ordering of Poisson streams

For any Poisson stream on $(0,T_0)$ with an integrable timedepending intensity $\lambda = (\lambda(t),\ t \in (0,T_0))$ an "age" function $a: (0,T_0) \longrightarrow [0,1]$ can be defined as

$$a(t) = \frac{\int_0^t \lambda(s)\,ds}{\int_0^{T_0} \lambda(s)\,ds} \quad .$$

Let $S_1,\dots,S_k$ be Poisson streams on $(0,T_0)$ with intensities λ_1, ..., λ_k and age functions $a_1,\dots,a_k$. Then $S_1,\dots,S_k$ are said chronologically ordered if $a_i \geqslant a_j$ for $i<j$, $i,j=1,\dots,k$. The ordering means that for any $i<j$ and any $t \in (0,T_0)$ the expected number of arrivals in the i-th stream up to the moment t forms a fraction of the expected number of arrivals of the i-th type in the interval $(0,T_0)$ larger than the respective fraction concerning the j-th stream.

It is intuitively clear that the "shift" between the chronologically ordered streams can be smaller or greater so that the ordering can be specified as weaker or stronger. To deal with this problem it is convenient to introduce for any sequence of intensities

$\lambda_1,\dots,\lambda_k$ a pair of random variables (X,T) taking values from $\{1,\dots,k\}$ and $(0,T_0)$, respectively, the distribution P of (X,T) being determined by $\lambda_1,\dots,\lambda_k$ in the following way: for any $i=1,\dots,k, t\in(0,T_0)$

$$P(X=i,T<t) = \frac{\int_0^t \lambda_i(s)ds}{\sum_{j=1}^k \left(\int_0^{T_0} \lambda_j(s)ds\right)} .$$

Consequently, for any $i=1,\dots,k$ the distribution of $T_{X=i}$, i.e. of the random variable T truncated to the set $\{X=i\}$, has the density equal $\lambda_i / \int_0^{T_0} \lambda_i(s)ds$.

It is easy to see that streams $S_1,\dots,S_k$ with intensities $\lambda_1,\dots,\lambda_k$ are chronologically ordered iff $T_{X=i} \underset{st}{\leq} T_{X=j}$ for $i<j$, $i,j=1,\dots,k$. Then using the terminology introduced in Lehmann (1968) $S_1,\dots,S_k$ are chronologically ordered iff T is positively regression dependent on X. This suggests that the chronological ordering of Poisson streams is closely connected with monotone dependence of two random variables and that measuring the intensity of the ordering can be replaced by measuring the intensity of monotone dependence of the corresponding variables.

However, real-valued measures are not sensitive enough for this purpose. It seems desirable to use a function-valued measure which will give the opportunity to follow the changes of intensity corresponding to the age p of the joint stream ($p\in(0,1)$). A natural candidate is the monotone dependence function of X on T, which will be denoted here by $\mu(\cdot;\lambda_1,\dots,\lambda_k)$.

It is easy to see that $\mu(\cdot;\lambda_1,\dots,\lambda_k)$ satisfies:

(3.1) For any $\lambda_1,\dots,\lambda_k$ $\mu(p;\lambda_1\dots,\lambda_k) \equiv 1$ if and only if for any $i, j = 1,\dots,k$ and any x and y belonging to the supports of λ_i and λ_j, respectively,

$$i<j \Rightarrow x<y.$$

(3.2) If $\lambda_1,\dots,\lambda_k$ are chronologically ordered then $\mu(\cdot;\lambda_1,\dots,\lambda_k)$ is nonnegative but not identically equal to zero.

(3.3) If the age functions are identical for all streams, then $\mu(p;\lambda_1,\dots,\lambda_k) \equiv 0$.

(3.4) For any $\lambda_1,\dots,\lambda_k$ $\mu(p;\lambda_1,\dots,\lambda_k) \equiv -\mu(p;\lambda_k,\dots,\lambda_1)$.

(3.5) For any $c>0$, $\lambda_1,\dots,\lambda_k$ and $p \in (0,1)$

$$\mu(p;c\lambda_1,\dots,c\lambda_k) = \mu(p;\lambda_1,\dots,\lambda_k).$$

(3.6) For any $\lambda_1,\dots,\lambda_k$ on $(0,T_o)$ and any $\Delta_1,\dots,\Delta_k$ satisfying $0 = \Delta_1 \leq \dots \leq \Delta_k$ let $\lambda_1',\dots,\lambda_k'$ be intensities on $(0,T_o+\Delta_k)$ such that for any $i=1,\dots,k$

$$\lambda_i'(t) = \begin{cases} \lambda_i(t-\Delta_i) & \text{when } \Delta_i < t < T_o + \Delta_i, \\ 0 & \text{for other } t. \end{cases}$$

Then for any $p \in (0,1)$

$$\mu(p;\lambda_1,\dots,\lambda_k) \leq \mu(p;\lambda_1',\dots,\lambda_k').$$

(3.7) Let $0=t_o<t_1<\dots<t_m=T_o$ and suppose that $S_1,\dots,S_k$ are independent Poisson streams with intensities $\lambda_1,\dots,\lambda_k$ defined on $(0,T_o)$ which are constant in the intervals (t_{j-1},t_j) for $j=1,\dots,m$. Let ν_{ijn} denote the number of arrivals in the interval (t_{j-1},t_j) which appear jointly in n independent realizations of the stream S_i. Let P denote the probability distribution, according to which one can generate independently infinite sequences of realizations of a set of Poisson streams $S_1,\dots,S_k$ on $(0,T_o)$. Putting

$$\hat{\lambda}_i^{(n)}(t) = \nu_{ijn}/n(t_j - t_{j-1}) \text{ for } t \in (t_{j-1},t_j),$$

we have for any $p \in (0,1)$

$$P(\lim_{n\to\infty} \mu(p;\ \hat{\lambda}_1^{(n)},\ldots,\hat{\lambda}_k^{(n)}) = \mu(p;\ \lambda_1,\ldots,\lambda_k) = 1.$$

The above properties show that $\mu(\cdot;\ \lambda_1,\ldots,\lambda_k)$ provides essential information about the intensities from the point of view of chronological ordering. Property (3.7) deals with a straightforward estimator of $\mu(\cdot;\ \lambda_1,\ldots,\lambda_k)$.

An inspiration for this research emerged from the contacts with the archeologists interested in a "chronological ordering" of various types of ceramics found in an ordered set of layers (cf. [8]).

4. Final remarks

It is hoped that this paper will provide some insight into our investigations on stochastic dependence. New methods of measuring dependence are treated by us from both theoretical and practical points of view. Some theoretical considerations on consistent systems of (nonnecessarily real-valued) measures of global, monotone and linear dependence are given in [1]. Connections between statistical decision theory and measures of dependence are investigated by Szczesny [10], special interest is paid to the problem of screening. Schemes of dependence in biological models are worked out by Wierzbowska. Statistical inference based on the shape of the monotone dependence functions is considered by Kowalczyk.

References

[1] D ą b r o w s k a, D., Regression-based measures of dependence. To be published (1978).

[2] K o w a l c z y k, T., General definition and sample counterparts of monotonic dependence functions of bivariate distribu-

tions. Math. Operationsforsch. Statist., Ser. Statistics, 8, 351--365 (1977).

[3] K o w a l c z y k, T., Shape of the monotone dependence function. To be published (1979).

[4] K o w a l c z y k, T., P l e s z c z y ń s k a, E., Monotonic dependence functions of bivariate distributions. Ann. Statist. 5, 1221-1227 (1977).

[5] K o w a l c z y k, T., Ć w i k, J., P l e s z c z y ń s k a, E., Chronological ordering of Poisson streams. Proceedings of Eighth Prague Conference on Information Theory, Statistical Decision Functions and Random Processes. Prague 1978, Vol. C (to appear).

[6] K o w a l c z y k, T., K o w a l s k i, A., M a t u s z e w s k i, A., P l e s z c z y ń s k a, E., Screening and monotonic dependence functions in the multivariate case. Ann. Statist. 7 (to appear) (1979).

[7] L e h m a n n, E. L., Some concepts of dependence. Ann. Math. Statist., 37, 1137-1153 (1966).

[8] M a e t z k e, G., P l e s z c z y ń s k a, E., T a b a c z y ń s k i, S., Problems of inference based on stratigraphic sequences: a tentative model. Archaeologia Polona XXI (to appear) (1979).

[9] N a l b a c h-L e n i e w s k a, A., Measures of dependence of the multivariate lognormal distribution. Math. Operationsforsch. Statist., Ser. Statistics 10 (to appear) (1979).

[10] S z c z e s n y, W., A decision theory approach to measures of dependence. To be published (1979).

LIFETESTING FOR MATCHED PAIRS

by

Kjell A. Doksum*
University of California, Berkeley

1. Introduction

Suppose a drug or chemical is to be tested for its effect on a group of lab animals. Because of the great variability between responses of lab animals it may be desirable to form matched pairs and give one animal in a given pair the treatment while the other animal serves as control. One simple way of obtaining matched pairs is described by Mantel et al. [10] where rats may be bought from a breeder in pairs of littermates of male (and female) rats.

In such experiments, what is measured is usually time-to-response, where the response is death or appearance of disease.

Plausible distributional models for such data are exponential, Weibull, gamma, lognormal, etc. In an experiment with moderate sample sizes, it is practically impossible to distinguish between these various models, i.e. a test for a Weibull hypothesis will have very low power for a gamma alternative. Thus we take the point of view that we can not determine a (approximate) parametric distributional model and turn to rank tests. These will have correct significance level. Moreover, it we derive the optimal rank test for one of the

* This research was supported by National Science Foundation grant MCS-78-01422.

plausible models above, we can expect it to have good power properties generally.

In Section 2, we consider the Mantel-Haenszel statistic and show that it is asymptotically equivalent to a rank exponential scores statistic based on the remnants (maxima) from pairs. In Section 3, we consider conditional rank tests based on the joint ranks of all responses, and give the locally most powerful conditional rank test for a parametric alternative. Sections 4 and 5 deal with two different classes of signed rank statistics, namely those based on treatment-control differences, and those based on log differences. In the case of log differences, we find that a tail-ordering on the distributions in the model induces an ordering on the power of rank tests.

2. A Mantel-Haenszel and exponential scores statistic

We let X_i and Y_i denote control and treatment responses in the i-th matched pair, respectively. It is assumed that $(X_1,Y_1),\dots,(X_N,Y_N)$ form a random sample from a population with a bivariate continuous distribution function $F(x,y)$, $x \geq 0$, $y \geq 0$. X_i and Y_i are assumed to be non-negative. The null-hypothesis H of bivariate symmetry is that (X_i,Y_i) and (Y_i,X_i) have the same distribution, $i = 1,\dots,N$. One rank statistic available for this hypothesis is the Mantel-Haenszel statistic given by Mantel et al, ([10], p.6) They proceed as follows. Let $M_i = \max\{X_i,Y_i\}$, $i = 1,\dots,N$ be called remnant responses (times) and let $L_i = \min\{X_i,Y_i\}$, $i = 1,\dots,N$ be called non-remnant responses. Order the N remnant times, and for the i-th remnant time in size form the 2×2 contingency table

	Died	Survived	At risk
Treatment	A_{1i}	$S-z_i$	$S - z_{i-1}$
Control	$1-A_{1i}$	$(N-S)-i+z_i$	$(N-S)-i+1+z_{i-1}$
Total	1	$N - i$	$N - i + 1$

where $A_{1i} = 1$ if the i-th remnant is a Y, $A_{1i} = 0$ otherwise, $z_j = \sum_{i=1}^{j} A_{1i}$, and $S = z_N$ = No. of treatment remnants = Sign statistic. Similarly, order the non-remnants times and for the i-th one compute the contingency table

	Died	Survived	At risk
Treatment	A_{2i}	$1 - A_{2i}$	1
Control	$1-A_{2i}$	A_{2i}	1
Total	1	1	2

where $A_{2i} = 1$ if the i-th non-remnant time is a Y, $A_{2i} = 0$ otherwise.

Let $e_{1i} = (S-z_{i-1})/(N-i+1)$, $v_{1i} = e_{1i}(1-e_{1i})$, $e_{2i} = \frac{1}{2}$ and $v_{ki} = e_{ki}(1-e_{ki})$, $k = 1,2$, then the Mantel-Haenszel statistic is

$$\frac{\left(\left|\sum_{k=1}^{2}\sum_{i=1}^{N}(A_{ki}-e_{ki})\right| - \frac{1}{2}\right)^2}{\sum_{k=1}^{2}\sum_{i=1}^{N} v_{ki}} .$$

Clearly,

$$\sum_{i=1}^{N}(A_{2i}-e_{2i}) = (N-S) - \tfrac{1}{2}N = \tfrac{1}{2}N - S.$$

Let $R_1,\ldots,R_S$ denote the ranks of the treatment (Y) remnants among $M_1,\ldots,M_N$, and let

$$J_0(k) = \sum_{j=N+k-1}^{N} (1/j), \quad k = 1,\ldots,N,$$

be the exponential scores. Then

$$T = \sum_{i=1}^{S} J_0(R_i)$$

is an exponential scores (Savage type) statistic. From Koziol and Petkau [8] we find,

$$\sum_{i=1}^{N} (A_{1i}-e_{1i}) = - \sum_{i=1}^{S} J_0(R_i) + S.$$

Thus

$$\sum_{k=1}^{2} \sum_{i=1}^{N} (A_{ki}-e_{ki}) = -T + \tfrac{1}{2}N$$

and the main part of the Mantel-Haenszel statistic is equivalent to the exponential scores statistic.

Even though $R_1,\ldots,R_S$ are not signed ranks (for signed ranks, see Section 4 and Lehmann ([9], p. 125)), their distribution under H is the same as that of signed ranks. Thus if $S_1<\ldots<S_S$ denotes $R_1,\ldots,R_S$ ordered, then

$$P_H(S_1 = s_1,\ldots,S_S = s_S,\ S=s) = \frac{1}{2^N}\ . \qquad (2.1)$$

To see this, let $M_i^* = M_i \operatorname{sign}(Y_i-X_i)$. Then $S_1<\ldots<S_S$ are the ordered signed ranks of $M_1^*,\ldots,M_N^*$. Moreover, since X_i and Y_i are interchangeable under the hypothesis, the distribution of M_i^* is symmetric about zero which is what is needed to conclude (2.1).

It follows easily that the expected value of T is

$$E_H(T) = \frac{1}{2} \sum_{k=1}^{N} J_0(k) = \frac{1}{2} N$$

while

$$\mathrm{Var}_H(T) = \frac{1}{4}\sum_{k=1}^{N} J_0^2(k) = \frac{1}{2}(N - \frac{1}{2}\sum_{k=1}^{N}\frac{1}{k}),$$

It can be shown that $\sum_{k=1}^{2}\sum_{i=1}^{N} v_{ki}$ converges in probability to $\mathrm{Var}_H(T)$ as $N \to \infty$ under H. Thus the Mantel-Haenszel statistic has approximately a χ_1^2 distribution under H and is asymptotically equivalent to $\left|T - \frac{1}{2}N\right|$.

The hypothesis distribution of T is given by Morita [11].

It should be noted that the Mantel-Haenszel statistic of Mantel et al. [10] can be applied not only to matched pairs but also to matched k-tuples, it can be applied to censored data as well, and it can handle ties.

3. Conditional rank tests

The tests of Section 2 do not use the ranks of the non-remnant responses. Moreover, the question of putting the ranks together in some optimal fashion is not considered. In this section, we consider ranks that take the non-remnant responses into account and from which the ranks of the previous section can be computed. We also give locally most powerful rank tests for certain alternatives.

Changing notation from Section 2, we let $R_1,\dots,R_N$ $(S_1,\dots,S_N)$ denote the ordered ranks of $X_1,\dots,X_N$ $(Y_1,\dots,Y_N)$ among $X_1, Y_1,\dots,X_N, Y_N$. These ranks arise naturally from invariance considerations when we test the hypothesis H_0 that (X,Y) has the same distribution as (Y,X) (bivariate symmetry) against the alternative that $P(X \leq h(Y)) \geq P(Y \leq h(X))$ for all $h \in \mathcal{K}$, where $\mathcal{K}$ is the class of continuous, increasing functions h from R onto R. This problem is invariant under transformations of the form $(X_i, Y_i) \to (h(X_i), h(Y_i))$, $i = 1,\dots,N$; $h \in \mathcal{K}$ and if we insist on tests invariant under these transformations, we are led to tests based on the above ranks. Some relevant references are Schaafsma [12,13], Bell and Haller [1], Yanagimoto and Sibuya [15] and Snijders [14].

Rank tests typically reject H_0 for large value of statistics of the form

$$T_a = \sum_{i=1}^{N} a(S_i)$$

for some increasing function $a(k)$ of $k \in \{1,\ldots,N\}$. The distribution of $S_1,\ldots,S_N$ under H depends on the distribution of (X_i,Y_i). In order to have distribution-free (similar) tests for the hypothesis H of bivariate symmetry of Section 2, we condition on the configuration of ranks $\underset{\sim}{r}$. That is, we compute probabilities conditionally given that the ranks in the pair (X_i,Y_i) are the observed ranks $\{r_i,s_i\}$, and under the hypothesis, r_i is equally likely to be the rank of X_i or the rank of Y_i. In other words, we use the permutation distribution of T obtained by assigning equal probability to all the 2^N possible values of T obtained by interchanging X's and Y's within pairs.

We consider conditional rank tests that are optimal for the following generalized scale alternative:

$$X_i = W_i + (\Delta -1)\,\varepsilon_i, \quad Y_i = \Delta W_i' + (\Delta-1)\,\varepsilon_i,$$

where W_i and W_i' independent identically distributed with continuous distribution function H satisfying $H(0)=0$ and the ε_i are independent of the W_i and have expected value $E(\varepsilon)$. Snijders [14], extending results of Hoeffding [7] , has given the locally most powerful similar tests for parametric alternatives.For our model,if H is exponential, or if H is Weibull and $E(\varepsilon) = 0$, this test reduces to the "conditional exponential scores or Savage test" which rejects H for large values of

$$T_1 = \sum_{i=1}^{N} J_0(S_i),$$

where $J_0(k) = \sum_{j=N+1-k}^{N} (1/j)$ as in Section 2.

Using the results of Snijders [14] we find that if we let

$$\hat{\sigma}^2 = \sum_{i=1}^{N} \left[J_0(S_i) - J_0(R_i) \right]^2 / 4N,$$

then $(T_1 - N)/\hat{\sigma}N^{1/2}$ is asymptotically standard normal under H.

In the case of ties, we use average scores as described by Hajek ([5], p. 134).

4. Signed rank tests

Let $Z_i = Y_i - X_i$ and, with a new change in notation, let $R_1 < \dots < R_s$ denote the ordered ranks of the positive Z_i among $|Z_1|, \dots, |Z_N|$. Tests based on these ranks are simple to use and are known to have good power properties [9].

We will consider rank tests that are optimal for the following generalized scale alternative:

$$X_i = W_i + \varepsilon_i, \quad Y_i = \Delta W_i' + \varepsilon_i$$

where $W_1, \dots, W_N$ and $W_1', \dots, W_N'$ are independent samples from a population with continuous distribution H satisfying $H(0) = 0$, and the ε_i are arbitrary random variables. Our hypothesis corresponds to "$\Delta = 1$", and the alternative, which corresponds to Y_i being stochastically larger than X_i, is $H_1: \Delta > 1$.

Scale alternatives make sense in reliability situations since a treatment that reduces wear at each instant a constant amount, leads to an overall percentage reduction in time to failure.

Hajek and Sidák [6] only deal with symmetric shift models, however, we can extend their results and find that in the important special case where H is the exponential distribution $1-\exp(-\lambda t)$, $t>0$, $\lambda > 0$, the locally most powerful rank test rejects H_0 for large values of the exponential scores statistic

$$T_2 = \sum_{i=1}^{s} J_0(r_i),$$

where $J_0(k)$ is the exponential (Savage) score $J_0(k) = \sum_{j=N+1-k}^{N} \frac{1}{j}$.

Under H_0, T_2 has mean $\frac{1}{2}N$, variance $\frac{1}{4}\sum_{k=1}^{N} J_0^2(k)$, its distribution is symmetric and (when standardized) tends to the normal distribution as $N \to \infty$. A table of the null-distribution is given by Morita [11].

T_2 looks very much like the statistic T of Section 2. The difference is that in Section 2, we used the treatment ranks of the remnants (maxima) while now we are using the treatment ranks of the differences. The hypothesis distributions of T and T_2 are the same.

5. Signed rank tests after a log transformation

In this section the notation is again different in that we let $Z_i = \log Y_i - \log X_i$ and let $R_1 < \ldots < R_s$ denote the ordered ranks of the positive Z_i among $|Z_1|, \ldots, |Z_N|$. In the two-sample case, the ranks are invariant under increasing transformations, but this is not the case here so the tests of this section are different from the ones in the preceding section.

We consider the following generalized log linear alternative

$$X_i = W_i\,\varepsilon_i, \quad Y_i = \Delta W_i'\,\varepsilon_i, \tag{5.1}$$

where W_i and W_i' are independent with common continuous distribution H satisfying $H(0) = 0$, the ε_i are arbitrary random variables, and $\Delta > 1$. The advantage of the log transformation is that it turns this generalized scale model into a symmetric shift model, namely, the distribution F of Z_i satisfies

$$F(z) = F_0(z-\Theta) \tag{5.2}$$

where $\Theta = \log \Delta$ and F_0 is the symmetric (about 0) distribution

$$F_0(z) = P(\log(Y/X) \leq z) = \int_0^\infty H(e^z w)\,dH(w). \tag{5.3}$$

Thus in this case, the results of Hájek and Sidák ([6] p.74, Theorem II.4.9) applies. In the important special case where H is the Weibull distribution, i.e., $H(x)=1-\exp\{-\lambda x^\alpha\}$, $x>0, \alpha>0, \lambda>0$, we find that the uniformly (in α, λ) locally most powerful rank tests is based on the signed rank Wilcoxon statistic $W = \sum_{i=1}^{s} R_i$. This follows by computing F_0 from (3.3) and observing that it is logistic, i.e., it equals

$$L(z) = 1/\left[1+ \exp(-x/\tau)\right], \quad \tau = 1/\alpha. \tag{5.4}$$

Having obtained the local optimality of W in an important class of parametric models, we turn to nonparametric properties. Note that F_0, as well as Θ, is unknown in the model (5.2). Thus we write the power of the test that rejects H for large values of W as $\beta(W;F_0,\Theta)$, and consider properties of $\beta(W;F_0,\Theta)$ as a function of F_0.

It would be reasonable to expect that the Wilcoxon test has higher power at alternatives $F_0(z-\Theta)$ with lighter tails than the logistic. This is the case; in fact, we can show that isotonic rank tests have isotonic power with respect to tail-ordering. A test φ is said to be <u>isotonic</u> (monotone) in $z = (z_1,\ldots,z_N)$ if $\varphi(\underset{\sim}{z}') \leq \varphi(\underset{\sim}{z})$ for each $\underset{\sim}{z}$, $\underset{\sim}{z}'$ with $z_i' \leq z_i$, $i = 1,\ldots,N$ (see [4]). If F_0 and G_0 are continuous distributions with median zero, we define: $F_0 <_t G_0$ (F_0 is tail ordered with respect to G_0 or F_0 has lighter tails than G_0) if $G_0^{-1}F_0(x) - x$ is nondecreasing on $\{x: 0 < F_0(x) < 1\}$ (see [2]). It is easy to see that "$F_0 <_t$ logistic" is equivalent to (see (5.4))

$$F_0(x)\left[1 - F_0(x)\right] \geqslant \tau f_0(x)$$

whenever F_0 is differentiable with derivative f_0. It can be shown that

Theorem. Suppose $F(z) = F_0(z-\Theta)$ and $G(z) = G_0(z-\Theta)$ where F_0 and G_0 are continuous and symmetric about zero and $F_0 <_t G_0$. If φ is an isotonic rank test, then $\beta(\varphi; G_0, \Theta) \leqslant \beta(\varphi; F_0, \Theta)$ for all $\Theta > 0$. If in addition, $F_0(x) < 1$ for each $x < \infty$ or $G_0(x) < 1$ for each $x < \infty$, then $\beta(\varphi; F_0, \Theta) \leqslant \beta(\varphi; G_0, \Theta)$ for each $\Theta \leqslant 0$.

It can be shown that isotonic non-rank tests do not necessarily have isotonic power. The t-test is not isotonic.

The above result is useful for deciding on the sample size N in an experiment. If we find $N = N_0(\varphi)$ such that the power is at least β for a certain Θ at the logistic distribution, it is at least β for all distributions with lighter tails. The Theorem shows that the smallest possible such $N_0(\varphi)$ for Θ small is obtained by using the Wilcoxon test φ_W. For this test the approximate power at the logistic distribution (5.4) is $\Phi(k_\alpha + \sqrt{N}\Theta/\sqrt{3}\tau)$, where $k_\alpha = \Phi^{-1}(\alpha)$ and Φ is the $\mathcal{N}(0,1)$ distribution.

As another application of the theorem, we note that φ_W is locally minimax (or maxmin) in the sense that it maximizes $\min_{F_0} \beta(\varphi; F_0, \Theta)$ for Θ in a neighborhood of zero, where the minimum is over the class of distributions with tails at least as heavy as the logistic. See also Doksum [3].

Acknowledgments. I am grateful to Odd Aalen and Peter Guttorp whose comments led to Section 2.

References

[1] Bell, C. B. and Haller, S. H., Bivariate symmetry tests: parametric and nonparametric. Ann. Math. Statist., 40, 259-269 (1969).

[2] Doksum, K. A., Starshaped transformations and the power of rank tests. Ann. Math. Statist., 40, 1167-1176 (1969).

[3] Doksum, K. A., Minimax results for IFRA scale alternatives. Ann. Math. Statist., 40, 1778-1783 (1969).

[4] Doksum, K. A. and Thompson, Rory, Power bounds and asymptotic minimax results for one-sample rank tests. Ann. Math. Statist., 42, 12-34 (1971).

[5] Hájek, J., Nonparametric Statistics. Holden Day, San Francisco 1969.

[6] Hájek, J. and Sidák, Z., Theory of Rank Tests. Academic Press, New York 1967.

[7] Hoeffding, W., "Optimum" nonparametric tests. In Proceedings of the 2nd Berkeley Symposium on Probability and Statistics, 83-92 (1951).

[8] Koziol, J. A. and Petkau, A. J., Sequential testing of the equality of two survival distributions using the modified Savage statistic. Biometrika, 65, 615-624 (1978).

[9] Lehmann, E. L., Nonparametrics: Statistical Methods Based on Ranks. Holden-Day, San Francisco 1975.

[10] Mantel, N., Bohidar, N. R. and Ciminera, J.L., Mantal-Haenszel analysis of litter-matched time-to-response data, with modifications for recovery of inter-litter information—an application to tumorigenicity testing. Technical report, Biostatistics Center, George Washington University (1978).

[11] Morita, J., Null distribution of the signed rank exponential scores statistic. Technical report, Statistics Department, University of California, Berkeley 1979.

[12] S c h a a f s m a, W., Testing Problems with Restricted Alternative. Noordhoff, Groningen, Netherlands 1966.

[13] S c h a a f s m a, W., Bivariate symmetry and asymmetry. Technical report, University of Groningen, Netherlands 1976.

[14] S n i j d e r s, T., Tests for the problem of bivariate symmetry. Technical report, University of Groningen, Netherlands 1976.

[15] Y a n a g i m o t o, T. and S i b u y a, M., Test of symmetry of a bivariate distribution. Sankya, Ser. A, Pt.2, 38 105-115 (1976).

D-OPTIMUM DESIGNS FOR THE INTERBLOCK-MODEL

by

N. Gaffke and O. Krafft
RWTH Aachen

Let a linear model with $u \geqslant 2$ treatments and $v \geqslant 2$ blocks be given. If there are no interactions between treatments and blocks, then by Yates [9] besides the customary intrablock estimators for the treatment effects the use of interblock estimators has been proposed- We will show in this paper that in set of all designs with constant block size k very simple designs are D-optimal, provided v is divisible by u.

In all what follows I_a denotes the identy matrix of order a, $1_{(a\times b)}$ the $(a\times b)$-matrix all elements of which are one and $E_a = I_a - \frac{1}{a} 1_{(a\times a)}$ the ortho-diagonal-projection.

1. The model

Let $N = (n_{ij})$, $n_{ij} \in \underline{N}_0$, $1 \leqslant i \leqslant u$, $1 \leqslant j \leqslant v$, be a design matrix with constant column sums $n_{.j} = k$, $1 \leqslant j \leqslant v$, and - to exclude trivialities - $n_{i.} \geqslant 1$, $1 \leqslant i \leqslant u$. Let $J_N = \{(i,j) : n_{ij} \geqslant 1\}$ and consider random variables

$$X_{ijl} = \mu + \alpha_i + \beta_j + e_{ijl}, \quad 1 \leqslant l \leqslant n_{ij} \quad (i,j) \in J_N. \tag{1}$$

Here μ is a general effect, α_i are treatment effects, β_j are random block effects with expectation zero and covariance $\sigma_b^2 I_v$, e_{ijl} are random errors with expectation zero and covariance $\sigma_e^2 I_n$,

$$n = \sum_{i=1}^{u} \sum_{j=1}^{v} n_{ij} = kv,$$

and uncorrelated with the block effects (cf. [7], p. 170, and [4], p. 233).

Ordering the observations by blocks and using $(n\times u)$- and $(n\times v)$-matrices B_1 and B_2 of zeros and ones, the system (1) can be written as vectors equation

$$\underset{\sim}{X} = \mu 1_{(n\times 1)} + B_1\underset{\sim}{\alpha} + B_2\underset{\sim}{\beta} + \underset{\sim}{e} . \tag{2}$$

According to the ordering of the components of X, B_1 and B_2 satisfy (cf. [6])

$$B_2 = 1_{(k\times 1)} \otimes I_v \quad (\otimes = \text{left Kronecker-product}), \tag{3}$$

$$B_2^T B_2 = kI_v, \tag{4}$$

$$B_2 1_{(v\times 1)} = 1_{(n\times 1)}, \quad B_2^T 1_{(n\times 1)} = k1_{(v\times 1)}, \tag{5}$$

$$B_1^T B_2 = N. \tag{6}$$

The interblock estimators are based on the block totals, i.e. on the random vector

$$\underset{\sim}{Y} = B_2^T \underset{\sim}{X}.$$

Using (2)-(6) one easily gets

$$E\underset{\sim}{Y} = \mu k1_{(v\times 1)} + N^T\underset{\sim}{\alpha}, \tag{7}$$

$$\operatorname{Cov} \underset{\sim}{Y} = k(k\sigma_b^2 + \sigma_e^2)I_v. \tag{8}$$

Putting $B_N = [k1_{(v\times 1)}, N^T]$, $\underset{\sim}{\gamma} = [\mu, \alpha^T]^T$, one obtains an ordinary linear model

$$\underset{\sim}{Y} = B_N \underset{\sim}{\gamma} + \underset{\sim}{e}', \tag{9}$$

with $\operatorname{Cov} \underset{\sim}{e}'$ given in (8). Least squares estimators $\hat{\underset{\sim}{\gamma}}$ for $\underset{\sim}{\gamma}$ are then given by

$$\hat{\underset{\sim}{\gamma}} = (B_N^T B_N)^- B_N^T \underset{\sim}{Y}, \tag{10}$$

where A^- denotes a g-inverse of the matrix A, i.e. a matrix satisfying $AA^-A = A$.

Using (3)-(6) and (49) in [8], p. 27, a g-inverse of $B_N^T B_N$ is given by

$$(B_N^T B_N)^- = \begin{bmatrix} \frac{1}{k^2 v} + \frac{1}{k^2 v^2}\, 1_{(1\times v)} N^T (NE_v N^T)^- N 1_{(v\times 1)}, & -\frac{1}{kv} 1_{(1\times v)} N^T (NE_v N^T)^- \\ -\frac{1}{kv}(NE_v N^T)^- N 1_{(v\times 1)} & (NE_v N^T)^- \end{bmatrix} \tag{11}$$

We are interested in designs N for which all treatment contrasts are estimable. If K is a $(u-1)\times u$-matrix whose rows are a basis of the space of all treatment contrasts, i.e. K satisfies rank K = u-1 and $K1_{(u\times 1)} = \underset{\sim}{0}$, then such designs can be characterized as follows:

<u>Lemma 1.</u> Let N be a design with constant block sizes k. Under the assumptions made above the following assertions are equivalent:

(a) In the model (9) all treatment contrasts are estimable.

(b) $K = K(NE_v N)^-(NE_v N^T)$ for some g-inverse $(NE_v N^T)^-$.

(c) rank $NE_v N^T = u-1$.

P r o o f: According to Krafft [5], p. 44, all treatment contrasts are estimable iff

$$\left[0_{(u-1)\times 1}, K\right] = \left[0_{(u-1)\times 1}, K\right] (B_N^T B_N)^- B_N^T B_N.$$

Using (11) we see that this condition is equivalent to (b).

Let (b) be satisfied. Then

$$u-1 = \text{rank } K = \text{rank } K(NE_v N^T)^- NE_v N^T \leq \text{rank } NE_v N^T.$$

Since we consider only designs such that

$$N^T 1_{(u\times 1)} = k1_{(v\times 1)} \tag{12}$$

and since $E_v 1_{(v\times 1)} = \underset{\sim}{0}$, we see that NE_vN^T has row- (and column-)sums zero. Hence rank $NE_vN^T \leq u-1$.

Let (c) be satisfied. Then NE_vN^T can be partitioned as

$$NE_vN^T = \begin{bmatrix} D & \underset{\sim}{d} \\ \underset{\sim}{d}^T & d \end{bmatrix},$$

where D is a regular $(u-1)\times(u-1)$-matrix and $\underset{\sim}{d}$ satisfies

$$\underset{\sim}{d} = -D1_{(u-1)\times 1}.$$

By A (3.6) in Krafft [5], p. 429, a g-inverse of NE_vN^T is given by

$$(NE_vN^T)^- = \begin{bmatrix} D^{-1} & \underset{\sim}{0} \\ \underset{\sim}{0}^T & 0 \end{bmatrix}.$$

Hence

$$(NE_vN^T)^- NE_vN^T = \begin{bmatrix} I_{u-1} & 1_{(u-1)\times 1} \\ \underset{\sim}{0}^T & 0 \end{bmatrix}.$$

Multiplying the last matrix from the left by K one gets K since $K1_{(u\times 1)} = \underset{\sim}{0}$. Thus (b) follows from (c).

<u>Remark.</u> Since rank $E_v = v-1$, we obtain from (c)

$$u-1 = \text{rank } NE_vN^T \leq \text{rank } E_v = v-1, \text{ i.e. } u \leq v.$$

2. D-optimum designs

Because of Lemma 1 we can restrict our attention to the class

$$\mathcal{N} = \left\{ N=(n_{ij}) : n_{ij} \in \underline{N}_0;\ n_{.j}=k, 1\leq j\leq v; n_{i.} \geq 1, 1\leq i\leq u, \text{ rank } NE_vN^T=u-1 \right\}.$$

D-optimum designs are defined as designs for which det Cov $K\hat{\underset{\sim}{\alpha}}$ is minimal. The estimator $K\hat{\underset{\sim}{\alpha}}$ can be obtained from (10) and, therefore, $N^* \in \mathcal{N}$ is D-optimum iff for all $N \in \mathcal{N}$

$$\det K(N^*E_vN^{*T})^-K^T \leq \det K(NE_vN^T)^-K^T. \tag{13}$$

By Lemma 2 in [3] this is equivalent to

$$\det KN^*E_vN^{*T}K^T \geq \det KNE_vN^TK^T \quad \text{for all } N \in \mathcal{N}. \tag{14}$$

Since

$$\det KNE_vN^TK^T \leq \left(\frac{1}{u-1}\, \mathrm{tr}KNE_vN^TK^T\right)^{u-1} = \left(\frac{1}{u-1}\mathrm{tr}NE_vN^TK^TK\right)^{u-1} \tag{15}$$

and since K can be so chosen that $K^TK = E_u$, we consider first the problem of maximizing

$$\mathrm{tr}NE_vN^TE_u = \mathrm{tr}\left(NE_vN^T - \frac{1}{u}NE_vN^T 1_{(u\times u)}\right) = \mathrm{tr}NE_vN^T = \sum_{i=1}^{u}\sum_{j=1}^{v} n_{ij}^2 - \frac{1}{v}\sum_{i=1}^{u} n_{i.}^2 .$$

(Here for the second equality (12) has been used).

Lemma 2. Let $v = au+b$, $a \geq 1$, $0 \leq b \leq u-1$ $(a,b \in \underline{N}_0)$. Then

$$\max\left\{\sum_{i=1}^{u}\sum_{j=1}^{v} n_{ij}^2 - \frac{1}{v}\sum_{i=1}^{u} n_{i.}^2 : N \in \mathcal{N}\right\} = \frac{k^2}{v}\left[(u-1)(ua^2+2ab)+b(b-1)\right]$$

and the maximum is attained exactly for those $N^* = (n^*_{ij}) \in \mathcal{N}$ for which $n^*_{ij} \in \{0,k\}$, $1 \leq i \leq u$, $1 \leq j \leq v$, and $n^*_{i.} \in \{ak,(a+1)k\}$, $1 \leq i \leq u$.

P r o o f: We consider first the set $\mathcal{N}'(\supset\mathcal{N})$ of those designs for which no restriction is made on rank NE_vN^T. Let N be maximal in $\mathcal{N}'$ and assume there is an index pair $(\mu,\nu) \in \{1,\dots,u\} \times \{1,\dots,v\}$ such that $1 \leq n_{\mu\nu} \leq k-1$. Let a $\rho \in \{1,\dots,u\}$ be defined by

$$n_{\rho.} - n_{\rho\nu} \leq n_{i.} - n_{i\nu},\ 1 \leq i \leq u,$$

and put

$$n'_{ij} = \begin{cases} n_{ij}, & j \neq \nu,\ 1 \leq i \leq u, \\ k, & j = \nu,\ i = \rho, \\ 0, & j = \nu,\ i \neq \rho. \end{cases}$$

Then for $N' = (n'_{ij}) \in \mathcal{N}'$ we obtain

$$\sum_{i=1}^{u}\sum_{j=1}^{v} n_{ij}'^2 - \frac{1}{v}\sum_{i=1}^{u} n_{i.}'^2$$

$$= k^2 + \sum_{\substack{j=1\\ j\neq\nu}}^{v}\sum_{i=1}^{u} n_{ij}^2 - \frac{1}{v}\left[(n_{\rho.}-n_{\rho\nu}+k)^2 + \sum_{\substack{i=1\\ i\neq\rho}}^{u}(n_{i.}-n_{i\nu})^2\right]$$

$$= \left(\sum_{i=1}^{u} n_{i\nu}\right)^2 + \sum_{\substack{j=1\\ j\neq\nu}}^{v}\sum_{i=1}^{n} n_{ij}^2 - \frac{1}{v}\left[n_{\rho.}^2 + n_{\rho\nu}^2 + k^2 - 2n_{\rho.}n_{\rho\nu} + 2kn_{\rho.}\right.$$

$$\left. - 2kn_{\rho\nu} + \sum_{\substack{i=1\\ i\neq\rho}}^{u}(n_{i.}^2 - 2n_{i\nu}n_{i.} + n_{i\nu}^2)\right]$$

$$= \sum_{i=1}^{u}\sum_{j=1}^{v} n_{ij}^2 - \frac{1}{v}\sum_{i=1}^{u} n_{i.}^2$$

$$+ 2\sum\sum_{1\leq i<l\leq u} n_{i\nu}n_{l\nu} - \frac{1}{v}\left[\sum_{i=1}^{u} n_{i\nu}^2 - 2\sum_{i=1}^{u} n_{i\nu}n_{i.} + k^2 + 2kn_{\rho.} - 2kn_{\rho\nu}\right]$$

$$= \sum_{i=1}^{u}\sum_{j=1}^{v} n_{ij}^2 - \frac{1}{v}\sum_{i=1}^{u} n_{i.}^2$$

$$+ 2\sum\sum_{1\leq i<l\leq u} n_{i\nu}\, n_{l\nu} - \frac{2}{v}\left[\sum_{i=1}^{u} n_{i\nu}^2 - \sum_{i=1}^{u} n_{i\nu}n_{i.} +\right.$$

$$\left. + \sum\sum_{1\leq i<l\leq u} n_{i\nu}\, n_{l\nu} + kn_{\rho.} - kn_{\rho\nu}\right].$$

Now, since $1\leq n_{\mu\nu}\leq k-1$ and $\sum_{i=1}^{u} n_{i\nu} = k$, we must have

$$(1-\frac{1}{v})\sum\sum_{1\leq i<l\leq u} n_{i\nu}n_{l\nu} > 0.$$

Furthermore,

$$n_{\rho .} - n_{\rho\nu} \leq n_{i.} - n_{i\nu}, \quad 1 \leq i \leq u,$$

implies

$$n_{\rho .} n_{i\nu} - n_{\rho\nu} n_{i\nu} \leq n_{i.}\, n_{i\nu} - n_{i\nu}^2, \quad 1 \leq i \leq u.$$

so that

$$kn_{\rho .} - kn_{\rho\nu} \leq \sum_{i=1}^{u} n_{i.}\, n_{i\nu} - \sum_{i=1}^{u} n_{i\nu}^2 .$$

It follows that

$$\sum_{i=1}^{u} \sum_{j=1}^{v} n_{ij}'^2 - \frac{1}{v} \sum_{i=1}^{u} n_{i.}'^2 > \sum_{i=1}^{u} \sum_{j=1}^{v} n_{ij}^2 - \frac{1}{v} \sum_{i=1}^{u} n_{i.}^2$$

in contradiction to the maximality of N. Therefore, we must have $n_{ij} \in \{0, k\}$, $1 \leq i \leq u$, $1 \leq j \leq v$.

The last condition shows that

$$\sum_{i=1}^{u} \sum_{j=1}^{v} n_{ij}^2 = vk^2 .$$

It remains to find $\min \sum_{i=1}^{n} n_{i.}^2$, where the $n_{i.}$ are all divisible by k, say $n_{i.} = kc_i$, $1 \leq i \leq u$, and $\sum_{i=1}^{u} n_{i.} = k \sum_{i=1}^{u} c_i = kv$. According to (21.3) in [5], p. 329, the problem

$$\min \left\{ \sum_{i=1}^{u} c_i^2 \; : \; \sum_{i=1}^{u} c_i = au + b; \; c_i \in \underline{N}_o, \; 1 \leq i \leq u \right\}$$

has exactly those $\underset{\sim}{c} = (c_1, \ldots, c_u)$ as solutions for which $c_i \in \{a, a+1\}$, $1 \leq i \leq u$.

Finally, with $M = \text{diag}\,(n_{1.}, \ldots, n_{u.})$ we have for $n_{ij} \in \{0, k\}$, $1 \leq i \leq u$, $1 \leq j \leq v$,

$$NE_vN^T = \frac{n}{v} M - \frac{1}{v} M1_{(u\times u)}M = (t_{il})_{1\leq i,l<u}.$$

(Here it is used that in every column of N there is exactly one non-zero element and that

$$NE_vN^T1_{(u\times 1)} = \underset{\sim}{0} \text{ implies } t_{ii} = -\sum_{\substack{l=1\\ l\neq i}}^{u} t_{il}.)$$

Since $t_{il} = -\frac{1}{v} n_{i.} n_{l.} < 0$, $i \neq l$, for $n_{i.} \in \{ak, (a+1)k\}$, $1 \leq i \leq u$, by Theorem (2.2) in Gaffke [2], p. 15, an N which is maximal in $\mathcal{N}'$ satisfies rank $NE_vN^T = u-1$ and is therefore also maximal in $\mathcal{N}$.

Remark. Lemma 2 can also be proved using convexity arguments similar to those in [2]. Theorem (2.5):

If one considers the functional $f(N) = trNE_vN^T$ on the larger set $\mathcal{N}_r$ of all real matrices $N = (n_{ij})$ with $n_{ij} \geq 0$, $1 \leq i \leq u$, $1 \leq j \leq v$, and $n_{.j} = k$ $(1 \leq j \leq v)$, then f is convex and hence the maximum of f over $\mathcal{N}_r$ is attained in an extreme point of $\mathcal{N}_r$. The extreme points of $\mathcal{N}_r$ can be shown to be given by the condition

$$n_{ij} \in \{0,k\}, \quad 1 \leq i \leq u, \ 1 \leq j \leq v.$$

Assume now that v is divisible by u, i.e. b = 0. Then, by Lemma 2, in the set of all designs with constant block size the equireplicate designs for which $n_{ij} \in \{0,k\}$ maximize $trNE_vN^T$. For those designs we have

$$NN^T = ak^2I_u \quad \text{and} \quad N1_{(v\times v)}N^T = a^2k^21_{(u\times u)}.$$

Hence

$$NE_vN^T = ak^2(I_u - \frac{a}{v}1_{(u\times u)}) = ak^2E_u.$$

This means that N is completely symmetric and that equality holds in (15). Thus we have proved

<u>Theorem.</u> Let v be divisible by u. Then in the set of all designs N of constant block size k, for which all treatment contrasts are estimable, the equireplicate designs for which $n_{ij} \in \{0,k\}$, $1 \leq i \leq u$, $1 \leq j \leq v$, are D-optimum.

<u>Remark.</u> It may be noted that in case where v is not divisible by u the $\mathcal{C}$-matrix NE_vN^T of designs N given by Lemma 2 is - up to permutations of rows and the same permutations for the columns - of the form

$$\frac{1}{k^2} NE_vN^T = \begin{bmatrix} (a+1)I_b - \frac{(a+1)^2}{v} 1_{(b\times b)}, & \frac{-a(a+1)}{v} 1_{(b\times(u-b))} \\ -\frac{a(a+1)}{v} 1_{((u-b)\times b)}, & aI_{u-b} - \frac{a^2}{v} 1_{((u-b)\times(u-b))} \end{bmatrix}. \tag{16}$$

This is slightly different from the $\mathcal{C}$-matrix given by Cheng [1], Lemma 2.1, for the intrablock-model. (16) has - in difference to Cheng's matrix - in general three different positive eigenvalues namely k^2a, $k^2a(1+\frac{u-b}{v})$ and $k^2(a+1)$ with multiplicities u-b-1, 1 and b-1, respectively. If v is not divisible by u and if $b > 1$, therefore the approach used by Cheng cannot be used to prove D-optimality in the interblock-model.

<u>Remark.</u> Let $v = au+b$. In the rather restricted subclass

$$\hat{\mathcal{N}} = \left\{ N \in \mathcal{N} : n_{i.} = (a+1)k,\ 1 \leq i \leq b;\ n_{i.} = ak,\ b+1 \leq i \leq u \right\}$$

of $\mathcal{N}$ designs $N^* \in \hat{\mathcal{N}}$ as derived in Lemma 2 are even U-optimal, i. e. $N^*E_vN^{*T} - NE_vN^T$ is non-negative definite for all $N \in \hat{\mathcal{N}}$.

This can be seen as follows:

Let $T = (t_{il}) = NE_vN^T$. Since E_v is symmetric and idempotent, we have $T = (NE_v)(NE_v)^T$, i. e. T is non-negative definite. Furtherfrom $T1_{(u\times 1)} = \underset{\sim}{0}$, cf. (12), it follows that

$$\underset{\sim}{x}^T T \underset{\sim}{x} = -\frac{1}{2} \sum_{\substack{1 \leq i,l \leq u \\ i \neq l}} t_{il}(x_i - x_l)^2.$$

For T^* we have $t^*_{il} = -\frac{1}{v} n^*_{i.} n^*_{l.}$, $i \neq l$, and for $N \in \hat{\mathcal{N}}$, $i \neq l$,

$$t_{il} = \sum_{j=1}^{v} n_{ij} n_{lj} - \frac{1}{v} n_{i.} n_{l.} \geq -\frac{1}{v} n_{i.} n_{l.} = -\frac{1}{v} n^*_{i.} n^*_{l.} = t^*_{il} .$$

The assertion then follows from (17).

References

[1] C h e n g, C h i n g-S h u i, Optimality of certain asymmetrical experimental designs. Ann. Statist., 6, 1239-1261 (1978).

[2] G a f f k e, N., Optimale Versuchsplanung für lineare Zwei-Faktor-Modelle. Ph. D.-Thesis, RWTH Aachen 1978.

[3] G a f f k e, N., K r a f f t, O., Optimum properties of latin square designs and a matrix inequality. Math. Operationsforsch. Statist., Ser. Statistics, 8, 345-350 (1977).

[4] J o h n, P.W.M., Statistical design and analysis of experiments, Macmillan, New York 1971.

[5] K r a f f t, O., Lineare statistische Modelle und optimale Versuchspläne, Vandenhoeck u. Ruprecht, Göttingen 1978.

[6] M i l l i k e n, G. A., A k d e n i z, F., A theorem on the difference of the generalized inverses of two nonnegative matrices, Commun. Statist.-Theor. Meth., A 6 (1), 73-79 (1977).

[7] S c h e f f é, H., The analysis of variance, Wiley, New York 1959.

[8] S e a r l e, S. R., Linear models, Wiley, New York 1971.

[9] Y a t e s, F., The recovery of inter-block information in balanced incomplete block designs, Ann. Eugenics 10, 317-325 (1940).

LOCALLY BEST LINEAR ESTIMATION IN EUCLIDEAN VECTOR SPACES

by

Stanisław Gnot

Polish Academy of Sciences, Wrocław

1. Summary and introduction

LaMotte [5] examined locally best estimators of the form y'Ay for variance components in a general linear model. He described these estimators in several classes, such as the class of unbiased estimators, the class of invariant estimators, and others. The purpose of this paper is to examine locally best estimators of the form y'Ay+ + a'y for variance components and compare their risks with the risks of LaMotte's estimators. A random vector z taking values in a finite dimensional vector space $\mathcal{X}$ endowed with an inner product $[\cdot,\cdot]$ is considered. The locally best estimator for expectation Ez in the class of linear estimators $\hat{\mathcal{G}}_0 = \{Tz: \mathcal{R}(T') \subset \mathcal{X}_0; T: \mathcal{X} \to \mathcal{X}\}$ is given for any subspace $\mathcal{X}_0$ of $\mathcal{X}$. The considerations are restricted to the quadratic loss function

$$\mathcal{L}_H\{Tz, Ez\} = [Tz-Ez,\ H(Tz-Ez)],$$

where H is a nonnegative definite linear operator mapping $\mathcal{X}$ into $\mathcal{X}$. The locally best linear plus quadratic estimators y'Ay + a'y for variance components are obtained by specifying $\mathcal{X}$, $\mathcal{X}_0$ and $[\cdot,\cdot]$.

2. Locally best estimation of the mean vector

Let $\mathcal{P}$ be a family of probability measures with an associated measurable space $\{\mathcal{U},\mathcal{F}\}$. Consider a random vector z (u), $u \in \mathcal{U}$, taking values in $\mathcal{X}$. Assume that the mean vector $E_P z$ and covariance operator $Cov_P z = \Sigma$ exist for each $P \in \mathcal{P}$ and that the Σ's are positive definite. We are interested in the problem of estimation of the mean vector Ez in the class $\hat{\mathcal{G}}_0$ of estimators under the quadratic loss $\mathcal{L}_H$. It is known [6], that for $\mathcal{X}_0 = \mathcal{X}$ the locally best estimator of Ez at (θ, Σ) has the form Tz, where

$$T = \theta \bar{\otimes} \theta \, (\Sigma + \theta \bar{\otimes} \theta)^{-1} = \gamma \theta \bar{\otimes} \Sigma^{-1} \theta, \tag{1}$$

while $\gamma = ([\Sigma^{-1} \theta, \theta] + 1)^{-1}$.

Here for any $a, b \in \mathcal{X}$ symbol $a \bar{\otimes} b$ denotes the linear operator from $\mathcal{X}$ to $\mathcal{X}$ defined for each $c \in \mathcal{X}$ by

$$(a \bar{\otimes} b) c = [b, c] a.$$

Now let $\mathcal{X}_0$ be any subspace of $\mathcal{X}$ and let Π project on $\Sigma(\mathcal{X}_0)$ along $\mathcal{X}_0^{\perp}$. Let H = I, where I stands for the identity operator. The following result can be established.

Theorem 1. Estimator T z is locally best for Ez at (θ, Σ) in $\hat{\mathcal{G}}_0$ under $\mathcal{L}_I$ iff

$$T = (\theta \bar{\otimes} \Pi\theta)(\Sigma + \theta \bar{\otimes} \Pi\theta)^{-1} \tag{2}$$

or, equivalently,

$$T = \gamma \theta \bar{\otimes} \Sigma^{-1} \Pi \theta,$$

where

$$\gamma = ([\Sigma^{-1} \Pi\theta, \theta] + 1)^{-1}.$$

Remark 1. It follows from the Shinozaki's lemma [7] that if Tz is locally best for Ez at (Θ, Σ) in $\hat{\mathcal{G}}_0$ under $\mathcal{L}_I$, then Tz is also locally best for Ez at (Θ, Σ) in $\hat{\mathcal{G}}_0$ under $\mathcal{L}_H$ with any nonnegative definite H.

3. Locally best estimation of parametric functions

Suppose that we are interested in estimation of parametric functions belonging to

$$g = \{[a, Ez]: a \in \mathcal{K}\},$$

within the class of estimators

$$\hat{g}_0 = \{[b, z]: b \in \mathcal{K}_0\}$$

under the loss function

$$l\{[b,z], [a,Ez]\} = ([b,z] - [a,Ez])^2.$$

Let $r\{b, a, Ez, \Sigma\} = E_{Ez,\Sigma}\, l\{[b,z], [a,Ez]\}$ and let $\mathcal{V}_0 = \{T: R(T') \subset \mathcal{K}_0\}$. Since $\{T' a: T \in \mathcal{V}_0\} = \mathcal{K}_0$ for all nonzero $a \in \mathcal{K}$ and since

$$r\{T'a, a, Ez, \Sigma\} = E_{Ez,\Sigma}\, \mathcal{L}_{a \bar{\otimes} a}\{Tz, Ez\},$$

Theorem 1 and Remark 1 yield the following result.

Lemma 1. If Tz is locally best for Ez at (Θ, Σ) in $\hat{\mathcal{G}}_0$ under $\mathcal{L}_I$ then $[T'a, z]$ is locally best for $[a, Ez]$ at (Θ, Σ) in $\hat{g}_0$ under l for each $a \in \mathcal{K}$. The risk of $[T'a, z]$ at (Θ, Σ) is equal to $\gamma[a, \Theta]^2$, where γ is defined as in Theorem 1,

4. The linear plus quadratic estimation of variance components

Let y be a n-dimensional, normally distributed random vector with expectation

$$Ey = X\beta,$$

where β is a vector of unknown parameters and X is a known matrix. The covariance matrix of y is assumed to be of the form

$$\text{Cov } y = V(\sigma) = \sum_{i=1}^{k} \sigma_i^2 V_i,$$

where $V_1, V_2, \ldots, V_k$ are known, nonnegative definite matrices, and $\sigma = (\sigma_1^2, \sigma_2^2, \ldots, \sigma_k^2)' \in \Omega$ is a vector of unknown parameters, called variance components. We assume that $V(\sigma)$ is positive definite for all $\sigma \in \Omega$. Locally best estimation of functions $c'\sigma$ in the class of all quadratic estimators

$$\hat{g} = \{y'Ay : A' = A\}$$

and in the class of invariant quadratic estimators

$$\hat{g}_0 = \{y'Ay : A' = A, AX=0\}$$

was considered extensively by LaMotte [5]. He has shown that the locally best estimators of $c'\sigma$ at a given point (σ_0, β_0) in the classes $\hat{g}$ and $\hat{g}_0$ are, respectively,

$$\frac{(c'\sigma_0)y'\left[(2d_0+1)V^{-1}(\sigma_0) - V^{-1}(\sigma_0)X\beta_0\beta_0'X'V^{-1}(\sigma_0)\right]y}{d_0^2 + (n+2)(2d_0+1)} \tag{3}$$

and

$$\frac{(c'\sigma_0)y'\left\{V^{-1}(\sigma_0) - V^{-1}(\sigma_0)X\left[X'V^{-1}(\sigma_0)X\right]^{+}X'V^{-1}(\sigma_0)\right\}y}{\varkappa+2} \tag{4}$$

where $d_0(\beta_0, \sigma_0) = \beta_0'X'V^{-1}(\sigma_0)X\beta_0$, and $\varkappa = \operatorname{tr} XX^{+}$. The risks of these estimators at (β_0, σ_0) are, respectively,

$$r_1 = \left\{2(2d_0+1)/\left[d_0^2 + (n+2)(2d_0+1)\right]\right\}(c'\sigma_0)^2$$

and

$$r_2 = \left[2/(\varkappa+2)\right](c'\sigma_0)^2.$$

Klonecki, Zmyślony and the author [1-3] have studied the problem of unbiased estimation of variance components in the above described model, when the choice of estimators is restricted to the class of linear plus quadratic estimators $y'Ay + a'y$. They have introduced a random element $\underline{z} = \{yy', y\}$ taking values in a vector space $\underline{\mathcal{X}} = \mathcal{X} \times \mathcal{R}^n$ with an inner product defined by

$$[\{A, a\}, \{B, b\}] = \text{tr}AB + a'b,$$

where $\mathcal{X}$ is the space of all symmetric matrices. Under assumption of normality of y the expectation and covariance operator of $\underline{z}$ are, respectively,

$$E\underline{z} = \underline{\Theta}(\beta, \sigma) = \{X\beta\beta'X' + V(\sigma), X\beta\},$$

$$\text{Cov}\,\underline{z} = \underline{\Sigma}(\beta, \sigma),$$

where for $a \in \mathcal{R}^n$ and $A \in \mathcal{X}$

$$\underline{\Sigma}(\beta,\sigma)(\{A, a\}) = \{2V(\sigma)AV(\sigma) + 2V(\sigma)AX\beta\beta'X' + 2X\beta\beta'X'AV(\sigma) +$$
$$+ V(\sigma)a\beta'X' + X\beta a'V(\sigma),\ 2V(\sigma)AX\beta + V(\sigma)a,$$

(compare also [4]). Using (1) and (2) we may find the locally best estimators of $E\underline{z}$ in the class of estimators $\hat{\underline{\mathcal{G}}}$ and $\hat{\mathcal{G}}_0$ and the locally best estimators for parametric functions of the form

$$[\{A, a\}, E\underline{z}] = \beta'X'AX\beta + \sum_{i=1}^{k} \sigma_i^2 \text{ tr } AV_i + a'X\beta,$$

in the class of estimators

$$\hat{\underline{g}}_0 = \{[\{B, b\}, \underline{z}] : \{B, b\} \in \underline{\mathcal{X}}_0\} = \{y'By + b'y;\ \{B, b\} \in \underline{\mathcal{X}}_0\}.$$

For $\underline{\mathcal{X}}_0 = \{\{B, b\} \in \underline{\mathcal{X}} : B \in \mathcal{R}(M \otimes M),\ b \in \mathcal{R}(M)\}$, where $M = I - XX^+$, $\hat{\underline{g}}_0$ is the class of all linear plus quadratic invariant estimators, that is the estimators satisfying the condition

$$y'By + b'y = (y - X\beta)'B(y - X\beta) + b'(y - X\beta)$$

for each β.

<u>Locally best estimators in case</u> $\underline{\mathcal{X}}_0 = \underline{\mathcal{X}}$. It can be easily checked that

$$\underline{\Sigma}^{-1}\underline{\Theta} = \left\{ (1/2)V^{-1}-(1/2)V^{-1}X\beta\beta'X'V^{-1},\ dV^{-1}X\beta \right\},$$

and

$$\left[\underline{\Sigma}^{-1}\underline{\Theta}, \underline{\Theta} \right] = (n+d)^2/2.$$

Hence it follows from (1) that the locally best estimator of $E\underline{z}$ at (β_0, σ_0) in $\hat{\underline{\mathcal{G}}}$ is

$$\frac{y' \left[V^{-1}(\sigma_0)-V^{-1}(\sigma_0)X\beta_0\beta_0'X'V^{-1}(\sigma_0) \right] y + (d_0'/2)\beta_0'X'V^{-1}(\sigma_0)\ y}{n+d^2+2}\ \underline{\Theta}(\beta_0,\sigma_0),$$

with the risk at (β_0, σ_0) equal to $\left[2/(n+d_0^2+2)\right]\left[\underline{\Theta}_0, \underline{\Theta}_0\right]$. For $\underline{A} = \{A, a\}$ the locally best estimator of $[\underline{A}, E\underline{z}]$ at (σ_0, β_0) is

$$\frac{[\underline{A}, E\underline{z}]\ y'\left[V^{-1}(\sigma_0)-V^{-1}(\sigma_0)X\beta_0\beta_0'X'V^{-1}(\sigma_0)\right]y + (d_0/2)\ \beta_0'X'V^{-1}(\sigma_0)y}{n+d_0^2+2}. \tag{5}$$

The risk of this estimator is

$$\underline{r}_1 = \left[2/(n+d_0^2+2)\right]\left[\underline{A}, E\underline{z}\right]^2.$$

In particular, if $X'AX = 0$ and $X'a = 0$, then $\left[\underline{A}, E\underline{z}\right] = c'\sigma$ and consequently (3) and (5) present estimators of the same function. The first one belongs to the class of quadratic estimators and the second one to the class of linear plus quadratic estimators. It is interesting to observe that the risk $\underline{r}_1$ of (5) is smaller than the risk r_1 of (3) at the point (β_0, σ_0), provided $c'\sigma_0 \neq 0$. It is so because

$$r_1/\underline{r}_1 = (n+2+d_0^2)/(n+2+a_0^2) > 1\ .$$

Here $a_0^2 = d_0^2/(2d_0^2+1) < d_0^2$, since $d_0^2 > 0$.

<u>Locally best estimators in case $\mathfrak{X}_0 = \{\{A, a\}: A \in \mathcal{R}\{M \otimes M\}, a \in \mathcal{R}(M)\}$.</u> The operator Π appearing in Theorem 1 is defined by

$$\Pi(\{A, a\}) = \left\{(I-G)A(I-G)', (I-G)a\right\},$$

where $G = X(X'V^{-1}X)^{+}X'V^{-1}$. From the above we have

$$\pi(\underline{\Theta}) = \left\{ V - X(X'V^{-1}X)^{+}X', 0 \right\},$$

and

$$\underline{\Sigma}^{-1}\pi(\underline{\Theta}) = \left\{ (1/2)V^{-1} - (1/2)V^{-1}X(X'V^{-1}X)^{+}X'V^{-1},\ 0 \right\}.$$

Now from Theorem 1 we find that in this case the locally best estimator for $E\underline{z}$ at (β_0, σ_0) is

$$\frac{y'\left\{V^{-1}(\sigma_0) - V^{-1}(\sigma_0)X\left[X'V^{-1}(\sigma_0)X\right]^{+} X'V^{-1}(\sigma_0)\right\}y}{\varkappa + 2}\ \underline{\Theta}(\beta_0, \sigma_0)$$

with the risk at (β_0, σ_0) equal to $\left[2/(\varkappa+2)\right]\left[\underline{\Theta}_0, \underline{\Theta}_0\right]$. Taking into account Lemma 1 we find that the locally best linear plus quadratic invariant estimator for $c'\sigma$ at (β_0, σ_0) is exactly the same as the locally best quadratic invariant estimator (4) at (β_0, σ_0).

References

[1] G n o t, S., K l o n e c k i, W., Z m y ś l o n y, R, Uniformly minimum variance unbiased estimation in Euclidean vector spaces. Bulletin de L'Academie Polonaise des Sciences, ser. Math. Ast. Phys., XXIV, 4, 281-286 (1976).

[2] G n o t, S., K l o n e c k i, W., Z m y ś l o n y, R.,Best linear plus quadratic unbiased estimation of parameters in mixed linear models. Applicationes Mathematicae, XV, 4,455-462(1977a).

[3] G n o t, S., K l o n e c k i, W., Z m y ś l o n y, R., Best unbiased linear estimation, a coordinate free approach. Institute of Mathematics Polish Academy of Sciences. Preprint 124 (1977b).

[4] K l e f f e, J., P i n c u s, R., Bayes and best quadratic unbiased estimation of variance components and heteroscedastic variances in linear models, Mathematische Operationsfor.Statist., 5, 147-159 (1974).

[5] L a M o t t e, L. R., Quadratic estimation of variance components. Biometrics, 29, 311-330 (1973).

[6] R a o, C. R., Unified theory of linear estimation. Sankhya ser., A, 33, 371-394 (1971).

[7] R a o, C. R., Estimation of parameters in a linear model. Ann. Statist. 4, 1023-1037 (1976).

ON STATISTICAL PROBLEMS OF STOCHASTIC PROCESSES WITH PENETRABLE BOUNDARIES

by

B. Grigelionis, R. Mikulevičius

Academy of Sciences of the Lithuanian SSR
and the University of Vilnius

Introduction

In paper [1] the structure of m-dimensional semimartingales, satisfying two-sided Wentzell's type boundary conditions on the set $\partial G = \{X = (X_1, \ldots, X_m) : X_1 = 0\}$ was considered. Conditions were found in terms of the defined system of local characteristics for absolute continuity of measures, corresponding to such processes, and formulas were given for Radon-Nikodym derivatives. In statistical estimation problems, where processes with penetrable boundaries arise as observation processes, it is important to derive the corresponding stochastic equations for non-linear filtering. In the paper the main attention is spared to this problem.

In § 1 a limit theorem illustrating the structure of stochastic processes with penetrable boundaries is proved. In § 2 a definition and properties of processes with penetrable boundaries are given. Following [2]-[4], stochastic equations for non-linear filtering are obtained in § 3 when the observation process is a semimartingale and, in particular, a process with penetrable boundaries.

1. A Limit Theorem

Let us consider a sequence of stochastic processes $\{X_n(t), t \geq 0, n \geq 1\}$ which are solutions to the K. Ito equations:

$$dX_n(t) = a_n \, \chi_{\left[-\frac{1}{n},\frac{1}{n}\right]}(X_n(t)\,dt + dW_t^n, \qquad (1)$$

$$X_n(0) = x, \quad t \geq 0, \; n = 1,2,\ldots,$$

where a_n are nonnegative constatnts and W^n is a standard process of Brownian motion.

By P_x^n denote the measure corresponding to the solution of (1) on the space $D = D_{R^1}[0, \infty)$ of right continuous with left limits functions with γ_1-topology of Skorochod and the standard family of σ-algebras $\{D_t, t \geq 0\}$. Let $P_x^{(a)}$ be a measure corresponding to the homogeneous Markov process, whose density of the transition probability function is the fundamental solution to the equation

$$\frac{\partial u(t,x)}{\partial t} = \frac{1}{2} \frac{\partial^2 u(t,x)}{\partial x^2}, \quad x \neq 0,$$

$$\gamma_+ \frac{\partial u^+(t,0)}{\partial x} = \gamma_- \frac{\partial u^-(t,0)}{\partial x}, \quad t > 0,$$

where

$$\gamma_+ = \frac{e^a}{e^a + e^{-a}}, \; \gamma_- = \frac{e^{-a}}{e^a + e^{-a}} \quad \text{for} \quad 0 \leq a < \infty,$$

$$\gamma_+ = 1, \; \gamma_- = 0 \quad \text{for} \quad a = \infty$$

$\partial u^{\pm}/\partial x$ denote one-sided derivatives.

Theorem 1. If $\frac{2a_n}{n} \to \infty$ as $n \to \infty$, then for all $x \in R^1$ $P_x^n \Rightarrow P_x^{(a)}$ as $n \to \infty$.

P r o o f: Denote $X_t(\omega) = \omega(t)$, $t \geq 0$, $\omega \in D$. From the results of paper [5] it follows that the measure $P_x^{(a)}$ can be characterized as a unique measure such that for all $t \geq 0$ $P_x^{(a)}$-a.e.

$$X_t = x + W_t^{(a)} + (\gamma_+ - \gamma_-)\varphi_t^{(a)} \tag{2}$$

and

$$X_t^+ = x^+ + \int_0^t \chi_{\{X_s \geq 0\}} \, dW_s^{(a)} + \gamma_+ \varphi_t^{(a)(*)}, \tag{3}$$

where $W^{(a)}$ is a standard process of Brownian motion and $\varphi^{(a)}$ is a continuous increasing process such that for all $t \geq 0$ $P_x^{(a)}$-a.e.

$$\int_0^t \chi_{\{0\}}(X_s)\, d\varphi_s^{(a)} = \varphi_t^{(a)} \quad \text{and} \quad \int_0^t \chi_{\{0\}}(X_s)\, ds = 0. \tag{4}$$

Thus it is enough to prove relative compactness of the sequence $\{P_x^n, n \geq 1\}$ for every $x \in R^1$ and to show that the process X enjoys the properties (2)-(4) for every measure which is the limiting measure of a weakly convergent subsequence of $\{P_x^n, n \geq 1\}$.

Set

$$A_t^n = \int_0^t a_n \, \chi_{\left[-\frac{1}{n}, \frac{1}{n}\right]}(X_s)\, ds, \quad t \geq 0, \quad n = 1, 2, \ldots$$

In view of corollary 2 in [6] $\{P_x^n, n \geq 1\}$ is relatively compact if for each $n \geq 1$ and $T > 0$ there exists a function $\beta_{n,T}(h)$ such that $\lim_{h \to 0} \overline{\lim_{n \to \infty}} \beta_{n,T}(h) = 0$, for all $0 \leq s \leq t \leq s + h \leq T$

$$E_x^n (A_t^n - A_s^n | D_s) \leq \beta_{n,T}(h) \tag{5}$$

and

$$\sup_{x,n} E_x^n (A_T^n)^2 < \infty.$$

Let us take $\Phi_n \in C_b^2(R^1)$ such that $\sup_x |\Phi_n(x)| \leq \frac{3}{n}$ and $\Phi_n'(x) = 1$ for $x \in \left[-\frac{1}{n}, \frac{1}{n}\right]$, $n \geq 1$. Then using the Ito formula we shall have

(*) $x^+ = x \vee 0$, $x^- = x \wedge 0$

$$E_x^n (A_t^n - A_s^n \mid D_s) = E_x^n (\Phi_n(X_t) - \Phi_n(X_s) \mid D_s) -$$

$$- E_x^n\left(\frac{1}{2}\int_s^t \Phi_n''(X_n)\,dn \,\Big|\, D_s\right) \leqslant C_1(n) + C_2(n)(t-s),$$

where $C_1(n) = \frac{6}{n}$ and $C_2(n) = \frac{1}{2} \sup_x \left| \Phi_n''(x) \right| < \infty$. If $h_n \downarrow 0$ and $C_2(n)h_n \downarrow 0$ as $n \to \infty$, then

$$\beta_{n,T}(h) = \begin{cases} C_1(n) + C_2(n)\, h_n & \text{for } h < h_n, \\ C_1(k) + C_2(k)\, h_k & \text{for } h_{k+1} \leqslant h < h_k,\ k = 1, \dots, n-1, \\ C_1(1) + C_2(1)\, T & \text{for } h \geqslant h_1 \end{cases}$$

satisfies (5). Further after some simple estimations we obtain with the help of the Ito formula for the function Φ_1, that

$$E_x^n A_T^2 \leqslant 2 \sup_x |\Phi_1(x)|^2 + \frac{3}{2} T^2 \sup_x |\Phi_1''(x)|^2 + 3T \sup_x |\Phi_1'(x)|^2 < \infty$$

and thus for each $T < \infty$

$$\sup_{x,n} E_x^n X_T^2 < \infty.$$

From [7] it follows that for all $t \geqslant 0$ P_x^n-a.e.

$$X_t^+ = x^+ + \int_0^t \chi_{\{X_s > 0\}} dW_s^n + C_t^n + \int_0^t \chi_{\{X_s > 0\}} dA_s^n,$$

$$X_t^- = x^- - \int_0^t \chi_{\{X_s < 0\}} d\, W_s^n + C_t^n - \int_0^t \chi_{\{X_s < 0\}} dA_s^n,$$

where C^n is an increasing process. Let

$$\Psi_\varepsilon(z) = \begin{cases} \dfrac{(\varepsilon - z)^3}{\varepsilon} & \text{for } z \leqslant \varepsilon, \\ 0 & \text{for } z \geqslant \varepsilon. \end{cases}$$

According to the Ito formula we have

$$\Psi_\varepsilon(|X_t|) - \Psi_\varepsilon(|X_0|) = \int_0^t \Psi_\varepsilon'(|X_s|)\, d|X_s| + \frac{1}{2}\int_0^t \Psi_\varepsilon''(|X_s|)\, ds.$$

Hence follows that for each $t \geqslant 0$ there exists $K(t) < \infty$ such that

$$\sup_{x,n} E_x^n\Big(\int_0^t \chi_{\{|X_n|<\frac{\varepsilon}{2}\}}\,du\Big) \leqslant 2\varepsilon^2 + K(t)\,\varepsilon . \tag{6}$$

By fixing x and considering that $P_x^n \Rightarrow P_x$ as $n\to\infty$, from (6) we find that for each $t\geqslant 0$ P_x - a.e.

$$\int_0^t \chi_{\{0\}}(X_s)\,ds = 0.$$

Define the measures μ_n^T on $[0,\infty)\times D$ for each $T>0$ by means of the equalities

$$\mu_n^T(Y) = E_x^n\Big(\int_0^T Y_s\, dA_s^n\Big).$$

Since the sequence $\{P_x^n,\ n\geqslant 1\}$ is relatively compact, for each $\varepsilon>0$ there exists a compact set $K\subset D$ such that $P_x^n(K^c)\leqslant\varepsilon$ and thus

$$\mu_n^T\big([0,T]\times K^c\big) \leqslant \big(P_x^n(K^c)\ E_x^n(A_T^n)^2\big)^{1/2},$$

i.e. $\{\mu_{n_j}^T,\ n\geqslant 1\}$ is a relatively compact sequence.

Let $\mu_{n_j}^T \Rightarrow \mu^T$ as $j\to\infty$. Hence and from the earlier obtained estimations for A_t^n it follows that $\mu^T(\{t\}\times D) = 0$ for each $t\geqslant 0$. If f is a continuous bounded function on D, then

$$\mu^T\big(\chi_{[0,t]}\,|f|\big) = \lim_{j\to\infty}\mu_{n_j}^T\big(\chi_{[0,t]}\,|f|\big)\leqslant$$

$$\leqslant \lim_{j\to\infty}\Big(E_x^{n_j}\ f^2\Big)^{1/2} C(t),\quad C(t)<\infty,$$

and therefore for all $t\geqslant 0$

$$\mu^T([0,\ t]\times\cdot) \ll P_x(\cdot).$$

Hence we find that μ^T is generated by some increasing process φ^T. Since we shall be interested only in the values of the measure μ^T on the σ-algebra of predictable sets with respect to the family $\{D_{t+},\ t\geqslant 0\}$ then we can suppose (see [8]) that φ^T is a predictable increasing process.

As well as for $n \geq m$

$$\Phi_m(X_{t\wedge T}) - \frac{1}{2}\int_0^{t\wedge T} \Phi_m''(X_s)\,ds - A^n_{t\wedge T}, \quad t \geq 0$$

is a martingale with respect to the measure P^n_x, then

$$\Phi_m(X_{t\wedge T}) - \frac{1}{2}\int_0^{t\wedge T} \Phi_m''(X_s)\,ds - \varphi^T_t, \quad t \geq 0,$$

is a martingale with respect to the measure P_x. Hence we find that if τ is a jump time of φ^T then $\mu^T([\tau]) \leq \frac{6}{m}$ for every m, i.e. φ^T is a P_x - a.e. continuous process.

Considering measures on $[0,m] \times D$ and properly choosing subsequences we obtain the P_x - a.e. continuous process φ such that φ and φ^m coincide on $[0, m]$. Since $X_t(\omega)$ is $d\varphi_t\, dP_x$ - a.e. continuous then for all $f \in C^2_b(R^1)$

$$f(X_t) - \frac{1}{2}\int_0^t f''(X_s)\,ds - \int_0^t f'(X_s)\,d\varphi_s, \quad t \geq 0,$$

is a martingale, i.e. with respect to the measure P_x

$$X_t = x + W_t + \varphi_t, \quad t \geq 0,$$

where W is a standard process of Brownian motion.

From [7] we also get the following decompositions:

$$X^+_t = x^+ + \int_0^t \chi_{\{X_s > 0\}}\,dW_s + C^+_t, \quad t \geq 0, \tag{7}$$

$$X^-_t = x^- + \int_0^t \chi_{\{X_s \leq 0\}}\,dW_s + C^+_t - \varphi_t, \tag{8}$$

where C^+ is an increasing process such that

$$\int_0^t \chi_{\{0\}}(X_s)\,dC^+_s = C^+_t, \quad t \geq 0 .$$

Denote

$$B_n(x) = \int_{-\infty}^x a_n\,\chi_{\left[-\frac{1}{n},\frac{1}{n}\right]}(y)\,dy, \quad f_n(x) = \int_0^x e^{-2B_n(y)}\,dy .$$

We have that

$$f_n(x) \longrightarrow f(x) \equiv -x^- + e^{-2a}x^+$$

uniformly on compact sets as $n \to \infty$.

Since $f_n(X_t)$, $t \geq 0$, is a martingale with respect to P^n_x, therefore $f(X_t)$, $t \geq 0$, is a martingale with respect to P_x. Hence and from (8)-(9) it follows that for $t \geq 0$ P_x-a.e.

$$\varphi_t - C^+_t + e^{-2a}C^+_t = 0.$$

Hence (2)-(4) hold with respect to P_x assuming $\varphi^{(a)} = (\gamma_+ - \gamma_-)^{-1}\varphi$, $a \neq 0$, $\varphi^{(0)} \equiv 0$. This gives us the equality $P_x = P^{(a)}_x$. Thus Theorem 1 is proved.

Remark 1. A close result to that of Theorem 1 is obtained using different arguments in [9] in the case when $0 \leq a < \infty$.

Note that the measure $P^{(\infty)}_x$ corresponds to the process coinciding with that of Brownian motion up to hitting time of the point 0, which acts as a boundary of reflecting to the right.

2. Stochastic Processes with Penetrable Boundaries

Let us consider a measurable space (Ω, F) with a right continuous $\underline{F} = \{F_t,\ t \geq 0\}$ of increasing σ-algebras. By $\mathcal{P}(\underline{F})$ denote σ-algebra of $\underline{F}$-predictable subsets of $[0,\infty) \times \Omega$, $E = R^m \setminus \{0\}$, $\mathcal{E} = \mathcal{B}(R^m \setminus \{0\})$, $\tilde{\mathcal{P}}(\underline{F}) = \mathcal{P}(\underline{F}) \otimes \mathcal{E}$. Let P be a probability measure on $\mathcal{F}$. By $\mathcal{M}_{loc}(P, \underline{F})$ denote a class of $(P, \underline{F})$-local martingales,

$$G = \{x \in R^m:\ x_1 \neq 0\},\quad G_+ = \{x \in R^m:\ x_1 > 0\},$$

$$G_- = \{x \in R^m:\ x_1 < 0\},\quad \partial G = \{x \in R^m:\ x_1 = 0\},$$

$\hat{C}^2(R^m)$ a class of continuous functions such that $D_j f$, $D^2_{jk} f$, $j, k = 2, \ldots, m$, exist and are continuous, $D_1 f, D^2_{1j} f$, $j = 1, \ldots, m$, exist and are continuous on the sets $G^{\pm} \cup \partial G$ where these derivatives $x \in \partial G$ are considered as one-sided, $\hat{C}_b^2(R^m)$ a class of bounded functions $f \in \hat{C}^2(R^m)$ with bounded mentioned derivatives.

Consider $\mathcal{P}(\underline{F})$-measurable functions

$$\gamma^+(t) \geq 0, \quad \gamma_-(t) \geq 0, \quad \delta(t) \geq 0, \quad \hat{\alpha}(t) = (\hat{\alpha}_1(t), \ldots, \hat{\alpha}_m(t)),$$
$$\hat{B}(t) = \| \hat{\beta}_{jk}(t) \|_1^m, \quad \hat{\pi}(t, \Gamma), \quad \tilde{\alpha}(t) = (\tilde{\alpha}_2(t), \ldots, \tilde{\alpha}_m(t)),$$
$$\tilde{B}(t) = \| \tilde{\beta}_{jk}(t) \|_2^m, \quad \tilde{\pi}(t, \Gamma), \quad t \geq 0, \quad \Gamma \in \mathcal{E},$$

and for $f \in \hat{C}_b^2(R^m)$ denote

$$\hat{A}(t) f(x) = \frac{1}{2} \sum_{j,k=1}^{m} \hat{\beta}_{jk}(t) D_{jk}^2 f(x) + \sum_{j=1}^{m} \hat{\alpha}_j(t) D_j f(x) +$$
$$+ \int_E \Big(f(x+y) - f(x) - \sum_{j=1}^{m} y_j D_j f(x) \chi_{\{|y| \leq 1\}}(y) \Big) \hat{\pi}(t, dy),$$

$$\tilde{A}(t) f(x) = \frac{1}{2} \sum_{j,k=2}^{m} \tilde{\beta}_{jk}(t) D_{jk}^2 f(x) + \sum_{j=2}^{m} \tilde{\alpha}_j(t) D_j f(x) + \int_E \Big(f(x+y) -$$
$$- f(x) - \sum_{j=2}^{m} y_j D_j f(x) \chi_{\{|y| \leq 1\}}(y) \Big) \tilde{\pi}(t, dy) + \gamma_+(t) D_1^+ f(x) -$$
$$- \gamma_-(t) D_1^- f(x).$$

In the case when $m=1$ we assume that $\tilde{\alpha} = 0$, $\tilde{B} = 0$ and

$$\tilde{A}(t) f(x) = \int_E \big(f(x+y) - f(x) \big) \tilde{\pi}(t, dy) + \gamma^+(t) D_1^+ f(x) - \gamma^-(t) D_1^- f(x).$$

Let us say that a m-dimensional right continuous with left limits adapted to the family $\underline{F}$ stochastic process $X = \{ X_t = (X_t^1, \ldots, \ldots, X_t^m), \ t \geq 0 \}$ is a process with a penetrable boundary ∂G and $(P, \underline{F})$-local characteristics $(\hat{\alpha}, \hat{B}, \hat{\pi}, \gamma_+, \gamma_-, \delta, \tilde{\alpha}, \tilde{B}, \tilde{\pi})$ if there exists an increasing continuous process φ, such that P -a.e.

$$\varphi_t=\int_0^t \chi_{\partial G}(X_s)\,d\varphi_s,\quad \int_0^t \chi_{\partial G}(X_s)\,ds=\int_0^t \delta(s)\,d\varphi_s,\qquad t\geq 0,$$

and for all $f\in \hat{C}_b^2(R^m)$, $M(f)\in \mathcal{M}_{loc}(\mathcal{P},\underline{F})$, where

$$M_t(f) = f(X_t) - \int_0^t \chi_G(X_s)\hat{A}(s)f(X_s)\,ds - \int_0^t \tilde{A}(s)f(X_s)\,d\varphi_s,\qquad t\geq 0.$$

Denote

$$\underline{F}^X = \left\{ F_t^X = \bigcap_{\varepsilon>0} \sigma(X_s,\ s\leq t+\varepsilon),\quad t\geq 0\right\}.$$

From the results of paper [1] we obtain that if the functions $\gamma_+, \gamma_-, \delta$ and $\tilde{B}$ are adapted to $\underline{F}^X$ and for all $t\geq 0$ -a.e.

$$\gamma_+(t) + \gamma_-(t) + \delta(t) + \mathrm{tr}\,\tilde{B}(t) > 0, \tag{9}$$

then φ, which is called the local time of the process X on the boundary ∂G, is defined uniquely and is adapted to $\underline{F}^X$. The process X has the local characteristics $(\hat{\alpha}^X, \hat{B}, \hat{\pi}^X, \gamma_+, \gamma_-, \delta, \tilde{\alpha}^X, \tilde{B}, \tilde{\pi}^X)$ and local time φ with respect to P and $\underline{F}^X$, where $\hat{\alpha}^X, \hat{\pi}^X, \tilde{\alpha}^X$ and $\tilde{\pi}^X$ are $(P,\underline{F}^X)$-predictable projections of $\hat{\alpha}, \hat{\pi}, \tilde{\alpha}$ and $\tilde{\pi}$ correspondingly.

Recall that the process X is a $(P,\underline{F})$-semimartingale with a triplet of characteristics (α, β, Π) if it has the following canonical form (see [10], [11]):

$$X_t = X_0 + \alpha_t + X_t^c + \int_0^t \int_{|x|\leq 1} x\, q(ds,dx) + \int_0^t \int_{|x|>1} x\, p(ds,dx),\quad t\geq 0,$$

where $p(dt,dx)$ is the jump measure of X, $q(dt,dx) = p(dt,dx) - \Pi(dt,dx)$, $\Pi(dt,dx)$ is the $(P, \underline{F})$-dual predictable projection of the measure p, α is a $\mathcal{P}(\underline{F})$-measurable process having P - a.e. finite variation on every finite time interval, X^{cj} are continuous local martingales,

$$B_t = \|\beta_{jk}(t)\|_1^m,\quad \beta_{jk}(t) = \langle X^{cj}, X^{ck}\rangle_t,\quad t\geq 0,\ j,k=1,\dots,m.$$

In [1] it is proved that a stochastic process X having the $(P,\underline{F})$-local characteristics $(\hat{\alpha},\hat{B},\hat{\pi},\gamma_+,\gamma_-,\delta,\tilde{\alpha},\tilde{B},\tilde{\pi})$ is a $(P,\underline{F})$-semimartingale, whose triplet of characteristic (α,B,Π) has the form:

$$\alpha_1(t)=\int_0^t \chi_G(X_s)\,\hat{\alpha}_1(s)\,ds+\int_0^t(\gamma_+(s)-\gamma_-(s))\,d\varphi_s+\int_0^t\int_{|x|\le 1} x_1\tilde{\pi}(s,dx)\,d\varphi_s\,,$$

$$\alpha_j(t)=\int_0^t \chi_G(X_s)\,\hat{\alpha}_j(s)\,ds+\int_0^t\tilde{\alpha}_j(s)\,d\varphi_s\,,\quad j=2,\dots,m,$$

$$\beta_{jk}(t)=\int_0^t \chi_G(X_s)\,\hat{\beta}_{1k}(s)\,ds,\quad k=1,\dots,m, \tag{10}$$

$$\beta_{jk}(t)=\int_0^t \chi_G(X_s)\,\hat{\beta}_{jk}(s)\,ds+\int_0^t\tilde{\beta}_{jk}(s)\,d\varphi_s\,,\quad j,k=2,\dots,m,$$

$$\Pi([0,t]\times\Gamma)=\int_0^t \chi_G(X_s)\,\hat{\pi}(s,\Gamma)\,ds+\int_0^t\tilde{\pi}(s,\Gamma)\,d\varphi_s\,,$$

moreover

$$\int_0^t\int_E |x_1|\wedge 1+|x|^2\wedge 1\ \ \tilde{\pi}(s,dx)\,d\varphi_s<\infty,\ \int_0^t\int_E(|x|^2\wedge 1\,\hat{\pi}(s,dx)\,ds<\infty,\quad t\geq 0.$$

Remark 2. It is easy to check, that the process X, considered in § 1, has the $(P^{(a)},\underline{F}^{x})$- local characteristics $(0,1,0,\gamma_+,\gamma_-,0,0,0,0)$ and is a Markov process with the density of the transition probability function

$$p(t,x,y)=\begin{cases}\varphi(t,x,y)+\operatorname{tha}\varphi(t,-|x|,y) & \text{for } y\geq 0,\\ \varphi(t,x,y)-\operatorname{tha}\varphi(t,|x|,y) & \text{for } y<0,\end{cases}$$

where

$$\varphi(t,x,y)=(2\pi t)^{-1/2}\exp\left\{-\frac{1}{2t}(x-y)^2\right\}.$$

A general class of Markov processes with penetrable boundaries and given local characteristics was constructed in [5].

3. Stochastic Equations for Non-Linear Filtering

Let us observe a quasileftcontinuous $(P,\underline{F})$-semimartingale X with the triplet of characteristics (α, B, Π).

(I) Suppose that

$$\alpha_t = \alpha'_t + \int_0^t h(s)\, dB_s + \int_0^t \int_{|x|\leq 1} [\rho(t,x)-1] x\, \Pi'(ds,dx),$$

$$\Pi(dt,dx) = \rho(t,x)\, \Pi'(dt,dx), \quad t \geq 0,\ x \in E,$$

where α' is a $\mathcal{P}(\underline{F}^X)$-measurable process having P -a.e. finite variation on every finite time interval, Π' is a $\underline{F}^X$ - predictable measure such that $\Pi'(dt,dx)dP$ is the $\tilde{\mathcal{P}}(\underline{F}^X)$- σ-finite measure, h is a $\mathcal{P}(\underline{F})$ - measurable process, ρ is a $\tilde{\mathcal{P}}(\underline{F})$-measurable function ($\rho > 0$ Π'-a.e) such that the processes

$$\int_0^t |h(s)|\, d\, \mathrm{tr}\, B_s, \quad t \geq 0,$$

and

$$\int_0^t \int_{|x|\leq 1} |\rho(s,x) - 1|\, |x|\, \Pi'(ds,dx), \quad t \geq 0,$$

are $(P,\underline{F}^X)$-locally summable.

Hence and from Theorem 2 in [1] we obtain that X is a $(P,\underline{F}^X)$-semimartingale having the triplet of characteristics $(\bar{\alpha}, B, \bar{\Pi})$, where

$$\bar{\Pi}(dt,dx) = \bar{\rho}(t,x)\, \Pi'(dt,dx),$$

$$\bar{\alpha}_t = \alpha'_t + \int_0^t \bar{h}(s)\, dB_s + \int_0^t \int_{|x|\leq 1} [\bar{\rho}(s,x) - 1] x\, \Pi'(ds,dx),$$

$\bar{\rho}$ and $\bar{h}$ are $(P,\underline{F}^X)$-predictable projections of ρ and h, correspondingly.

The canonical form of the $(P,\underline{F}^X)$-semimartingale X is the following:

$$X_t = X_0 + \bar{\alpha}_t + \bar{X}_t^c + \int_0^t \int_{|x|\le 1} x \ \bar{q}(ds, dx) + \int_0^t \int_{|x|>1} x \ p(ds, dx), \quad t \geq 0,$$

where

$$\bar{X}_t^c = X_t^c + \int_0^t \big(h(s) - \bar{h}(s)\big)\, dB_s, \qquad t \geq 0,$$

and

$$\bar{q}(dt, dx) = p(dt, dx) - \bar{\Pi}(dt, dx).$$

(II) Let an estimable stochastic process Z be quasileftcontinuous and have the form:

$$Z_t = M_t^z + A_t^z, \qquad t \geq 0,$$

where for all $t \geq 0$ $E|Z_t| < \infty$, M^z is a $(P, \underline{F})$-martingale and A^z is a continuous $\underline{F}$-adapted process, $E|A^z|_t < \infty$, $t \geq 0$; $|A^z|_t$ denotes the variation of A^z on the interval $[0, t]$.

Denote

$$L^2_{loc}(B, P, F) = \Big\{ g = (g_1, \dots, g_m):$$

g is $\mathcal{P}(\underline{F})$-measurable and for all $t \geq 0$ P - a.e. $\int_0^t \big(g(s), g(s)\, B'(s)\big)\, d\, tr\, B_s < \infty \Big\}$(*), $G_{loc}(\Pi, P, F) = \Big\{ \eta = \eta(t,x) : \eta$ is $\tilde{\mathcal{P}}(\underline{F})$-measurable and $\int_0^t \int_E \frac{\eta^2(s,x)}{1 + |\eta(s,x)|} \Pi(ds, dx) < \infty$ P - a.e. for all $t \geq 0 \Big\}$.

By the standard way we define stochastic integrals

$$X_t^c(g) = \sum_{j=1}^m \int_0^t g_j(s)\, dX_s^{cj}, \qquad g \in L^2_{loc}(B, P, \underline{F}),$$

(*) $B'(t) = \| \beta'_{jk}(t) \|_1^m$, $\quad \beta'_{jk}(t) = \dfrac{d\beta_{jk}(t)}{d\, tr\, B_t}$, $\quad j,k = 1, \dots, m.$

$$\bar{X}_t^c(g) = \sum_{j=1}^{m} \int_0^t g_j(s)\, d\bar{X}_s^{cj}, \quad g \in L^2_{loc}(B, P, \underline{F}^x),$$

$$Q_t(\eta) = \int_0^t \int_E \eta(s,x)\, q(ds,dx), \; \eta \in G_{loc}(\Pi, P, \underline{F}),$$

$$\bar{Q}_t(\eta) = \int_0^t \int_E \eta(s,x)\, \bar{q}(ds,dx), \; \eta \in G_{loc}(\bar{\Pi}, P, \underline{F}^x).$$

(III) Assume that each $M \in M_{loc}(P, \underline{F}^X)$ has the form

$$M_t = M_o + \bar{X}_t^c(g) + \bar{Q}_t(\eta), \quad t \geq 0,$$

for some $g \in L^2_{loc}(B, P, \underline{F}^x)$ and $\eta \in G_{loc}(\bar{\Pi}, P, \underline{F}^x)$ (see [12]).

Define $\underline{D}^z \in L^2_{loc}(B, P, \underline{F})$ and $F^z \in G_{loc}(\Pi, P, \underline{F})$ by means of the following unique decomposition:

$$M_t^z = M_o^z + X_t^c(D^z) + Q_t(F^z) + \hat{M}_t^z, \quad t \geq 0,$$

where $\hat{M}^z \perp X^c(g)$ and $\hat{M}^z \perp Q(\eta)$ for all $g \in L^2_{loc}(B, P, \underline{F}), \eta \in G_{loc}(\Pi, P, \underline{F})$ such that $Q(\eta)$ is locally bounded.

(IV) Assume that the measure

$$dJ_t\, d\mathcal{P} \equiv \left| Z_{t-}(h(t) - \bar{h}(t)) + D^z(t) \right| dt\, trB_t\, d\mathcal{P}$$

is $\mathcal{P}(\underline{F}^x)$-$\sigma$-finite and the measure

$$\tilde{J}(dt,dx)dP \equiv \frac{\left| Z_{t-}(\rho(t,x) - \bar{\rho}(t,x)) + F^z(t,x)\, \rho(t,x) \right|}{\bar{\rho}(t,x)} \bar{\Pi}(dt,dx)\, dP$$

is $\mathcal{P}(\underline{F}^x)$-$\sigma$-finite.

<u>Theorem 2.</u> Under the assumptions I-IV the following equality (the stochastic equation for non-linear filtering) holds for all $t \geq 0$ P- a.e.

$$E(Z_t \mid F_t^x) = E(Z_o \mid F_o^x) + \bar{A}_t^z + \bar{X}_t^c(\bar{g}^z) + \bar{Q}_t(\bar{\eta}^z),$$

where

$$g^z(t) = Z_{t-}\big(h(t) - \bar{h}(t)\big) + D^z(t),$$

$$\eta^z(t,x) = \Big[Z_{t-}(\rho(t,x) - \bar{\rho}(t,x)) + F^z(t,x)\,\rho(t,x)\Big]\,(\bar{\rho}\,(t,x))^{-1},$$

$\bar{A}^z$ is a $(P,\underline{F}^x)$-dual predictable projection of A^z and $\bar{g}^z$, $\bar{\eta}^z$ are $(P,\underline{F}^x)$-predictable projections of g^z and η^z respectively.

P r o o f: Let

$$\bar{M}^z_t = E\big[Z_t \,|\, F^x_t\big] - \bar{A}^z_t, \qquad t \geq 0.$$

Since $\bar{M}^z_t = E\big[M^z_t \,|\, F^x_t\big] + E\big[A^z_t - \bar{A}^z_t \,|\, F^x_t\big]$, $t \geq 0$, then $\bar{M}^z$ is a $(P,\underline{F}^x)$-martingale. According to the assumption (III)

$$\bar{M}^z_t = \bar{M}^z_0 + \bar{X}^c_t\,(\hat{g}) + \bar{Q}_t(\hat{\eta}), \quad t \geq 0, \tag{11}$$

for some $g \in L^2_{loc}(B,P,\underline{F}^x)$ and $\hat{\eta} \in G_{loc}\,(\bar{\Pi},P,\underline{F}^x)$.

Now it is enough to show that $\hat{g} = \bar{g}^z$ and $\hat{\eta} = \bar{\eta}^z$. Let us choose the increasing sequences of subsets $A_r \in \mathcal{P}(\underline{F}^x)$, $A_r \uparrow [0,\infty)\times\Omega$, $\tilde{A}_r \in \tilde{\mathcal{P}}(\underline{F}^x)$, $\tilde{A}_r \uparrow [0,\infty)\times\Omega\times E$ such that

$$E\Big[\int_0^\infty \chi_{A_r}\, dJ_t\Big] < \infty \quad \text{and} \quad E\Big[\int_0^\infty\int_E \chi_{\tilde{A}_r}\,\tilde{J}(dt,dx)\Big] < \infty.$$

Let $U_n = [0,\infty)\times\Omega\times\{x : |x| > \frac{1}{n}\}$. Take $g \in L^2_{loc}(B,P,\underline{F}^x)$ and $\eta \in G_{loc}\,(\bar{\Pi},P,\underline{F}^x)$ such that g, η are bounded, $\eta = \chi_{U_n \cap \tilde{A}_r}\,\eta$, $g = \chi_{A_r} g$ and the processes $\bar{X}^c(g)$ and $\bar{Q}(\eta)$ are square integrable $(P,\underline{F}^x)$-martingales. Setting $\bar{N} = \bar{X}^c(g) + \bar{Q}(\eta)$ by means of (I) we obtain that

$$\bar{N}_t = X^c_t(g) + Q_t(\eta) + C_t - \bar{C}_t, \quad t \geq 0,$$

where

$$C_t = \int_0^t \big(g(s),\; h(s)\,B'(s)\big)\; d\,\mathrm{tr}\,B_s + \int_0^t\int_E \eta(s,x)\;\rho(s,x)\;\Pi'(ds,dx),$$

$\bar{C}$ is the $(P,\underline{F}^x)$-dual predictable projection of C. If we set $N = X^c(g) + Q(\eta)$ then we obtain

$$\bar{M}_t^z \bar{N}_t = E\left[M_t^z \bar{N}_t + \left(A_t^z - \bar{A}_t^z\right)\bar{N}_t \middle| F_t^x\right], \quad t \geq 0. \tag{12}$$

However

$$M_t^z \bar{N}_t = M_t^z N_t + \int_0^t (C_u - \bar{C}_u)\, dM_u^z + \int_0^t M_{u-}^z\, d(C_u - \bar{C}_u) \tag{13}$$

and

$$\left(A_t^z - \bar{A}_t^z\right)\bar{N}_t = A_t^z\left(N_t + C_t - \bar{C}_t\right) - \bar{A}_t^z \bar{N}_t =$$

$$= \int_0^t \bar{A}_u^z\, dN_u + \int_0^t A_u^z\, d(C_u - \bar{C}_u) - \int_0^t \bar{A}_u^z\, d\,N_u +$$

$$+ \int_0^t \bar{N}_{u-}\, d\left(A_u^z - \bar{A}_u^z\right).$$

Therefore supposing $L = M^z\bar{N} + (A^z - \bar{A}^z)\,\bar{N}$, from (12)-(13) we find

$$L_t = M_t^z N_t - \langle M^z, N\rangle_t + \int_0^t (C_u - \bar{C}_u)\, dM_u^z +$$

$$+ \int_0^t \bar{A}_u^z\, d\,N_u - \int_0^t \bar{A}_u^z\, d\bar{N}_u + \int_0^t \bar{N}_{u-}\, d(A_u^z - \bar{A}_u^z) +$$

$$+ \int_0^t Z_{u-}\, d\,(C_u - \bar{C}_u) + \langle M^z, N\rangle_t\,.$$

Let $\tau_p \uparrow \infty$ be a sequence of $\underline{F}^x$-stopping times such that

$$E\Bigg[\int_0^{\tau_p} |\hat{g}(s)|^2\, d\operatorname{tr} B_s + \int_0^{\tau_p}\int_{|\eta(s,x)|\leq 1} |\eta(s,x)|^2\, \bar{\Pi}(ds,dx) +$$

$$+ \int_0^{\tau_p}\int_{|\eta(s,x)|>1} |\eta(s,x)|\, \bar{\Pi}(ds,dx)\Bigg] < \infty, \quad \bar{M}^z_{\cdot\wedge\tau_p -} \leq p$$

and

$$\tau_n' = \inf\left\{t : |\bar{N}_t| \geq n \text{ or } |\bar{A}^z|_t \geq n\right\}.$$

Then

$$E\left[\bar{M}^z_{t\wedge\tau_p}\, \bar{N}_{t\wedge\tau_n'}\right] = E\Bigg[\int_0^{t\wedge\tau_p\wedge\tau_n'} (\hat{g}(s), g(s)B'(s))\, d\operatorname{tr} B_s +$$

$$+ \int_0^{t\wedge\tau_p\wedge\tau_n'}\int_E \hat{\eta}(s,x)\,\eta(s,x)\,\bar{\Pi}(ds,dx)\Bigg] = E\left[\bar{M}_t^z\, \bar{N}_{t\wedge\tau_n'\wedge\tau_p}\right] =$$

$$= \lim_{k\to\infty} E\left[L_{t\wedge\tau_n'\wedge\tau_p\wedge T_k}\right] =$$

$$= E\Bigg[\int_0^{t\wedge\tau_p\wedge\tau_n'} (\bar{g}^z(s)\, g(s)\, B'(s))\, d\operatorname{tr} B_s +$$

$$+ \int_0^{t\wedge\tau_p\wedge\tau_n'} \int_E \eta(s,x) \quad \overline{\eta}^z(s,x) \quad \overline{\Pi}(ds,dx) \Big],$$

where $T_k \uparrow \infty$ is a sequence of $\underline{F}$ -stopping times such that

$$M^z_{\cdot\wedge T_k} \in H^1(P,\underline{F}), \quad X^c_{\cdot\wedge T_k} \quad (D^z) \in H^1(P,\underline{F}), \quad Q_{\cdot\wedge T_k} \quad (F^z) \in H^1(P,\underline{F}),$$

$$M^z_{\cdot\wedge T_k -} \leqslant k, \quad X^c_{\cdot\wedge T_k} \quad (g) \leqslant k, \quad Q_{\wedge T_k -} \leqslant k$$

and

$$\int_0^{t\wedge T_k} A^z_u \, dN_u, \quad t \geqslant 0, \quad \int_0^{t\wedge T_k} (C_u - \bar{C}_u) \; d\, M^z_u, \quad t \geqslant 0,$$

are martingales.

Recall that

$$H^1(P,\underline{F}) = \Big\{ M: \; M \in M_{loc}(P,\underline{F}), \; E\Big[\sup_t |M_t|\Big] < \infty \Big\}.$$

Therefore we have the equality

$$E\Big[\int_0^{t\wedge\tau_p\wedge\tau_n'} (\hat{g}(s), g(s) \; B'(s)) \; dtrB_s + \int_0^{t\wedge\tau_p\wedge\tau_n'} \int_E \hat{\eta}(s,x)\eta(s,x) \; \overline{\Pi}(ds,dx) \Big] =$$

$$= E\Big[\int_0^{t\wedge\tau_p\wedge\tau_n'} (\overline{g}^z(s), g(s)B'(s)) \; d \; tr \; B_s + \int_0^{t\wedge\tau_p\wedge\tau_n'} \int_E \overline{\eta}^z(s,x)\,\eta(s,x)\overline{\Pi}(ds,dx) \Big].$$

By passing to the limit as $n \to \infty$ and varying g and η hence we get that $\hat{g} = \overline{g}^z$ and $\hat{\eta} = \overline{\eta}^z$. Theorem 2 is proved.

Remark 3. Stochastic equations for non-linear filtering when the observation process is a semimartingale were derived in [13], [14], under different form of assumptions.

Assume now that the observation process X is a process with the penetrable boundary ∂G and the $(P,\underline{F})$-local characteristics $(\hat{\alpha}, \hat{B}, \hat{\pi}, \gamma_+, \gamma_-, \delta, \tilde{\alpha}, \tilde{B}, \tilde{\pi})$. Assume that $\gamma_+, \gamma_-, \delta$ and $\tilde{B}$ are adapted to $\underline{F}^X$ and (9) is fulfilled.

Since in this case the local time φ is adapted to $\underline{F}^X$, then assuming that

$$\hat{\alpha}(t) = \hat{\alpha}'(t) + \hat{h}(t)\hat{B}(t) + \int_{|x|\le 1} \left[\hat{\rho}(t,x) - 1\right] x \hat{\pi}'(t,dx),$$

$$\tilde{\alpha}(t) = \tilde{\alpha}'(t) + \tilde{h}(t)\tilde{B}(t) + \int_{|x|\le 1} \left[\tilde{\rho}(t,x) - 1\right] \tilde{x}\, \tilde{\pi}'(t,dx),^{(*)}$$

$$\hat{\pi}(t,dx) = \hat{\rho}(t,x)\, \hat{\pi}'(t,dx),$$

$$\tilde{\pi}(t,dx) = \tilde{\rho}(t,x)\, \tilde{\pi}'(t,dx),$$

and using equalities (10), evidently we can find the analogs of the assumptions (I)-(IV) and the formulation of Theorem 2. We shall omit the details here.

References

[1] G r i g e l i o n i s, B., M i k u l e v i č i u s, R., On stochastic processes with penetrable boundaries. Liet. matem. rink., XX, 2 (1980).

[2] F u j i s a k i, M., K a l l i a n p u r, G., K u n i t a, H., Stochastic differential equations for the non-linear filtering problem. Osaka J. of Math., 9, 1, 19-40 (1972).

[3] G r i g e l i o n i s, B., On stochastic non-linear filtering equations of stochastic processes. Liet. matem. rink., XII, 4, 37--51 (1972).

[4] G r i g e l i o n i s, B., On statistical problems of stochastic processes with boundary conditions. Liet. matem. rink., XVI, 1, 63-87 (1976).

[5] M i k u l e v i č i u s, R., On existence and uniqueness of solutions of martingale problem on branched manifold. Liet. matem. rink., XX (1980).

(*) $\tilde{x} = (x_2, \ldots, x_m)$.

[6] G r i g e l i o n i s, B., On relative compactness of the sets of probability measures on $D_{[0,\infty)}(X)$. Liet.matem.rink., XIII,4, 89-96 (1973).

[7] M e y e r, P. A., Un cours sur les integrales stochastiques. Sém. Probab. X, Lecture Notes in Math., 511, Springer, 245-400 (1976).

[8] D e l l a c h e r i e, C., Capacités et processus stochastiques Springer, Verlag: Berlin-Heidelberg-New York 1972.

[9] P o r t e n k o, N. I., Generalized diffusion processes. Dissertation, Kiev 1977.

[10] J a c o d, J., M é m i n, J., Charactéristiques locales et conditions de continuité absolue pour les semimartingales.Z.Wahrscheinlichkeitstheorie verw. Geb., 35, 1-37 (1976).

[11] G r i g e l i o n i s, B., On the martingale characterization of random processes with independent increments. Liet. matem. rink., XVII, 1, 75-86 (1977).

[12] J a c o d, J., A general theorem of representation for martingales. Proc. of Symp. in Pure Math., 31, 37-53 (1977).

[13] B r e m a u d, P., Y o r, M., Changes of filtrations and of probability measures. Z. Wahrscheinlichkeitstheorie verw. Geb., 45, 4, 269-295 (1978).

[14] S z p i r g l a s, J., M a z z i o t t o, G., Modéle général de filtrage non linéaire et équations différentielles stochastiques associées. C. R. Acad.Sci.Paris, Sér. A, 286, 1067-1070, (1978).

ON TWO-SIDED NONPARAMETRIC TESTS FOR THE TWO-SAMPLE PROBLEM

by

Piotr Hellmann

University of Warsaw

1. Introduction

Let $X_1,\dots,X_n$ and $Y_1,\dots,Y_m$ be two independent random samples from distributions dominated by the Lebesgue measure, and let their distribution functions be, respectively, F and G. Consider the hypothesis H_0: F = G and the alternatives

$$H_1: F \neq G,$$

$$H_2: P(X_i > Y_j) < 1/2 \quad \text{or} \quad P(Y_j > X_i) < 1/2,$$

$$H_3: (F \geqslant G \quad \text{or} \quad G \geqslant F) \text{ and } F \neq G.$$

It is clear that $H_3 \dot{\subset} H_2 \subset H_1$.

There exist some tests that are proposed against these alternatives but even their unbiasedness is not known yet (cf. [5], p. 240). The one exception is the Lehmann test which is unbiased by the construction [4].

In the sequel, we shall discuss the tests that are commonly employed for the problem (cf. e.g. [2]). At first consider the tests with the critical region of the shape

$$\left|\sum_{j=1}^{m} h(t_j) - \sum_{i=1}^{n} h(r_i)\right| > c,$$

where r_i and t_j are the ranks of elements of the first and the second sample respectively, c is a constant and h some function. By choice of h, the following tests may be obtained.

1. The Wilcoxon test - by putting $h(r_i) = r_i$ and $h(t_j) = t_j$.

2. The Fisher-Yates test - by putting $h(r_i) = E(V^{(r_i)})$ and $h(t_j) = E(V^{(t_j)})$, where $V^{(1)} < \dots < V^{(n+m)}$ is an order sample from the standardized normal distribution.

3. The van der Waerden test - by putting

$$h(r_i) = \psi^{-1}\left(\frac{r_i}{n+m+1}\right) \quad \text{and} \quad h(t_j) = \psi^{-1}\left(\frac{t_j}{n+m+1}\right),$$

where ψ^{-1} is an inverse of standardized normal distribution functions.

Then consider the other tests.

4. The Cramer - von Misses test which rejects H_0 if

$$\sum_{i=1}^{n}\sum_{j=1}^{m} (r_i - t_j)^2 > c.$$

5. The Kolmogorov-Smirnov test which rejects H_0 if

$$\sup_z \left| S_{(x_1,\dots,x_n)}(z) - S_{(y_1,\dots,y_m)}(z) \right| > c,$$

where $S_{(x_1,\dots,x_n)}(z)$ and $S_{(y_1,\dots,y_m)}(z)$ are the sample distribution functions of the two samples.

6. The Haga test which rejects H_0 if

$$| A + B - A' - B' | > c,$$

where A and B' denote the number of observations among X's larger than $\max_j Y_j$ or smaller than $\min_j Y_j$, respectively, and where A' and B denote the number of observations among Y's larger than $\max_i X_i$ or smaller than $\min_i X_i$.

7. The Wald-Wolfowitz run test which rejects H_0 if the number of runs in the combined sample is too small.

8. The Lehmann test, which we are going to detail. According [5], p. 256, let X, X' and Y, Y' be independent samples of size two

from continuous distributions F and G, respectively. Then

$$p = P\{\max(X,X') < \min(Y,Y') \text{ or } \max(Y,Y') < \min(X,X')\} = \tag{1}$$
$$= 1/3 + 2\Delta ,$$

where

$$\Delta = \int (F - G)^2 d(F + G)/2.$$

From the continuity of F and G it follows that $\Delta = 0$ iff $F = G$. Let now X_i, X_i' and Y_i, Y_i' $(i = 1,\dots,n)$ be independently distributed, the X's with distribution F, the Y's with distribution G, and let $V_i = 1$ if $\max(X_i,X_i') < \min(Y_i,Y_i')$ or $\max(Y_i,Y_i') < \min(X_i,X_i')$ and $V_i=0$ otherwise. Then $\sum V_i$ has a binomial distribution with probability p and the problem reduces to that of testing $p = 1/3$ against $p > 1/3$. Unbiasedness of the test is obvious.

In the sequel, we shall consider the family of the rank tests of H_0 invariant with respect to the renumeration of the samples. The considerations shall be restricted to the case of samples of size two.

2. Distribution of symmetric ranks

Let Re denote the real line, B the σ-field of Lebesgue measurable sets and Φ the family of all distribution functions dominated by the Lebesgue measure. In (Re^4, B^4) we denote by Z the σ-ideal of the null sets. Define statistical structure

$$\left(Re^4,\ B^4,\ \left\{P_{(F,G)} : (F,G) \in \Phi^2\right\}\right), \tag{2}$$

where the distribution function of $P_{(F,G)}$ is equal to $F(\cdot)F(\cdot)G(\cdot)G(\cdot)$.

The restriction to the rank tests reduces the space (Re^4, B^4) to $(\overline{Re}^4,\ \overline{A}_R')$, where $\overline{Re}^4$ is derived from Re^4 by rejection of all points with at least two coordinates equal and $\overline{A}_R'$ is the field generated by the vector of ranks in the pooled sample. The vector of ordered ranks of one of the samples, say of the first one, is a sufficient statistic for the family of distributions on $\overline{A}_R'$. Therefore we can consider the space $\left(\overline{Re}^4, \overline{A}_R\right)$, where $\overline{A}_R$ is generated by the vector of ordered ranks of the first sample. Restriction to the tests, that are invariant with respect to the renumeration of the samples, reduces finally the initial measurable space to $\left(\overline{Re}^4, \overline{A}_{SR}\right)$, where $\overline{A}_{SR}$ is generated by the three atoms

$$\overline{A}_1 = \left\{x \in Re^4 :\ x_i < x_{1-i} < y_j < y_{1-j} \text{ or } y_j < y_{1-j} < x_i < x_{1-i}\right\},$$

$$\overline{A}_2 = \left\{x \in Re^4 :\ x_i < y_j < y_{1-j} < x_{1-i} \text{ or } y_j < x_i < x_{1-i} < y_{1-j}\right\},$$

and

$$\overline{A}_3 = \left\{x \in Re^4 :\ x_i < y_j < x_{1-i} < y_{1-j} \text{ or } y_j < x_i < y_{1-j} < x_{1-i}\right\},$$

where $i, j \in \{0, 1\}$.

We can extend $\left(\overline{Re}^4, \overline{A}_{SR}\right)$ defining $(Re^4, \overline{A}_{SR} \cup Z)$. Now, there exists a partition of Re^4 into subsets A_1, A_2, A_3, such that $A_i \approx \overline{A}_i$ mod Z and $P(A_i) = P(\overline{A}_i)$ for $i = 1, 2, 3$ and for all distributions vanishing on Z ([1], p. 144). The sets A_1, A_2, A_3 (called symmetric ranks) generate the subfield denoted by A_{SR}.

Finally, we define statistical structure

$$\left(Re^4,\ A_{SR},\ \{P_p :\ p \in S\}\right), \tag{3}$$

where S is a unit simplex in Re^3 and, for any p, P_p is defined by $P_p(A_i) = p_i$, $i = 1, 2, 3$. Let M be the mapping $\Phi^2 \to S$ such that $P_{M(F,G)}$ is equal to $P_{(F,G)}$ restricted to A_{SR}. In the sequel, we shall examine images of the hypotheses (preserving the notation $H_0, \ldots, H_3$). In order to do so, we define the following parameter functions of structure (2):

$$\Delta_1^2(F,G) = \left(\int (F - G)\,dF\right)^2,$$

$$\Delta_2^2(F,G) = \int (F - G)^2\,dF.$$

<u>Theorem 1.</u> The mapping M is given by the formula:

$$p_1 = 1/3 + 2\Delta_2^2 ,$$

$$p_2 = 1/3 + 2\left(\Delta_2^2 - 2\Delta_1^2\right), \tag{4}$$

$$p_3 = 1/3 - 4\left(\Delta_2^2 - \Delta_1^2\right).$$

The images of the hypotheses are of the shape:

$$H_0 = \left\{ p \in S: \Delta_1^2 = \Delta_2^2 = 0 \right\},$$

$$H_1 = \left\{ p \in S: 0 \leqslant \Delta_1^2 < \Delta_2^2 \right\},$$

$$H_2 = \left\{ p \in S: 0 < \Delta_1^2 < \Delta_2^2 \right\}.$$

P r o o f. The result for p_1 is obtained from (1) by noting that $\Delta_2^2 = \Delta$ while, in view of Savage, Sobel and Woodworth ([6] p. 110) $p_3 = 2\left(\int F\, dG\right)^2 + 2\left(\int G\, dF\right)^2 - 2p_1$. Moreover, the Schwartz inequality and continuity of F and G imply $\Delta_2^2 \geqslant \Delta_1^2$ with the equality holding iff F = G. The proof is completed by noting that $\Delta_1^2 = 0$ iff $P(X > Y) = \int G dF = 1/2 = \int F dG = P(Y > X)$.

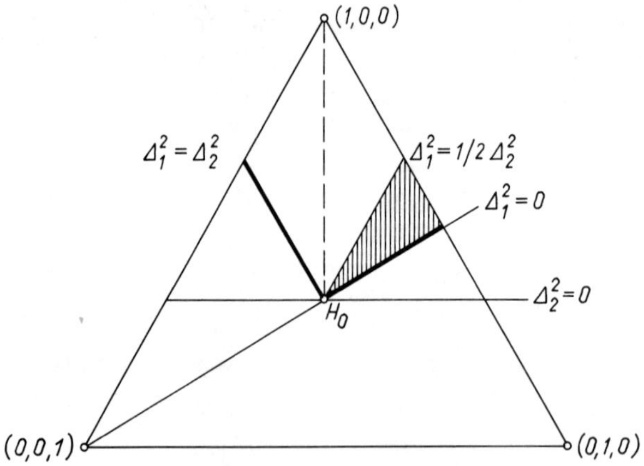

Fig. 1

Now, let us introduce the local coordinates $\left(\Delta_1^2, \Delta_2^2\right)$ in S. Some lines of importance and their equations, which are obtained from (4), are shown in Fig. 1, so that it is easy to situate H_0, H_1 and H_2. The image of H_3 was not found, but, for further considerations, it is enough to notice that the intersection of H_3 and the shaded area in Fig. 1 $\left(0 \leqslant \Delta_1^2 < 1/2\,\Delta_2^2\right)$ is not empty. Indeed, if, for instance, F and G correspond to the uniform distributions on (0,1) and $(0, 1/2) \cup \cup (1, 3/2)$ respectively, then one can compute that $1/64 = \Delta_1^2 < 1/2\,\Delta_2^2 = = 1/48$.

3. Properties of two-sided tests

Any test γ of H_0, defined on statistical structure (3), is of the shape $\gamma(x) = \gamma_i$ if $x \in A_i$ (where $x \in Re^4$, $0 \leqslant \gamma_i \leqslant 1$ and $i = 1, 2, 3$). It is useful for further considerations to visualize the power of γ, denoted by β_γ, as a plane over the triangle presented in Fig. 1. Denote by Γ_α the family of the tests of size α which satisfy the conditions $\gamma_2 = \gamma_3 = 0$ if $\alpha \leqslant 1/3$ and $\gamma_1 = 1$ otherwise.

Theorem 2. For every α, the family Γ_α is the minimal essentially complete class for testing H_0 against the alternatives H_1, H_2 and H_3.

P r o o f: It is enough to consider H_1. For all $(F,G) \in H_1$, $p_1 \geqslant p_2$ and $p_1 \geqslant p_3$. If $\alpha \leqslant 1/3$, then Γ_α contains only one test which is uniformly most powerful. Suppose now that $\alpha > 1/3$. Then, for any test γ' of size α, there exists a test γ in Γ_α such that $\gamma_i = \gamma_i' - c_i$ ($i = 2, 3$), where c_i's are some constants such that $c_2 + c_3 = 1 - \gamma_1'$. It is clear that γ is uniformly better than γ' and therefore, Γ_α is a complete class. The proof is completed by noting that for all $\gamma \in \Gamma_\alpha$, β_γ is independent from γ_2 and γ_3 over the line $p_2 = p_3$.

Theorem 3. For $0 \leqslant \alpha \leqslant 1$ $(0 < \alpha < 1)$ any test $\gamma \in \Gamma_\alpha$ is unbiased as a test of H_0 against H_1 (strictly unbiased as a test of H_0 against H_2) iff $\gamma_3 \leqslant \alpha$.

P r o o f: It is enough to consider H_1. If $\alpha \leqslant 1/3$ or $\alpha = 1$, then the only test $\gamma \in \Gamma_\alpha$ is unbiased. Let $1/3 < \alpha < 1$. For every test $\gamma \in \Gamma_\alpha$

$$\beta_\gamma - \alpha = (1 + \gamma_2 - 2\gamma_3)\Delta_2^2 - 2(\gamma_2 - \gamma_3)\Delta_1^2 \,. \qquad (5)$$

Assume first that $\gamma_3 \leqslant \alpha$. By (5), the claim is immediate if we note that the assumption is equivalent to $1 + \gamma_2 - 2\gamma_3 \geqslant 0$. On the other hand, if $\gamma_3 > \alpha$, then it is enough to put $\Delta_1^2 = 0$ and, in view of (5) the proof is completed.

It is easy to see from (5) that, for the test $\gamma \in \Gamma_{2/3}$ such that $\gamma_1 = \gamma_3 = 1$ and $\gamma_2 = 0$, the power β_γ is smaller than 2/3 on the entire area shaded in Fig. 1. Therefore, γ is biased against all the alternatives (including H_3).

Finally, let us consider the properties of the tests mentioned in the introduction.

Table

Test	1	2	3	4	5	6	7
A_1	4	2.65	2.19	18	1	4	2
A_2	0	0	0	10	1/2	0	3
A_3	2	1.46	1.17	12	1/2	2	4

The table contains the values of the tests statistics on symmetric ranks A_1, A_2, A_3. It is easy to see that tests 1, 2, 3, 4, 6 are biased against H_1 and H_2 for $\alpha > 1/2$ and against H_3 at least for $\alpha = 2/3$. Tests 5 and 7 are unbiased at least for the samples of size two and, therefore, their unbiasedness is an open problem.

I would like to point out that the geometrical approach, presented by me in this paper, was inspired by most remarkable book written by Czenzov [1].

References

[1] C z e n z o v, N. N., Statisticzeskije reszajuszczije prowila i optimalnyje wywody. Nauka, Moscow 1972.

[2] H á j e k, I. and Š i d á k, Z., Theory of rank tests. Academia, Prague 1967.

[3] H e l l m a n n, P., Unbiasedness of two-sided nonparametric tests in the two-sample problem (to be published) (1979).

[4] L e h m a n n, E. L. Consistency and unbiasedness of certain nonparametric tests. Ann. Math. Statist., 22, 165-179 (1951).

[5] L e h m a n n, E. L., Testing statistical hypotheses. Wiley. New York 1959.

[6] S a v a g e, I. R., S o b e l, M. and W o o d w o r t h, G., Fine structure of the ordering of probabilities of rank orders in the two-sample case. Ann. Math. Statist., 37, 98-112 (1966).

ON LIMIT THEOREMS FOR SUMS OF DEPENDENT HILBERT SPACE VALUED RANDOM VARIABLES

by

Adam Jakubowski
Nicolaus Copernicus University, Toruń

1. Introduction

Let $\{X_{nk}\}$, $k = 1,2,\ldots,k_n$; $n = 1,2,\ldots$, be an array of random variables defined on a common probability space $(\Omega, \mathcal{F}, P)$. If $\{X_{nk}\}$ are row-wise independent, then there exists a quite satisfactory theory of the weak convergence of sums $S_n' = \sum_{k=1}^{k_n} X_{nk}$. One of the most reasonable trends in the analogous theory for dependent random variables is initiated by papers of Brown [2] and Dvoretzky [4], [5].

This new successful approach (see [3], [6] for generalizations of [2], [5]) can be described very briefly: To obtain limit theorems for dependent random variables one has to replace usual expectations in classical theorems for independent random variables by conditional expectations with respect to a suitably chosen family of σ-subfields of $\mathcal{F}$ and the convergence of numbers by the convergence in probability. This procedure can be observed most explicitly in Theorem C (Section 3) - the Hilbert space version of the Brown's Theorem.

The present paper contains generalizations of theorems of such form for the case when X_{nk} are random variables taking values in real separable Hilbert space. Their proofs are new even in the finite dimensional case and are based on the technics of the regular conditional distributions. Such an approach gives possibility for the use of the Varadhan's theory for weak convergence of convolutions in Hilbert space (see [8], also [7], Chapter VI).

Basic Theorems A and B, which can be treated as modified accompanying laws, are contained in Section 2. In particular, Theorem A is a sufficient tool for quick proofs in the finite dimensional case. In Section 3 it is shown, how to obtain from Theorem B the required results: the Brown's Theorem (Theorem C) and the Hilbert space generalization of theorem of Kłopotowski (Theorem D). More detailed proofs of Theorems B and D will be published elsewhere.

2. Main Theorems

Let H be a real separable Hilbert space with the inner product $(\cdot,\cdot)$ and let $\mathcal{B}_H$ be the σ-field of Borel subsets of H. All H-valued random variables considered in this paper are defined on fixed probability space $(\Omega, \mathcal{F}, P)$. Let $\{X_{nk}\}$, $k = 1,2,\ldots,k_n$; $n = 1,2,\ldots$, be an array of random variables and $\{\mathcal{F}_{nk}\}$, $k = 0,1,\ldots,k_n$; $n = 1,2,\ldots$, be an array of row-wise increasing σ-subfields of $\mathcal{F}$ (i.e. $\mathcal{F}_{nk} \subset \mathcal{F}_{n,k+1}$ for n fixed and $k = 0,1,\ldots,k_n-1$). The array $\{X_{nk}\}$ is said to be adapted to $\{\mathcal{F}_{nk}\}$ if every X_{nk} is $\mathcal{F}_{nk}$ - measurable.

For $\{X_{nk}\}$ adapted to $\{\mathcal{F}_{nk}\}$ we can define an array $\{\mu_{nk}\}$ of regular random measures by choosing for every (n,k) a regular version of the conditional distribution of X_{nk} given $\mathcal{F}_{n,k-1}$. In other words, for every n,k

$$\mu_{nk} : \mathcal{B}_H \times \Omega \longrightarrow [0,1]$$

is a function such that for every $\omega \in \Omega$ $\mu_{nk}(\cdot,\omega)$ is a probability measure on $\mathcal{B}_H$ and for every $A \in \mathcal{B}_H$ $\mu_{nk}(A,\cdot)$ is a version of $P(X_{nk} \in A \mid \mathcal{F}_{n,k-1})$ (hence $\mathcal{F}_{n,k-1}$-measurable). For some properties of the regular conditional distribution and for the proof of its existence see [1], Chapter 4.

In the sequel we will deal with the arrays $\{X_{nk}\}$ adapted to $\{\mathcal{F}_{nk}\}$ and the regular random measures μ_{nk} defined above and the definitions will not be repeated in the theorems.

Now we can formulate Theorem A, which is sufficient for applications in the finite dimensional case.

<u>Theorem A.</u> Let μ be a distribution on $\mathcal{B}_H$ with the non-vanishing characteristic functional:

$$\forall y \in H \qquad \hat{\mu}(y) := \int e^{i(y,x)} \mu(dx) \neq 0.$$

If for almost all $\omega \in \Omega$, the convolutions

$$\mu_n(\cdot,\omega) := \mu_{n1}(\cdot,\omega) * \mu_{n2}(\cdot,\omega) * \dots \mu_{nk_n}(\cdot,\omega)$$

are weakly convergent to μ ($\mu_n \Rightarrow \mu$ a.s.) then the characteristic functionals of $S_n = \sum_{k=1}^{k_n} X_{nk}$ are pointwise convergent to $\hat{\mu}$:

$$\forall y \in H \qquad E\, e^{i(y,S_n)} \to \hat{\mu}(y).$$

P r o o f: Denoting $\hat{\mu}_{nk}(y,\omega) := \widehat{\mu_{nk}(\cdot,\omega)}(y)$ we have

$$\forall y \in H \qquad \prod_{k=1}^{k_n} \hat{\mu}_{nk}(y,\omega) \to \hat{\mu}(y) \text{ a.s.}$$

For fixed $y \in H$ the set

$$A_{nk} := \left\{\omega ; \prod_{j=1}^{k} |\hat{\mu}_{nj}(y,\omega)| \geq \tfrac{1}{2}|\hat{\mu}(y)|\right\}$$

is $\mathcal{F}_{n,k-1}$-measurable $\langle$since $\hat{\mu}_{nj}(y,\cdot) = E(e^{i(y,X_{nj})} \mid \mathcal{F}_{n,j-1})\rangle$. Moreover $A_{n,k+1} \subset A_{nk}$ and $P\left(\bigcup_{m=1}^{\infty} \bigcap_{n \geq m} A_{nk_n}\right) = 1$.

Putting

$$X^*_{nk} := I_{A_{nk}} X_{nk}, \qquad S^*_n := \sum_{k=1}^{k_n} X^*_{nk},$$

we obtain $S_n - S^*_n \xrightarrow{P} 0$; hence $E\, e^{i(y,S_n)} \to \hat{\mu}(y)$ if and only if $E\, e^{i(y,S^*_n)} \to \hat{\mu}(y)$. But as a regular version of the conditional distribution of X^*_{nk} given $\mathcal{F}_{n,k-1}$ we can choose

$$I_{A^c_{nk}}\, \delta_0(\cdot) + I_{A_{nk}}\, \mu_{nk}(\cdot)$$

so that

$$\prod_{k=1}^{k_n} \left| E(e^{i(y,X^*_{nk})} \mid \mathcal{F}_{n,k-1}) \right| \geq \tfrac{1}{2} \, |\hat{\mu}(y)| \quad \text{a.s.}$$

Hence the following computation

$$H_n := E\, e^{i(y,S^*_n)} \left[\prod_{k=1}^{k_n} E(e^{i(y,X^*_{nk})} \mid \mathcal{F}_{n,k-1}) \right]^{-1} = 1$$

is true.

By the estimation

$$\left| E\, e^{i(y,S^*_n)} - \hat{\mu}(y) \right| = \left| E\, e^{i(y,S^*_n)} - \hat{\mu}(y)\, H_n \right| \leq$$

$$\leq E \left| (\hat{\mu}(y))^{-1} - \left[\prod_{k=1}^{k_n} E(e^{i(y,X^*_{nk})} \mid \mathcal{F}_{n,k-1}) \right]^{-1} \right|$$

and the fact, that $\prod_{k=1}^{k_n} E(e^{i(y,X^*_{nk})} \mid \mathcal{F}_{n,k-1}) \to \hat{\mu}(y)$ a.s. the proof is completed.

The next theorem gives the conditions for the weak convergence in the infinite dimensional case.

Theorem B. If μ is an infinitely divisible distribution (for the definition and some properties see [7]), then the following conditions

B1. $\mu_n = \mu_{n1} * \mu_{n2} * \dots * \mu_{nk_n} \Rightarrow \mu$ a.s.

B2. $\forall \varepsilon > 0 \ \max_{1 \leq k \leq k_n} \mu_{nk}(\|x\| > \varepsilon) \to 0$ a.s.

imply the weak convergence of distributions P_{S_n} of sums S_n to μ: $P_{S_n} \Rightarrow \mu$.

We give only a sketch of the proof.

Since μ is infinitely divisible its characteristic functional is non-vanishing. So B1 together with Theorem A imply $E\, e^{i(y,S_n)} \to \hat{\mu}(y)$ for every $y \in H$. By Lemma 2.10, Chapter VI, [7] it is sufficient to prove that $\{P_{S_n}\}$ form a conditionally compact set of measures.

To accomplish it, let us define:

$$a_{nk} = a_{nk}(\omega) := \int_{[\|x\| \leq 1]} x\, \mu_{nk}(dx, \omega) = E(X_{nk} I(\|X_{nk}\| \leq 1) | \mathcal{F}_{n,k-1}) \text{ a.s.},$$

$$\Theta_{nk} := \mu_{nk} * (-a_{nk}),$$

$$\lambda_n := e\Big(\sum_{k=1}^{k_n} \Theta_{nk}\Big) * \Big(\sum_{k=1}^{k_n} a_{nk}\Big),$$

where $e(F)$ is defined for a finite measure F by the formula

$$e(F) := e^{-F(H)} \sum_{n=0}^{\infty} F^{*n}/n!.$$

Under condition B2 by the accompanying laws (Corollary 6.1. Chapter VI [7]) condition B1 is equivalent to B1', $\lambda_n \Rightarrow \mu$ a.s.

We have introduced λ_n because for such measures we have good criteria of compactness (see [7], paragraph 5, Chapter VI).

Now let us define

$$Z_{nk} := X_{nk} - a_{nk},$$

$$U_{nk} := Z_{nk} I(\| Z_{nk} \| \leq t) - E(Z_{nk} I(\| Z_{nk} \| \leq t) | \mathcal{F}_{n,k-1}),$$

$$V_{nk} := Z_{nk} I(\| Z_{nk} \| > t),$$

$$W_{nk} := a_{nk} + E(Z_{nk} I(\| Z_{nk} \| \leq t) | \mathcal{F}_{n,k-1}),$$

where $t > 0$ is a fixed real number such that $M(\| x \| = t) = 0$ (M is the measure in the Levy's representation of μ, see Section 3 of this paper).

Due to the equality

$$S_n = \sum_{k=1}^{k_n} X_{nk} = \sum_{k=1}^{k_n} U_{nk} + \sum_{k=1}^{k_n} V_{nk} + \sum_{k=1}^{k_n} W_{nk} =$$

$$=: U_n + V_n + W_n$$

the conditional compactness of $\{P_{S_n}\}$ follows from the conditional compactness of the sets $\{P_{U_n}\}$, $\{P_{V_n}\}$, $\{P_{W_n}\}$. For each of the mentioned sets we use the criteria of compactness given by B1:In the proof of the conditional compactness of $\{P_{V_n}\}$ we use the following lemma:

<u>Lemma.</u> Let $\{F_n\}$ be a sequence of finite regular random measures on $\mathcal{B}_X \times \Omega$, where X is a complete separable metric space. If for almost every $\omega \in \Omega$ the family $\{F_n(\cdot, \omega);\ n \in N\}$ is uniformly tight, then for every $\delta > 0$ there exists a set A_δ with the properties

(a) $P(A_\delta) > 1 - \delta$,

(b) the set of measures $\{F_n(\cdot, \omega);\ n \in N,\ \omega \in A_\delta\}$ is uniformly tight.

<u>Remark 1.</u> If $\{F_n\}$ is a.s. conditionally compact, then the set A_δ can be chosen in such a way, that $\{F_n(\cdot, \omega);\ n \in N,\ \omega \in A_\delta\}$ is conditionally compact.

The above lemma can be proved analogously as the well known Egorov's Theorem.

3. Consequences

In this section we will give two applications of Theorem B.

First let us remind the Levy's representation of infinitely divisible laws. As in the real case, an infinitely divisible law μ has a unique representation $\mu = l(a, S, M)$ given by the formula

$$\hat{\mu}(y) = \exp\left[i(a,y) - \frac{1}{2}(Sy,y) + \int \left(e^{i(y,x)} - 1 - \frac{i(y,x)}{1+\|x\|^2}\right) M(dx)\right],$$

where $a \in H$, S is an S-operator (i.e. positive and hermitian with the finite trace $\mathrm{tr}\, S = \sum_{i=1}^{\infty} (Se_i, e_i)$) and M is a σ- finite measure on $\mathcal{B}_H$, which is finite outside every neighbourhood of 0 and has the following properties $M(\{0\}) = 0$, $\int_{[\|x\| \leq 1]} \|x\|^2 M(dx) < +\infty$. IF $M \equiv 0$, then $\mu = G(a, S)$ is called the Gaussian distribution with mean a and covariance operator S.

We need also the notion of martingale difference array (MDA). An array $\{X_{nk}\}$ is called MDA with respect to $\{\mathcal{F}_{nk}\}$ if $\{X_{nk}\}$ is adapted to $\{\mathcal{F}_{nk}\}$, $E\|X_{nk}\|^2 < +\infty$ and $E(X_{nk} \mid \mathcal{F}_{n,k-1}) = 0$.

<u>Theorem C.</u> Let $\{X_{nk}\}$ be MDA with respect to $\{\mathcal{F}_{nk}\}$ and $G(0,S)$ be the Gaussian distribution.

If the following conditions hold:

C1. $\sum_k E(\|X_{nk}\|^2 \mid \mathcal{F}_{n,k-1}) \xrightarrow{P} \mathrm{tr}\, S$,

C2. $\sum_k E(\|X_{nk}\|^2 I(\|X_{nk}\| > \varepsilon) \mid \mathcal{F}_{n,k-1}) \xrightarrow{P} 0$ for every $\varepsilon > 0$,

C3. $\sum_k E((X_{nk}, e_i)(X_{nk}, e_j) \mid \mathcal{F}_{n,k-1}) \xrightarrow{P} (Se_i, e_j)$ for some orthonormal basis $\{e_l\}$ in H and $i, j \in N$, then $P_{S_n} \Longrightarrow G(0, S)$.

(Here and in the sequel we use the convention $\sum_k \equiv \sum_{k=1}^{k_n}$).

Remark 2. For the stronger conditions see [9]. Theorem 2.

Remark 3. For discussion of the equivalence of the Condition C2 to the Lindeberg Condition see [2]:

$$\text{(LC)} \quad \sum_k E\|X_{nk}\|^2 \, I(\|X_{nk}\| > \varepsilon) \to 0, \qquad \varepsilon > 0.$$

P r o o f: Let us choose and fix regular version μ_{nk} of the conditional distributions of X_{nk} given $\mathcal{F}_{n,k-1}$.

Now we can rewrite conditions C1 - C3 in the equivalent form:

$$C1'. \quad \sum_k \int \|x\|^2 \, \mu_{nk}(dx) \longrightarrow \operatorname{tr} S,$$

$$C2'. \quad \sum_k \int_{[\|x\|>q]} \|x\|^2 \, \mu_{nk}(dx) \longrightarrow 0 \text{ for positive rationals } q,$$

$$C3'. \quad \sum_k \int (x,e_i)(x,e_j) \, \mu_{nk}(dx) \longrightarrow (Se_i,e_j) \text{ for } i,j \in N.$$

Since we have only countable number of conditions, by the diagonalization procedure from every subsequence $\{S_{n_k}\}$ of $\{S_n\}$ we can choose a further subsequence $\{S_{n_{k_l}}\}$, for which the convergence in probability in conditions C1 - C3′ is replaced by the a.s. convergence. Hence there exists a set of probability one Ω' such that for $\omega \in \Omega'$ the convolutions

$$\mu_{n_{k_l}}(\cdot,\omega) := \mu_{n_{k_l}1}(\cdot,\omega) * \mu_{n_{k_l}2}(\cdot,\omega) * \ldots$$

are weakly convergent to $G(0,S)$ by the Central Limit Theorem. Therefore by Theorem B (the condition B2 is implied by C2′) the distributions of $S_{n_{k_l}}$ weakly converge to $G(0,S)$.

Suppose that $P_{S_n} \not\Rightarrow G(0,S)$. Then we can choose a subsequence $\{P_{S_{n_k}}\}$, any subsequence of which is not weakly convergent to $G(0,S)$. But we have already shown that this is impossible.

Using the same method the following generalization of Theorem 4.2 [12] of A. Kłopotowski can be proved:

<u>Theorem D.</u> If $\mu = 1(a,S,M)$ is an infinitely divisible distribution, then the following conditions imply the weak convergence of distributions P_{S_n} to μ:

D1. $\sum_k \mu_{nk}(A) \xrightarrow[n\to\infty]{P} M(A)$

for every $A \in \mathcal{B}_H$ such that $\bar{A} \not\ni 0$ and $M(\partial A) = 0$.

D2. there exists a.s. finite real random variable $C(\omega)$ such that

$$P(\operatorname{tr} T_n^1 > C) \xrightarrow[n\to\infty]{} 0$$

D3. $$\sup_n \sum_{i=N}^{\infty} (T_n^1 e_i, e_i) \xrightarrow[N\to\infty]{P} 0$$

for some orthonormal basis $\{e_l\}$ in H,

D4. $$(T_n^{\varepsilon} e_i, e_j) \xrightarrow[n\to\infty]{P} (Se_i, e_j) + \int_{[\|x\| \leqslant \varepsilon]} (x, e_i)(x, e_j) M(dx)$$

for the mentioned basis $\{e_l\}$, $i,j \in N$ and every $\varepsilon > 0$ with $M(\|x\| = \varepsilon) = 0$

D5. $$\sum_k \left(a_{nk} + \int \frac{x - a_{nk}}{1 + \|x - a_{nk}\|^2} \mu_{nk}(dx)\right) \xrightarrow[n\to\infty]{P} a$$

D6. $$\max_{1\leqslant k\leqslant k_n} \mu_{nk}(\|x\| > \varepsilon) \xrightarrow[n\to\infty]{P} 0$$

for every $\varepsilon > 0$.

where a_{nk} is defined by $a_{nk} := \int_{[\|x\| \leq 1]} x \; \mu_{nk}(dx)$ and for every $t>0$ $\{T_n^t = T_n^t(\omega)\}$ is a set of random S-operators defined by the formulas

$$(T_n^t y, y) := \sum_k \int_{[\|x - a_{nk}\| \leq t]} (y, x - a_{nk})^2 \; \mu_{nk}(dx)$$

Remark 4. Conditions D1 - D6 can be translated into the language of conditional expectations.

References

[1] B r e i m a n, L., Probability, Addison - Wesely, London 1968.

[2] B r o w n, B. M., Martingale central limit theorems, Ann. Math. Statist., 42, 59-66 (1971).

[3] B r o w n, B. M., E a g l e s o n, G. K., Martingale convergence to infinitely divisible laws with finite variances, Trans. Amer. Math. Soc., 162, 449-453 (1971).

[4] D v o r e t z k y, A., The central limit theorems for dependent random variables, Proc. of the Int. Congress of Math. Nice 1970.

[5] D v o r e t z k y, A., Asymptotic normality for sums of dependent random variables, Proc. 6th Berkeley Sympos. Math. Statist. Probab. Univ. Calif., 513-535 (1971).

[6] K ł o p o t o w s k i, A., Limit theorems for sums of dependent, random vectors in R^d, Dissert. Math. CLI, 1-55 (1977).

[7] P a r t h a s a r a t h y, K. R., Probability measures on metric spaces, Academic Press, New York - London 1967.

[8] V a r a d h a n, S. R. S., Limit theorems for sums of independent random variables with values in a Hilbert space, Sankhya 2.4, 213-238 (1962).

[9] W a l k, H., An invariance principle for the Robbins-Monro process in a Hilbert space, Z. Wahrscheinlichkeitstheorie verw. Geb., 39, 135-150 (1977).

C. R. RAO'S MINQUE FOR REPLICATED AND MULTIVARIATE OBSERVATIONS

by

J. Kleffe

Academy of Sciences of the German Democ ratic Republic, Berlin

Summary

Assuming a basic variance-covariance components model we derive C. R. Rao's MINQUE for its m-fold replicated and its multivariate version. Both extensions do not essentially increase the extent of necessary calculations and our formula for replicated observations gives some new light on the asymptotic behaviour of MINQUE. Simultaneously, all formulae derived present you with Minimum Bias Minimum Norm Quadratic Estimates if MINQUE does not exist.

Introduction

C. R. Rao's MINQUE principle introduced a new idea to statistical theory. It provides some kind of optimality and does not refer to specific distributional assumptions. It is also going to be used in practice now. Ahrens [2] and Swallow and Searle [9] gave computable explicite expressions for MINQUE under the unbalanced one-way random model. These applications rise the question how to extend such results to multivariate or replicated models.

Consider a random N-vector y which follows a linear model

$$y = X\beta + U\xi\ , \quad E\xi = 0, \ E\xi\xi' = F(\Theta) = \sum_{i=1}^{p} \Theta_i F_i \tag{1}$$

with unknown k-vector β and unknown p-vector $\Theta = (\Theta_1, \ldots, \Theta_p)'$. U, X and $F_1, \ldots, F_p$ are known matrices of appropriate orders. In what follows we think of (1) as representing a basic experiment and call it the basic model. The matrices $V_i = UF_iU'$ $i = 1, \ldots, p$ are assumed to be linearly independent. The m-fold replicated version of model (1) is defined by

$$\underline{y} = (1_m \otimes X)\beta + (I_m \otimes U)\underline{\xi}, \quad E\underline{\xi}\underline{\xi}' = \sum_{i=1}^{p} \Theta_i (I_m \otimes F_i), \tag{2}$$

where $\underline{y} = (y_1', \ldots, y_m')'$ is an mN-dimensional vector of observations and $\underline{\xi}$ is a high-dimensional error vector. The symbols I_m and 1_m are used to denote the identity matrix of order m and the m-vector of ones.

Similarly, we define the n-variate version of model (1) as

$$\underline{y} = (X \otimes I_n)\underline{\beta} + (U \otimes I_n)\underline{\xi}\ , \quad E\underline{\xi}\underline{\xi}' = \sum_{i=1}^{p} (F_i \otimes \Theta_i), \tag{3}$$

where $\underline{y}$ is a nN-vector of observations, $\underline{\beta}$ is now of dimension kn and $\Theta_1, \ldots, \Theta_p$ are unknown symmetric $n \times n$ matrices. The connection between (1) and (3) generalizes that kind of correspondence which is normally observed between univariate and multivariate ANOVA models. More commonly, such models are presented in matrix from as

$$Y = XB + U\Xi, \tag{4}$$

where Y is obtained by writing the frist n components of $\underline{y}$ into the first row of Y, the second n observations into the second line and so on. B and Ξ result from $\underline{\beta}$ and $\underline{\xi}$ just as Y does from $\underline{y}$. A more detailed motivation of (3) and (4) has been given in Kleffe [5].

The m-fold replicated as well as the multivariate version of (1) are special cases of model (1) such that MINQUE theory may be routinely applied to (2) or (3). All we contribute in this paper is a somewhat elegant way to do so which yields easily interpretable results and offers convenient formulae for calculating MINQUE.

To end up we study models like

$$y_j = X_j \beta_j + U_j \xi_j, \quad E \xi_j \xi_j' = \sum_i \Theta_i F_{ji}, \tag{5}$$

where y_j $(j = 1,\ldots,m)$ are independently distributed random N-vectors. Such models allow for an extremly simple analysis and have been introduced by J. Focke and G. Dewess [4]. A well-known example of model (5) is the two way nested classification mixed model in analysis of variance.

Let us use following notations throughout the paper: Frequently we have to deal with Gramian p×p-matrices $\mathcal{S}_{A,B}$ which are given by the terms

$$S_{ij}^{A,B} = \mathrm{tr}\left[V_i A V_j B\right]. \tag{6}$$

A and B are given symmetric matrices and "tr" stands for trace. For short we write simply S_A if A = B and in case $A = B = I_N$ we write S. The symbols $R(\cdot)$ and $N(\cdot)$ are used for column space and nullspace of matrices and the capitals P and M remain reserved to denote the orthogonal projections onto R(X) or N(X'), respectively. The superscripts $^+$ and $'$ denote Moore-Penrose-inverse and transpose of matrices. Sp (A,B,C,...) is used for the linear span of all matrices within the parenthesis.

1. MINQUE under model (1)

MINQUE theory is concerned with estimating linear functions $\gamma = f'\Theta$ of Θ by quadratic forms in y. This problem has been extensively studied for heteroscedastic variances and variance-covariance components models by Rao [7, 8]. Theorem 1 summarizes what we know about MINQUE under model (1).

Theorem 1

(i) MINQUE of $\gamma = f'\Theta$ exists iff $f \in R(S_M)$.

(ii) MINQUE of $\gamma = f'\Theta$ is uniquely given by $\hat{\gamma} = y'Ay$, where

$$A = \sum_{i=1}^{p} \lambda_i (MVM)^+ V_i (MVM)^+ \quad \text{and } V = UU', \tag{1.1}$$

and the vector of coefficients $\lambda = (\lambda_1,\ldots,\lambda_p)'$ may be taken as $\lambda = S^-_{(MVM)^+} f$ for any g-inverse of $S_{(MVM)^+}$.

(iii) Under normality of y, the sample variances of MINQUE are

$$\text{var}_\Theta\ \hat{\gamma} = 2\Theta' S_A \Theta$$

with A as given by (1.6).

A proof of Theorem 1, as general as it is presented here,has not been published yet, but is a rather straightforward extension of earlier results by Rao. Additionally, choice of $\lambda = S^+_{(MVM)^+} f$ yields the Minimum Bias Minimum Norm Quadratic Estimate of γ as it was introduced by Pukelsheim [6]. The matrix A given by (1.1) is frequently called to be the MINQUE-matrix for estimating γ. It is unique iff the columns of X and U together span the entire Euclidain N-space. Otherwise we have several matrices serving the same purpose but their quadratic forms coincide on the linear subspace R(MU) in which My = MU takes values. Statement (iii)of Theorem 1 is a trivial consequence of the well-known formula

$$\operatorname{cov}_{\Theta}(y'Ay, y'By) = 2 \operatorname{tr}\left[A\ V(\Theta) B\ V(\Theta)\right] = 2\Theta' S_{A,B}\Theta \qquad (1.2)$$

which holds for every pair of invariant quadratic forms and normally distributed y. Moreover, $V(\Theta)$ stands for $\sum \Theta_i V_i$.

2. Replicated model

Existence of MINQUE under model (2) is basically answered by part (i) of Theorem 1 but may become alternatively clear by observing that for $m > 2$

$$\hat{\Sigma} = \frac{1}{m-1} \sum_{i=1}^{m} (y_i - \bar{y}_.)(y_i - \bar{y}_.)', \quad \bar{y}_. = \frac{1}{m} \sum y_i, \qquad (2.1)$$

is an unbiased estimate of $V(\Theta)$ from which invariant quadratic unbiased estimates of $\gamma = f'\Theta$ may be derived by decomposition of $\hat{\Sigma}$.

<u>Theorem 2.</u> Consider model (2) with $m \geq 2$

(i) MINQUE exists for every linear function of Θ

(ii) MINQUE of $\gamma = f'\Theta$ is uniquely given by

$$\hat{\gamma} = \operatorname{tr}\left[G_{(m)} \hat{\Sigma}\right] + \frac{m}{m-1} \bar{y}_.' A_{(m)} \bar{y}_., \qquad (2.2)$$

where

$$G_{(m)} = \sum \lambda_{i(m)} V^+ V_i V^+,$$

$$A_{(m)} = \sum \lambda_{i(m)} (MVM)^+ V_i (MVM)^+$$

and the vector of coefficients $\lambda_{(m)} = (\lambda_{1(m)}, \ldots, \lambda_{p(m)})'$ is

$$\lambda_{(m)} = \left(S_{V^+} + \frac{1}{m-1} S_{(MVM)^+}\right)^{-1} f. \qquad (2.3)$$

(iii) Under the normal assumption MINQUE has sample variances

$$\operatorname{var}_{\Theta} \hat{\gamma} = \frac{2}{m-1} \Theta' (S_{G_{(m)}} + \frac{1}{m-1} S_{A_{(m)}}) \Theta. \qquad (2.4)$$

P r o o f: The proof requires calculation of the terms appearing in Theorem 1 under model (2). For example the projection matrix onto the nullspace of $(1'_m \otimes X')$ denoted by $M^{(m)}$ is

$$M^{(m)} = I_{Nm} - \left(\frac{1}{m} 1_m 1'_m \otimes P\right). \tag{2.5}$$

This implies the orthogonal decomposition

$$M^{(m)}(I \otimes V)M^{(m)} = (I_m - \frac{1}{m} 1_m 1'_m \otimes V) + (\frac{1}{m} 1_m 1'_m \otimes MVM) \tag{2.6}$$

which is the basic tool for proving all statements, Letting $V = I$ in (2.6) yields the matrix S_M for model (2) which we now denote by $S_M^{(m)}$,

$$S_M^{(m)} = (m-1)S + S_M.$$

Linear independence of $V_1,\ldots,V_p$ implies invertibility of S and shows $R(S_M^{(m)})$ to be the entire Euclidian p-space if $m \geqslant 2$.

Statement (ii) follows from

$$(M^{(m)}(I \otimes V)M^{(m)})^+ = (I_m - \frac{1}{m} 1_m 1'_m \otimes V^+) + (\frac{1}{m} 1_m 1'_m \otimes (MVM)^+) \tag{2.7}$$

and

$$\gamma = \sum_{i=1}^{p} \lambda_i^{(m)} (\underline{y}'(I_m - \frac{1}{m} 1_m 1'_m \otimes V^+V_iV^+)\underline{y} +$$

$$+ \underline{y}'(\frac{1}{m} 1_m 1'_m \otimes (MVM)^+V_i(MVM)^+)\underline{y}).$$

after substituting $\lambda_{i(m)}$ for $\lambda_i^{(m)}(m-1)$ and some straightforward simplifications. In the same way we find the matrix $S_{(MVM)^+}$ for model (2) as

$$S^{(m)}_{(MVM)^+} = (m-1)\, S_{V^+} + S_{(MVM)^+}.$$

The variance of the first term of (2.2) becomes by using (1.2)

$$\mathrm{var}_{\Theta}\,\mathrm{tr}\left[G_{(m)}\hat{\sum}\right] = \frac{2}{m-1}\,\mathrm{tr}\left[G_{(m)}\,V(\Theta)\,G_{(m)}\,V(\Theta)\right].$$

The variance of the second term of (2.2) is

$$\mathrm{var}_{\Theta}\,\frac{m}{m-1}\,\bar{y}_.'A_{(m)}\bar{y}_. = \frac{2}{(m-1)^2}\,\mathrm{tr}\left[A_{(m)}V(\Theta)A_{(m)}V(\Theta)\right].$$

and follows from statement (iii) of Theorem 1 by observing that $\bar{y}_.$ has dispersion matrix $\frac{1}{m}\,V(\Theta)$.

Independence of $\bar{y}_.$ and $\hat{\sum}$ implies (2.4).

Formula (2.2) is in fact very convenient to handle with. It is based on $\bar{y}_.$ and $\hat{\sum}$ only and the effort to calculate $A_{(m)}$ and $G_{(m)}$ is as much as necessary for the basic model. Or in other words, if MINQUE is known under the basic model it makes only little effort to extend these results to replicated models, too. Obviously, (2.4) also shows consistency of MINQUE as m tends to infinity. An alternative proof of this fact given by Brown [3] does not take advantage of the explicit expression (2.2) and requires great efforts. MINQUE is also asymptotically equivalent to the estimate

$$\hat{\gamma}^*_{(m)} = \mathrm{tr}\left[G_{(\infty)}\sum\right] = \frac{1}{m-1}\sum_{i=1}^{n}(y_i - \bar{y}_.)'\,G_{(\infty)}(y_i - \bar{y}_.),$$

where $G_{(\infty)}$ is obtained from (1.1) by assuming $X = 0$. This means that MINQUE tends to ignore all information about Ey.

3. Multivariate observations

The nice relations between ANOVA estimates in univariate and multivariate ANOVA models are well-known and frequently used. The subject of our next theorem is to show the same for MINQUE of

$$\gamma = \sum_{i=1}^{p} \mathrm{tr}\left[C_i \Theta_i\right], \tag{3.1}$$

with arbitrary symmetric $n \times n$-matrices $C_1,\ldots,C_p$.

Theorem 3: Assume a multivariate model (3) to be given. Then

(i) MINQUE of (3.1) exists iff

$$\sum b_i C_i = 0 \text{ for all } b = (b_1,\ldots,b_p)' \in N(S_M) \tag{3.2}$$

(ii) MINQUE of (3.1) is uniquely given by

$$\hat{\gamma} = \sum_{i=1}^{p} \mathrm{tr}\left[\Lambda_i Y'(MVM)^+ V_i (MVM)^+ Y\right], \tag{3.3}$$

where

$$\begin{pmatrix} \Lambda_1 \\ \cdot \\ \cdot \\ \cdot \\ \Lambda_p \end{pmatrix} = (S^-_{(MVM)^+} \otimes I) \begin{pmatrix} C_1 \\ \cdot \\ \cdot \\ \cdot \\ C_p \end{pmatrix}$$

for an arbitrary g-inverse of $S_{(MVM)^+}$.

(iii) If S_M is of full rank and $\hat{\Theta}_r = y'A_r y$ is MINQUE of Θ_r under model (1), then we have under normality of $\underline{y}$

$$\mathrm{var}_{\Theta_1,\ldots,\Theta_p}\, \hat{\gamma} = 2 \sum_{\kappa\lambda} \sum_{rs} \mathrm{tr}\left[A_r V_\kappa A_s V_\lambda\right] \mathrm{tr}\left[C_r \Theta_\kappa C_s \Theta_\lambda\right]. \tag{3.4}$$

P r o o f: An obvious interpretation of statement (i) of Theorem 1 is that MINQUE of $\gamma = f'\Theta$ exists iff there is a symmetric matrix $A \in sp(MV_1M,\ldots,MV_pM)$ such that $Ey'Ay = \gamma$.

Investigating this under model (3) yields (3.2).

Due to statement (ii) of Theorem 1 MINQUE may be obtained by searching for a matrix $G \in sp(V_1,\ldots,V_p)$ such that $E\, y'(MVM)^+G(MVM)^+y$

$= \gamma$. Here G is not necessarily unique. Applying this approach to model (3), MINQUE of (3.1) is obtained by equating the expectation of

$$\hat{\gamma} = \sum_{i=1}^{p} \underline{y}' \left((MVM)^{+} V_i (MVM)^{+} \otimes \Lambda_i \right) \underline{y} \tag{3.5}$$

to (3.1). It leads to the estimating equations

$$\sum_i s_{ij}^{(MVM)^{+}} \Lambda_i = C_j \quad j=1,\ldots,p$$

which possess a solution iff (3.2) holds. One class of solutions is given by

$$\Lambda_i = \sum_j s^{ij}_{(MVM)^{+}} C_j \quad i=1,\ldots,p, \tag{3.6}$$

where $s^{ij}_{(MVM)^{+}}$ are the elements of an arbitrary g-inverse of $S_{(MVM)^{+}}$.

The proof of statement (ii) is completed by observing that (3.3) is just an alternative formulation of (3.5).

Statement (iii) follows from

$$\mathrm{cov}_{\Theta_1,\ldots,\Theta_p} \left(\mathrm{tr}\left[CY'AY\right], \mathrm{tr}\left[DY'BY\right] \right)$$

$$= 2 \sum_{k\lambda} \mathrm{tr}\left[AV_k BV_\lambda \right] \mathrm{tr}\left[C\, \Theta_k\, D\, \Theta_\lambda \right],$$

what holds for arbitrary symmetric $N \times N$ matrices A and B with $AX = 0$ and $BX = 0$. Its application to (3.3) yields

$$\mathrm{var}_{\Theta_1,\ldots,\Theta_p} \hat{\gamma} =$$

$$= 2 \sum_{ijk\lambda} \mathrm{tr}\left[(MVM)^{+} V_i (MVM)^{+} V_k (MVM)^{+} V_j (MVM)^{+} V_\lambda \right] \mathrm{tr}\left[\Lambda_i \Theta_k \Lambda_j \Theta_\lambda \right]. \tag{3.7}$$

Using the explicit expressions (3.6) we arrive at (3.4) where

$$A_r = \sum_i s^{ri}_{(MVM)^{+}} (MVM)^{+} V_i (MVM)^{+}. \tag{3.8}$$

Nonsingularity of S_M is equivalent to those of $S_{(MVM)^+}$ and calculation of $Ey'A_r y$ shows A_r to be the MINQUE-matrix for estimating Θ_r under model (1).

Theorem 3 shows that multidimensionality of model (3) does not pose additional difficulties for calculation of MINQUE. All terms that appear in Theorem 3 are known from model (1). There is also obvious similarity between (3.3) and (1.1) as well as between the ways in which the coefficients λ_i or Λ_i may be found. This becomes even more clear if only a so called elementary parametric function

$$\gamma = \sum f_i \operatorname{tr} [C\Theta]_i$$

is to be estimated. Then (3.3) reduces to

$$\hat{\gamma} = \operatorname{tr} [CY'AY], \tag{3.9}$$

where A is the MINQUE-matrix for estimating $\gamma = f'\Theta$ under model (1). Therefore $\hat{\Gamma} = Y'AY$ might be considered as MINQUE of the matrix-valued parametric function $\Gamma = \sum f_i \Theta_i$.

This is in full accordance to known relations between univariate and multivariate ANOVA estimates. The sample variances of (3.9) are given by

$$\operatorname{var}_{\Theta_1,\ldots,\Theta_p} \hat{\gamma} = 2 \sum_{ij} s_{ij}^A \operatorname{tr} \left[C\Theta_i \, C\Theta_j \right], \tag{3.10}$$

where s_{ij}^A are the terms of S_A as it appears in Theorem 1. This connection between (3.10) and statement (iii) of Theorem 1 is of considerable importance and was discovered by Ahrens [1] for ANOVA estimates. But in general we have to use the more complicated expression in Theorem 3.

The regularity assumption made in statement (iii) of Theorem 3 is not very important. It serves to give an interpretation to the matrices (3.8) which otherwise are not MINQUE-matrices for estimating Θ_r. But by consequently making use of the Moore-Penrose g-inverse of $S_{(MVM)^+}$ throughout all derivations $\hat{\Theta}_r = y'A_r y$ may easily seen to be minimum bias minimum norm quadratic estimator of Θ_r under model (1) and so is $\hat{\gamma}$ for model (3) independent of whether condition (i) is satisfied or not.

Sometimes expression (3.7) seems to be more advantageous than (3.4). It only requires knowledge of such terms which are necessary to calculate any way.

4. Models with a block structure

Let us finally assume to have a model (5) consisting of m submodels. Now we omit proofs because of their straight-forwardness and similarity to those of the preceding sections.

Theorem 4.

(i) MINQUE of $\gamma = f'\Theta$ exist iff $f \in R\left(\sum_{j=1}^{m} S_M^{(j)}\right)$, where $S_M^{(j)}$ is defined as S_M for the j-th submodel of (5).

(ii) MINQUE of $\gamma = f'\Theta$ is given by

$$\hat{\gamma} = \sum_{i=1}^{p} \lambda_i \left(\sum_{j=1}^{m} y_j'(M_jV_jM_j)^+ V_{ji}(M_jV_jM_j)^+ y_j \right), \tag{4.1}$$

where the vector of coefficients may be taken as

$$\lambda = \left(\sum_{j=1}^{m} S^{(j)}_{(MVM)^+} \right)^- f \tag{4.2}$$

for any choice of g-inverse. M_j, V_j, V_{ji} and $S^{(j)}_{(MVM)^+}$ are the matrices M, V, V_i and $S_{(MVM)^+}$ for the j-th submodel of (5).

(iii) The sample variance of MINQUE becomes under normality of y_j

$$\operatorname{var}_{\Theta} \hat{\gamma} = 2\,\Theta' \left(\sum_{j=1}^{m} S_{A_j} \right) \Theta,$$

where

$$A_j = \sum_{i=1}^{p} \lambda_i (M_j V_j M_j)^+ V_{ji} (M_j V_j M_j)^+ .$$

Also here the effort needed for calculating MINQUE is reduced to that which is necessary to calculate MINQUE for the single submodels of (5). In fact only one vector of coefficients λ_i has to be found.

Combining e.g. Theorem 4 with the results by Ahrens [2] or Swallow and Searle [9] we arrive at easily computable formulae for MINQUE under the unbalanced two way nested classification mixed model without interactions.

In case of identical submodels (4.1) reduces to the average of the m MINQUE estimates based on the separate observations y_j , j = = 1,...,m. But in general such result cannot be expected, for the separate MINQUEs may have different variances and a weighted average seems to be more appropriate. But even this will not be true except of a small number of cases.

Lemma: Let $\omega_1,\dots,\omega_m$ be weights which sum up to 1 and $\hat{\gamma}_1,\dots,\hat{\gamma}_m$ be MINQUEs of $\gamma = f'\Theta$ based on the separate submodels of (5).

Then, MINQUE of γ under model (5) is given by $\hat{\gamma} = \sum_{j=1}^{m} \omega_j \hat{\gamma}_j$ iff there exists λ such that

$$S^{(j)}_{(MVM)^+} \lambda = \omega_j f \qquad j = 1,\dots,m. \tag{4.3}$$

A vector λ as required by (4.3) exists for all $f \in R\left(\sum_{j=1}^{m} S_M^{(j)}\right)$ iff the matrices $S^{(j)}_{(MVM)^+}$ happen to be proportional to each other. This is of course a very restrictive property.

The minimum bias minimum norm quadratic estimate of γ is again obtained by using the Moore-Penrose g-inverse in (4.2).

References

[1] A h r e n s, H., An invariance property for first and second order moments of estimated variance-covariance-components. Biom. J., 19, 7, 485-496 (1977).

[2] A h r e n s, H., MINQUE and ANOVA estimator for one-way classification - a risk comparision. Biom. J., 20, 6, 535-556 (1978).

[3] B r o w n, K. G., Asymptotic behaviour of MINQUE-type estimators of variance components. Ann. Statist., 4, 746-754 (1976).

[4] F o c k e, J. and D e w e s s, G., Über die Schätzmethode MINQUE von C. R. Rao und ihre Verallgemeinerung. Math. Operationsforsch. u. Statist., 3, 129-143 (1972).

[5] K l e f f e, J., Optimal estimation of variance components - a survey. Sankhya, 39, Ser. B. (1977).

[6] P u k e l s h e i m, F., Schätzen von Mittelwert und Streuungsmatrix in Gauss-Markoff-Modellen. Diplomarbeit. Inst. f. Math. Stochastik der Univ. Freiburg (1974).

[7] R a o, C. R., Estimation of heteroscedastic variance in linear models. J. Am. Statist. Assoc., 65, 161-172 (1970).

[8] R a o, C. R., Estimation of variance and covariance components - MINQUE theory. J. Multivariate Analysis, 1, 257-275 (1971).

[9] S w a l l o w, W. H. and S e a r l e S. R., Minimum Variance Quadratic Unbiased Estimation of Variance Components, Technometrics, 10, 3, 265-272 (1978).

INVARIANT QUADRATIC UNBIASED ESTIMATION FOR VARIANCE COMPONENTS

by

Witold Klonecki

Polish Academy of Sciences, Wrocław

1. Introduction and summary

In their recent paper Olsen, Seely and Birkes [8] have given a characterization of the class of admissible invariant, unbiased and quadratic estimators for two variance components in a mixed linear model. In this paper our goal is to characterize the class of admissible, invariant and unbiased estimators for any number of variance components. The method suggested by Olsen et al. [8] is generally not applicable to more than two variance components, because as noted by I. Wistuba [9], the conditions of their basic theorem (see Corollary 3.7 in [8]) may not be met. Since we think that the problem of a full characterization of admissible estimators for mixed linear models is important (compare also [4]), an effort has been made in this paper to extend the method suggested in [8].

The developments in this paper are based on a work of S. Gnot, R. Zmyślony and myself [2]. The main results are Theorems 2.3 and 4.1. Theorem 2.3 gives sufficient conditions for an estimator to be admissible within a general linear model framework. Theorem 4.1 shows that these conditions are necessary when the underlying model is a mixed linear model as defined in Section 3 and the covariance operators commute. We give explicitly the minimal complete class of estimators for three variance components in a mixed linear model. Finally we show that the class of admissible estimators is closed.

Our terminology is generally consistent with that introduced in [8] except we use the expression best locally at Σ in place of Σ-best. For the sake of completness and convenience of the reader we reproduce in the paper most of the results for two variance components established by Olsen et al. [8].

2. Admissible estimators in general linear models

Let $\mathcal{K}$ stand for an Euclidean vector space with an inner product denoted by $[\cdot,\cdot]$ and let $\mathcal{L}$ stand for the space of linear mappings of $\mathcal{K}$ into itself. For every $\Theta \in \Omega$, where Ω is the set of parameters considered, let $\{U, \mathcal{A}, P_\Theta\}$ be a probability space, and let $Y : U \to \mathcal{K}$ be a random vector. Assume that the expectation $E_\Theta Y$ and covariance operator $\mathrm{Cov}_\Theta Y$ exist for each $\Theta \in \Omega$. Moreover, let $\mathcal{E} = \mathrm{span}\{E_\Theta Y \mid \Theta \in \Omega\} \subset \mathcal{K}$ and let $\circleddash = \{\mathrm{cov}_\Theta Y \mid \Theta \in \Omega\} \subset \mathcal{L}$. A model having the above described structure will be denote by $\mathcal{M}(\mathcal{E}, \circleddash)$ and called a general linear model. Now let $\mathcal{K}_0$ be a subspace of $\mathcal{K}$. Function $g : \Omega \to \mathcal{R}$ is said to be $\mathcal{K}_0$-estimable if there exists a vector $A \in \mathcal{K}_0$ such that $E[A, Y] = g$, and then A is said to be a $\mathcal{K}_0$-unbiased vector for g. Given that g is $\mathcal{K}_0$-estimable, a $\mathcal{K}_0$-unbiased vector A of g is said to be $\mathcal{K}_0$-best locally at $\Gamma \in \circleddash$, if $[A, \Gamma A] \leq [B, \Gamma B]$ for all $\mathcal{K}_0$-unbiased vectors B of g. As known [2], vector $A \in \mathcal{K}_0$ is $\mathcal{K}_0$-best locally at Γ iff $\Gamma A \in \mathcal{E} + \mathcal{K}_0^{\perp}$, where $\mathcal{K}_0^{\perp}$ stands for the set of vectors orthogonal to $\mathcal{K}_0$. If there exists an uniformly $\mathcal{K}_0$-best estimator for each $\mathcal{K}_0$-estimable function, then model $\mathcal{M}(\mathcal{E}, \circleddash)$ is said to be $\mathcal{K}_0$-regular. Throughout the paper let g be a nonzero $\mathcal{K}_0$-estimable function, and let $\mathcal{B}_0 = \{A \in \mathcal{K}_0 \mid E[A, Y] = g\}$. Our concern in the paper is to investigate admissibility in this general linear model context. More precisely we will compare the estimators $[A, Y]$ according to their possible variance $[A, \Gamma A], \Gamma \in \Theta$. For $B, A \in \mathcal{B}_0$ we say B is as good as A if $[B, \Gamma B] \leq [A, \Gamma A]$ for all $\Gamma \in \Theta$; B is better than A if B is as good as A and $[B, \Gamma B] < [A, \Gamma A]$ for at least one $\Gamma \in \Theta$; B is admissible within $\mathcal{B}_0$ if no vector in $\mathcal{B}_0$ is better than B. A vector B that is admissible within $\mathcal{B} = \{A \in \mathcal{K} \mid E[A, Y] = q\}$ is called admissible. A subset $\mathcal{C}_0$ of $\mathcal{B}_0$ is said to be a complete class within $\mathcal{B}_0$ if for every vector $B \in \mathcal{B}_0$, which is not in $\mathcal{C}_0$, there exists a vector in $\mathcal{C}_0$ which is better than B. A complete class $\mathcal{C}_0$ is said to be minimal if no proper subset of $\mathcal{C}_0$ is a complete class.

The following theorem is an obvious extension of Proposition 3.3 in [8].

<u>Theorem 2.1.</u> The set of admissible vectors in $\mathcal{B}_0$ is a minimal complete class within $\mathcal{B}_0$.

To formulate the next theorem we need to introduce additional notation. For any subset $\mathcal{L}_0$ of the space $\mathcal{L}$ of linear mappings of $\mathcal{K}$ into itself, symbol $[\mathcal{L}_0]$ denotes the smallest closed convex cone in $\mathcal{L}$ containing $\mathcal{L}_0$. We will say that $\mathcal{L}_0 \in \mathcal{L}$ generates $[\Theta]$, if $[\mathcal{L}_0] = [\Theta]$. As known (see Lemma 3.5 in [8]) there exists a compact convex set $\mathcal{V} \subset \mathcal{L}$, not containing the zero vector, which generates $[\Theta]$ such that every nonzero vector of $[\Theta]$ is a positive multiple of an element of $\mathcal{V}$. Note that a vector A is admissible within $\mathcal{B}_0$ with respect to model $\mathcal{M}(\mathcal{E}, \Theta)$ iff A is admissible within $\mathcal{B}_0$ with respect to model $\mathcal{M}(\mathcal{E}, [\Theta])$.

With minor and fairly obvious modifications, the arguments used by Olsen et al. [8] to show that an admissible estimator is locally best at some point, may be used to prove the following result.

<u>Theorem 2.2.</u> If A is admissible within $\mathcal{B}_0$, then A is $\mathcal{K}_0$-best locally at some $\Gamma_0 \in \mathcal{V}$.

In the remaining part of the paper we assume that the elements of $\mathcal{V}$ commute, e.i. that $\Gamma\Lambda = \Lambda\Gamma$ for all $\Gamma, \Lambda \in \mathcal{V}$. Under this assumption we establish the following simple lemma, which plays an important role in our considerations. Let $\mathcal{R}(\Gamma), \mathcal{N}(\Gamma), \Gamma^*, \Gamma^+$ denote, respectively, the range, the null space, the adjoint and the Moore-Penrose inverse of $\Gamma \in \mathcal{L}$. Moreover, let g be a $\mathcal{K}_0$-estimable parametric function.

Lemma 2.1. Suppose that the elements of $\mathcal{V}$ commute and suppose that A is admissible within the set $\mathcal{B}_0 = \{ A \in \mathcal{K}_0 \mid E[A,Y] = g \}$. If A may be decomposed into A = B + C, where $B \in \mathcal{R}(\Gamma_0)$ and $C \in \mathcal{K}_0 \cap \mathcal{N}(\Gamma_0)$, while $\Gamma_0 \in \mathcal{V}$, then C is admissible within $\mathcal{B}_1 = \{ A \in \mathcal{K}_0 \cap \mathcal{N}(\Gamma_0) \mid E[A,Y] = g_1 \}$, where $g_1 = E[A-B, Y]$.

P r o o f: From the adopted assumptions it follows that for any $\Gamma \in \mathcal{V}$

$$[A, \Gamma A] = [B, \Gamma B] + [C, \Gamma C],$$

because $[B, \Gamma C] = [\Gamma_0 D, \Gamma C] = [\Gamma D, \Gamma_0 C] = 0$. Suppose that C is not admissible within $\mathcal{B}_1$ and let C_0 be better than C. In this situation $B + C_0$ is better than A, and this contradiction concludes the proof.

For $A_0 \in \mathcal{K}$ and nonnegative-definite operators $\Gamma_1, \ldots, \Gamma_{r+1}$ let $A_1, \ldots, A_r$ be defined by the recurrence formula $A_i = (I - \Gamma_i \Gamma_i^+) A_{i-1}$, $i = 1, \ldots, r$. Using this notation A_0 may be decomposed into

$$A_0 = \sum_{i=1}^{r} \Gamma_i \Gamma_i^+ A_{i-1} + A_r.$$

The following theorem gives a sufficient condition for A_0 to be admissible (compare Corollary 3.7 in [8]).

Theorem 2.3. Suppose that $\Gamma \Lambda = \Lambda \Gamma$ for all $\Gamma, \Lambda \in \Theta$. If there exists a sequence of operators $\Gamma_1, \ldots, \Gamma_{r+1} \in \Theta$ such that

(i) $\Gamma_1, \ldots, \Gamma_r$ are noninvertible, Γ_{r+1} is invertible,

(ii) A_0 is best locally at Γ_1, and for $i = 1, \ldots, r$

(iii) A_i is $\mathcal{N}(\Gamma_1) \cap \ldots \cap \mathcal{N}(\Gamma_i)$-best locally at Γ_{i+1}, then A_0 is admissible.

P r o o f: For r=0 the assertion follows from Corollary 3.7 in [8]. To establish it for r=1 we proceed as follows.

(i) It is assumed that there exist a singular operator Γ_1 and an invertible operator Γ_2 such that

$$A_o \text{ is best locally at } \Gamma_1$$

and

$$A_1 \text{ is } \mathcal{N}(\Gamma_1)\text{-best locally at } \Gamma_2.$$

Suppose that A_o is not admissible within $\mathcal{B}$ and that B_o is better than A_o. Because B_o must be best locally at Γ_1 too, we may write that $\Gamma_1(A_o - B_o) \in \mathcal{E}$. This combined with $A_o - B_o \in \mathcal{E}^{\perp}$ leads to $\Gamma_1 A_o = \Gamma_1 B_o$. Hence we may write

$$B_o = \Gamma_1 \Gamma_1^{+} A_o + B_1,$$

where $B \in \mathcal{N}(\Gamma_1)$. Vector B_1 must be $\mathcal{N}(\Gamma_1)$-best locally at Γ_2 so that $\Gamma_2(A_1 - B_1) \in \mathcal{E} + \mathcal{R}(\Gamma_1)$. Combining this with $A_1 - B_1 \in \mathcal{E}^{\perp} \cap \mathcal{N}(\Gamma_1)$ we conclude that $A_1 = B_1$ because Γ_2 is invertible by assumption. Thus $A_o = B_o$ contrary to the assumption that B_o is better than A_o.

(ii) For r = 2 the proof is accomplished similarly as for r = 1. The assumptions of the theorem amount to the following: there exist operators Γ_1 and Γ_2, which are non-invertible, and Γ_3, which is invertible such that

$$A_o \text{ is best locally at } \Gamma_1,$$

$$A_1 \text{ is } \mathcal{N}(\Gamma_1)\text{-best locally at } \Gamma_2$$

and

$$A_2 \text{ is } \mathcal{N}(\Gamma_1) \cap \mathcal{N}(\Gamma_2)\text{-best locally at } \Gamma_3.$$

Suppose that A_o is not admissible within $\mathcal{B}$ and that B_o is better than A_o. As for r = 1 we easily establish that

$$B_o = \Gamma_1 \Gamma_1^{+} A_o + B_1.$$

where $B_1 \in \{A \in \mathcal{N}(\Gamma_1) | E[A,Y] = g_1\}$, while $g_1 = E[B_o - \Gamma_1 \Gamma_1^+ A_o, Y]$. Since $A_1 - B_1 \in \mathcal{E}^\perp \cap \mathcal{N}(\Gamma_1)$ and since $\Gamma_2(A_1 - B_1) \in \mathcal{E} + \mathcal{R}(\Gamma_1)$ we obtain that $\Gamma_2 A_1 = \Gamma_2 B_1$. Thus we may decompose B_1 into

$$B_1 = \Gamma_2 \Gamma_2^+ A_1 + B_2,$$

where $B_2 \in \{A \in \mathcal{N}(\Gamma_1) \cap \mathcal{N}(\Gamma_2) | E[A,Y] = g_2\}$, while $g_2 = E[B_1 - \Gamma_2 \Gamma_2^+ A_1, Y]$. Noting that $A_2 - B_2 \in \mathcal{E}^\perp \cap \mathcal{N}(\Gamma_1) \cap \mathcal{N}(\Gamma_2)$ and that B_2 must be $\mathcal{N}(\Gamma_1) \cap \mathcal{N}(\Gamma_2)$-best locally at Γ_3 we conclude that $A_2 = B_2$. Comparing the above obtained results we may infer $A_o = B_o$. This contradicts the assumption that B_o is better than A_o. Thus the proof of Theorem 2.3 is completed for $r = 2$.

It should be obvious that using the same reasoning as for $r = 1$ and $r = 2$ we may prove Theorem 2.3 for $r \geq 3$.

Remark. Example 3 in Section 4 shows that there may not exist a sequence of nonnegative-definite operators in Θ fulfilling conditions (i)-(iii) of Theorem 2.3.

3. The mixed linear model

Let Y be a random vector distributed according to an m-dimensional normal distribution with mean vector $X\beta$, where X is a known $m \times p$ matrix of rank p and β is a $p \times 1$ vector of unknown parameters. The covariance matrix of Y is taken to be $V = \sum_{i=1}^{k} \sigma_i V_i$, where $V_1, \ldots, V_k$ are known linearly independent, nonnegative difinite matrices, while $\sigma = (\sigma_1, \ldots, \sigma_k)'$ is a vector of unknown parameters called variance components. Finally we assume that the V's commute, that $V_1 = I$, $|V_k| = 0$ and that β and σ range over $\mathcal{R}^p$ and $\Omega = \{\sigma | \sigma_1 > 0, \sigma_2 \geq 0, \ldots, \sigma_k \geq 0\}$, respectively.

As known, invariant quadratic unbiased estimators of linear parametric functions $c'\sigma$, $c \in R^k$, may be conveniently investigated within the framework of the theory of the general linear model (see Section 2). Take $\mathcal{K}$ to be space of all $m \times m$ matrices with the usual trace inner product denoted by $[\cdot,\cdot]$. The class of invariant quadratic unbiased estimators of estimable functions $c'\sigma$ coincides then with the class of unbiased estimators of $c'\sigma$ within model $\mathcal{M}(\mathcal{E},\Theta)$ defined as follows. Let Q be a $m \times (m-p)$ matrix whose columns form an orthogonal basis for $\mathcal{N}(X')$ so that $\mathcal{R}(Q) = \mathcal{N}(X')$. Next form the random matrix $Z = Q'YY'Q$ and take $\mathcal{E}$ to be the space generated by $E\,Z$ and Θ to be the set of covariance operators of Z. This means that

$$\mathcal{E} = \operatorname{span}\{W_1,\dots,W_k\}$$

and

$$\Theta = \left\{ W_\sigma \otimes W_\sigma \;\middle|\; W_\sigma = \sigma_1 W_1 + \dots \sigma_k W_k,\ \sigma \in \Omega \right\},$$

where $W_i = Q'V_iQ$ for $i = 1,\dots,k$. Symbol $A \otimes B$ is defined by $(A \otimes B)C = ACB'$ for all A, B and $C \in \mathcal{K}$. For simplicity of notation assume that all matrices $W_1,\dots,W_k$ are diagonal.

The following alternative formula for Θ will be useful in latter considerations. Let

$$S = \left\{ \delta \;\middle|\; \sum_i \delta_i + \sum_{i<j} \delta_{ij} = 1,\ \ \delta_{ij} = 2\sqrt{\delta_i}\sqrt{\delta_j},\ i,j = 1,\dots,k,\ i \neq j \right\},$$

where $\delta = (\delta_1,\dots,\delta_k,\ \delta_{12},\dots,\delta_{k-1,k})' \in \mathcal{R}^{\frac{(k+1)k}{2}}$,

and let

$$\Gamma_\delta = \sum_i \delta_i W_i \otimes W_i + \frac{1}{2} \sum_{i<j} \delta_{ij} (W_i \otimes W_j + W_j \otimes W_i).$$

With this notation

$$\Theta = \{\kappa \Gamma_\delta \mid \kappa \in \mathcal{R}^+, \delta \in S\}.$$

Moreover, observe that $[\Theta] = [\mathcal{V}]$, where $\mathcal{V} = \{\Gamma_\delta \mid \delta \in H(S)\}$, while

$$H(S) = \Big\{\delta \Big| \sum_i \delta_i + \sum_{i<j} \delta_{ij} = 1 \leq \sum_i \delta_i,\ 0 \leq \delta_{ij} \leq 2\sqrt{\delta_i}\ \sqrt{\delta_j},$$

$$i,j = 1,\dots,k, i \neq j\Big\}$$

is the smallest convex hull in $\mathcal{R}^{\frac{(k+1)k}{2}}$ containing S. Note that $\Gamma_\delta \Gamma_\xi = \Gamma_\xi \Gamma_\delta$ for all $\delta, \xi \in H(S)$.

Our concern in the next sections will be to describe within the above specified model the minimal complete class for the problem of invariant, unbiased and quadratic estimation for a given estimable fixed nonzero function $g = c'\sigma$.

We conclude this section by adding a lemma due to R. Zmyślony [10] which is basic in further considerations. It states that within the model framework considered in this section the problem of unbiased estimation may be reduced to considering estimators in the smallest quadratic subspace containing $\mathcal{E}$.

Lemma 2.2. Let $\mathcal{F}$ be the smallest quadratic subspace containing $\mathcal{E}$. If A is admissible within $\mathcal{B}$, then $A \in \mathcal{F}$.

4. The minimal complete class for the mixed linear model

Theorem 4.1 below asserts that within the context of a mixed linear model the conditions appearing in Theorem 2.3 are also necessary. In fact using the same notation as in Theorem 2.3 we have the following result.

<u>Theorem 4.1.</u> In order that A_o be admissible it is necessary and sufficient that there exists a sequence of covariance operators $\Gamma_1, \Gamma_2, \ldots, \Gamma_{r+1} \in \mathcal{V}$ such that:

(i) $\Gamma_1, \ldots, \Gamma_r$ are noninvertible, while Γ_{r+1} is invertible.

(ii) A_o is best locally at Γ_1 and, for $i = 0,1,\ldots,r$,

(iii) A_i is $\mathcal{N}(\Gamma_0) \cap \mathcal{N}(\Gamma_1) \cap \ldots \cap \mathcal{N}(\Gamma_i)$-best locally at Γ_{i+1}.

P r o o f: As mentioned above the sufficiency of these conditions is granted by Theorem 2.3.

The necessity will be shown for $k = 1$, 2 and 3. For $k > 3$ it may be proved in a similar manner.

In case $k = 1$ the theorem is evidently true, because $\mathcal{V} = (I \otimes I)$. For $k = 2$ we proceed as follows. Suppose that A_o is admissible. From Lemma 2.2 it then follows that $A \in \mathcal{F}$. Thus A is a diagonal matrix. Moreover, by Theorem 2.2 vector A_o is best locally at some $\Gamma_1 \in \mathcal{V}$. When Γ_1 is invertible, the necessity of the conditions of Theorem 4.1 is obvious. When Γ_1 is not invertible, e.i. when $\Gamma_1 = W_2 \otimes W_2$ we decompose A_o into $A_o = \Gamma_1 \Gamma_1^+ A_o + A_1$. From Lemma 2.1 it then follows that A_1 is admissible within $\mathcal{B}_1 = \left\{ A \in \mathcal{N}(W_2 \otimes W_2) \,|\, E[A, Y] = g_1 \right\}$, where $g_1 = E\left[(I - \Gamma_1 \Gamma_1^+) A_o, Y\right]$. Now observe that $\Gamma_\delta A = \delta_1 A$ for each $\Gamma_\delta \in \mathcal{V}$, $\Gamma_\delta \neq \Gamma_1$, and for each $A \in \mathcal{B}_1$ because $AW_2 = 0$. In view of this fact it should be clear that as far as we are concerned with admissibility within $\mathcal{B}_1$ we may for a moment, for the sole reason of the proof, consider model $\mathcal{M}(\varepsilon, \Theta_1)$ with $\Theta_1 = [\{I \otimes I\}]$. With reference to this reduced model Theorem 2.2 implies that A_1 is $\mathcal{N}(W_2 \otimes W_2)$-best locally at $\Gamma_2 = I \otimes I$. Now it should be obvious that the operators Γ_1, Γ_2 and vectors A_o and A_1 meet the conditions of Theorem 4.1 with $r = 1$.

For $k = 3$ we show the necessity of the conditions of Theorem 4.1 in a similar manner. Suppose that A_o is admissible and suppose that A_o is best locally at $\Gamma_1 = \delta_2 W_2 \otimes W_2 + \delta_3 W_3 \otimes W_3 + \delta_{23} W_2 \otimes W_3 \in \mathcal{V}$ which is noninvertible. Otherwise, when Γ_1 is invertible, the assertion follows with r = 0. First suppose that $\delta_2 > 0$ and $\delta_3 > 0$. Decomposing A_o into $A_o = \Gamma_1 \Gamma_1^+ A_0 + A_1$, we note that A_1 is admissible within

$$\mathcal{B}_1 = \left\{ A \in \mathcal{N}(\Gamma_1) \,|\, E[A,Y] = g_1 \right\},$$

where $g_1 = E\left[(I - \Gamma_1 \Gamma_1^+)A_o,\ Y\right]$.

Notice that for each $\Gamma_\delta \in \mathcal{V}$, $\Gamma_\delta \neq \Gamma_1$, and for each $A \in \mathcal{B}_1$ we have $\Gamma_\delta A = \delta_1 A$. Hence we may use the same argument as above to conclude from Theorem 2.2 that A_1 is $\mathcal{N}(\Gamma_1)$-best locally at $\Gamma_2 = I \otimes I$. This shows that the assertion holds with r = 1.

Next suppose that $\delta_2 > 0$ and that $\delta_3 = 0$. In this case $\Gamma = W_2 \otimes W_2$, and, consequently, for each $\Gamma_\delta \in \mathcal{V}$, $\Gamma_\delta \neq \Gamma_1$, and for each $A \in \mathcal{B}_1$ we have

$$\Gamma_\delta A = \delta_1 A + \delta_3 W_3^2 A + \delta_{13} W_3 A.$$

Introducing model $\mathcal{M}(\varepsilon, \Theta_2)$, where $\Theta_2 = \left[\left\{ \delta_1\ I \times I + \delta_3 W_3 \otimes W_3 + \right.\right.$ $\left.\left. + \delta_{13}\ I \otimes W_3 \,|\, \delta_1 + \delta_3 + \delta_{13} = 1, 0 \leq \delta_{13} \leq 2 \sqrt{\delta_1}\sqrt{\delta_3} \right\} \right]$, we conclude from Theorem 2.2 that A_1 is $\mathcal{N}(\Gamma_1)$-best locally at some $\Gamma_2 \in \Theta_2$. When Γ_2 is invertible the assertion follows with r = 1. Otherwise, when $\Gamma_2 = W_3 \otimes W_3$, we decompose A_1 into

$$A_1 = \Gamma_2 \Gamma_2^+ A_1 + A_2,$$

where $A_2 \in \mathcal{B}_2 = \left\{ A \in \mathcal{N}(\Gamma_1) \cap \mathcal{N}(\Gamma_2) \,|\, E[A,Y] = g_2 \right\}$, while $g_2 = E\left[(I - \right.$ $\left. - \Gamma_2 \Gamma_2^+) A_1, Y\right]$. Now observe that $\Gamma_\delta A = \delta_1 A$ for $\Gamma_\delta \in \mathcal{V}$, $\Gamma_\delta \neq \Gamma_1, \Gamma_2$, and for each $A \in \mathcal{B}_2$. Using again the same argument as above we infer that A_2 is $\mathcal{N}(\Gamma_1) \cap \mathcal{N}(\Gamma_2)$-best locally at $\Gamma_3 = I \otimes I$. This establishes the necessity with r = 2.

When $\delta_2 = 0$ and $\delta_3 > 0$ or when $\delta_2 = 0$ and $\delta_3 = 0$ the necessity of the conditions of Theorem 4.1 may be demonstrated in exactly the same manner. We shall omit the details. Examples 1 and 2 present admissible estimators for two mixed linear models, with two and with three variance components, respectively. In these examples we make use of Theorems 2.1 and 4.1.

Example 1. For the mixed linear model with $k = 2$ and with W_2 being noninvertible the subset of vector in $\mathscr{B}$ given by

$$A = \rho_1 \Gamma_\delta^{-1}(I) + \rho_2 \Gamma_\delta^{-1}(I)W_2, \tag{4.1}$$

where $\delta = (\delta_1, \delta_2, \delta_{12})'$ ranges over H(S) with $\delta_1 > 0$, and by

$$A = \rho_1 W_2^+ + \rho_2(I - W_2W_2^+) \tag{4.2}$$

is the minimal complete class. Here ρ_1 and ρ_2 have to be chosen so that $E[A,Z] = g$. Note that the estimator given by (4.1)is best locally at the invertible covariance operator $\Gamma_\delta \in \Theta$. Also note that (4.2) is best locally at $W_2 \otimes W_2 \in \Theta$ and that $I - W_2W_2^+$ is $\mathscr{N}(W_2 \otimes W_2)$--best locally at each $\Gamma_\delta, \delta \in H(S)$.

Example 2. For $k = 3$ we give the minimal complete class when $\mathscr{R}(W_2) \neq \mathscr{R}(W_3)$ and $\mathscr{R}(W_2 + W_3) \neq \mathscr{R}^m$. Under these assumptions the subset of vectors in $\mathscr{B}$ given by (4.3)-(4.8) below is the minimal complete class:

$$A = \rho_1\Gamma_\delta^{-1}(I) + \rho_2\Gamma_\delta^{-1}(I)W_2 + \rho_3\Gamma_\delta^{-1}(I)W_3, \tag{4.3}$$

where $\delta = (\delta_1, \delta_2, \delta_3, \delta_{12}, \delta_{13}, \delta_{23})'$ ranges over H(S) with $\delta_1 > 0$;

$$A = \rho_1\Gamma_\delta^+(I)W_2 + \rho_2\Gamma_\delta^+(I)W_3 + \rho_3[I - (W_2 + W_3)(W_2 + W_3)^+], \tag{4.4}$$

where $\delta = (0, \delta_2, \delta_3, 0, 0, \delta_{23})'$ ranges over H(S) with $\delta_2 > 0$ and $\delta_3 > 0$;

$$A = \rho_1 W_2^+ + \rho_2\Gamma_\delta^{-1}(I)W_3(I - W_2W_2^+) + \rho_3\Gamma_\delta^{-1}(I)(I - W_2W_2^+), \tag{4.5}$$

where $\delta = (\delta_1, 0, \delta_3, 0, \delta_{13}, 0)'$ ranges over H(S) with $\delta_1 > 0$;

$$A = \rho_1 W_3^+ + \rho_2\Gamma_\delta^{-1}(I)W_2(I - W_3W_3^+) + \rho_3\Gamma_\delta^{-1}(I)(I - W_3W_3^+), \tag{4.6}$$

where $\delta = (\delta_1, \delta_2, 0, \delta_{12}, 0, 0)'$ ranges over H(S) with $\delta_1 > 0$;

$$A = \rho_1 W_2^+ + \rho_2 W_3^+(I - W_2W_2^+) + \rho_3(I - W_2W_2^+)(I - W_3W_3^+); \qquad (4.7)$$

$$A = \rho_1 W_3^+ + \rho_2 W_2^+(I - W_3W_3^+) + \rho_3(I - W_2W_2^+)(I - W_3W_3^+). \qquad (4.8)$$

As above ρ_1, ρ_2 and ρ_3 are uniquely determined by $\mathrm{tr}A = c_1$, $\mathrm{tr}\,A\,W_2 = c_2$ and $\mathrm{tr}\,A\,W_3 = c_3$. Notice that (4.3) is best locally at Γ_δ, which is an invertible operator; vector (4.4) is best locally at Γ_δ, which is not invertible, and, moreover, $I - (W_2 + W_3)(W_2 + W_3)^+$ is $\mathcal{N}(\Gamma_\delta)$-best locally at each Γ_τ, $\tau \in H(S)$; the vector given by (4.5) is best locally at $W_2 \otimes W_2$ and $\Gamma_\delta^{-1}(I)(\rho_3 I + \rho_2 W_3)(I - W_2W_2^+)$ is $\mathcal{N}(W_3 \otimes W_3)$-best locally at any invertible Γ_τ such that $\tau = (\delta_1, \tau_2, \delta_3, \tau_{12}, \delta_{13}, \tau_{23})' \in H(S)$; vector (4.6) is best locally at $W_3 \otimes W_3$ and $\Gamma_\delta^{-1}(I)(\rho_3 I + \rho_2 W_2)(I - W_3W_3^+)$ is $\mathcal{N}(W_2 \otimes W_2)$-best locally at any invertible Γ_τ such that $\tau = (\delta_1, \delta_2, \tau_3, \delta_{12}, \tau_{13}, \tau_{23})' \in H(S)$. Finally (4.7) is best locally at $W_2 \otimes W_2$, $\left[\rho_2 W_3^+ + \rho_3(I - W_3W_3^+)\right](I - W_2W_2^+)$ is $\mathcal{N}(W_2 \otimes W_2)$-best locally at $W_3 \otimes W_3$ and $(I - W_2W_2^+)(I - W_3W_3^+)$ is $\mathcal{N}(W_2 \otimes W_2) \cap \mathcal{N}(W_3 \otimes W_3)$-best locally at $I \times I$. Analogously (4.8) is best locally at $W_3 \otimes W_3$, $\left[\rho_2 W_2^+ + \rho_3(I - W_2W_2^+)\right](I - W_3W_3^+)$ is $\mathcal{N}(W_3 \otimes W_3)$-best locally at $W_2 \otimes W_2$ and $(I - W_2W_2^+)(I - W_3W_3^+)$ is $\mathcal{N}(W_2 \otimes W_2) \cap$ $\cap \mathcal{N}(W_3 \otimes W_3)$ - best locally at $I \otimes I$.

The following example shows that the "only if" part of Theorem 4.1 may not be true for a general linear model even when the covariance operators commute.

Example 3. Put $W_1 = I$, $W_2 = \mathrm{diag}\,(1,2,0,0,0,0)$ and $W_3 = \mathrm{diag}\,(1, 0,1,2,3,0)$. Consider model $\mathcal{M}_1 = \mathcal{M}(\mathcal{E}, \Theta_1)$, where $\mathcal{E} = \mathrm{span}\left\{I, W_2, W_3\right\}$, while $\Theta = \left\{W_\sigma \otimes W_\sigma \mid W_\sigma = \sigma_1 I + \sigma_2 W + \sigma_3 W, \sigma \in \Omega\right\}$. Note that $\mathcal{M} = \mathcal{M}(\mathcal{E}, \Theta)$ is not regular, because $\mathcal{E}$ is not a quadratic subspace. Letting $\delta =$ $= (1/2, 0, 1/2, 0, 0, 0)'$ we notice (see Example 2) that

$$A = \Gamma_\delta^{-1}(I) W_3(I - W_2W_2^+) = \frac{1}{5}\,\mathrm{diag}\,(0,0,5,4,3,0)$$

is admissible with reference to model $\mathcal{M}$. Vector A is best locally at $W_2 \otimes W_2$ only. Moreover, A is $\mathcal{N}(W_2 \otimes W_2)$-best loccaly at Γ_δ.

Now define $H(\mathcal{T})$ as a closed convex portion of an equilaterial triangle (see Figure 1) whose boundary consists of the line segment joining $(1,0,0)'$ and $(0,1,0)'$ and curve $\mathcal{T}$. Assume that the line segment joining $(0,1,0)'$ and $(1/2,0,1/2)'$ is tangent to $\mathcal{T}$ at $(0,1,0)'$ and that this is the only common point.

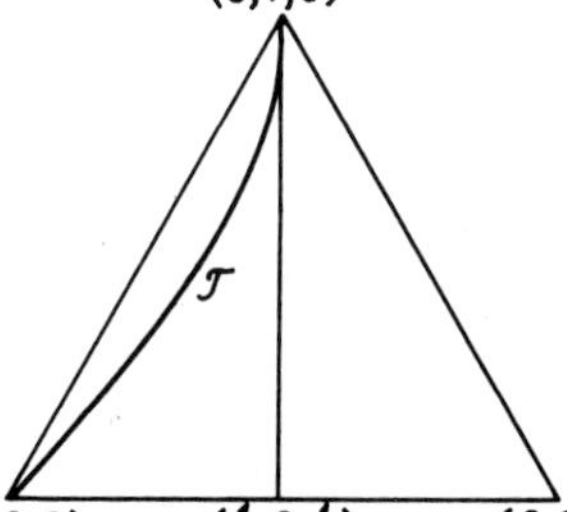

Fig. 1

In addition to $\mathcal{M}$ we consider model $\mathcal{M}_1 = \mathcal{M}(\varepsilon, \Theta_1)$, where Θ_1 is a subset of Θ and is defined by

$$\Theta_1 = \left[\left\{\Gamma_\delta \in \mathcal{V} \mid \delta \in H(S), (\delta_1, \delta_2, \delta_3)' \in H(\mathcal{T})\right\}\right].$$

We shall establish that A is admisible with reference to model $\mathcal{M}_1$. Suppose that it is not so, and let B be better than A. Clearly B must be then best locally at $W_2 \otimes W_2$. But in such a case $B \in \mathcal{N}(W_2 \otimes W_2)$. Moreover, B is $N(W_2 \otimes W_2)$-best locally at Γ_δ since B is as good as A within model $\mathcal{M}_1$. This may be shown as follows. Take a sequence $\{\xi_n\}$, where $\xi_n = (\xi_1^n, \xi_2^n, \xi_3^n)' \in \mathcal{T}$, such that $\xi_n \neq (0,1,0)'$ for each $n = 1,2,\ldots$ and $\xi_n \to (0,1,0)'$ as $n \to \infty$. Now define the sequence $\{\eta_n\}$, where $\eta_n = \alpha_n(\xi_1^n, 0\ \xi_3^n, 0,0,0)' \in H(S)$, while $\alpha_n = (\xi_1^n + \xi_3^n)^{-1}$. Notice that $\Gamma_{\eta_n}(A) = \alpha_n \Gamma_{\xi_n}(A)$ and $\Gamma_{\eta_n}(B) = \alpha_n \Gamma_{\xi_n}(B)$. Because B is better than A within model $\mathcal{M}_1$, therefore

$$[B, \Gamma_{\eta_n}(B)] \leq [A, \Gamma_{\eta_n}(A)]$$

for each $n = 1,2,\ldots$ This in turn yields $[B, \Gamma_{\delta}(B)] \leq [A, \Gamma_{\delta}(A)]$, because by assumption $\eta_n \to \delta = (\frac{1}{2},0,\frac{1}{2},0,0,0)'$ as $n \to \infty$. Recalling that A is $\mathcal{N}(W_2 \otimes W_2)$-best locally at Γ_{δ} within model $\mathcal{M}$, it is now clear that B must be $\mathcal{N}(W_2 \otimes W_2)$-best locally at Γ_{δ} as asserted. In view of the above $A - B \in \mathcal{E}^{\perp} \cap (W_2 \otimes W_2)$ and $\Gamma_{\delta}(A - B) \in \mathcal{E} + \mathcal{R}(W_2 \otimes W_2)$. However this leads to a contradictory conclusion A = B. Thus A is admissible within $\mathcal{M}_1$.

Now observe that if we restrict attention to non-zero convariance operators $\Gamma \in \Theta_1$ then A is best locally at $W_2 \otimes W_2$ only and also $\mathcal{N}(W_2 \otimes W_2)$-best locally at $W_2 \otimes W_2$ only.

Thus conditions given in Theorem 2.3 are sufficient but not necessary for a vector to be admissible.

5. Descriptions of the minimal class under additional assumptions

In some cases, as observed by Olsen et al. [8], the entire set H(S) is not needed to characterize the minimal complete class. In particular a reduction may be accomplished when $\dim(\mathcal{F}) - \dim(\mathcal{E}) = 1$, where $\mathcal{F}$ is the smallest quadratic subspace containing $\mathcal{E}$. Clearly, if $\dim(\mathcal{F}) - \dim(\mathcal{E}) = 0$, e.i. if the considered model is regular(see[2]), then the minimal complete class consists of a single vector, the uniquely determined uniformly best estimator. For simplicity let $\Delta = \dim(\mathcal{F}) - \dim(\mathcal{E})$.

Theorem 5.1. Suppose that $\Delta = 1$. Then the minimal complete class form vectors that are admissible and best locally at some Γ_{ξ}, where $\xi \in S$.

P r o o f: By Lemma 2.2 we may take as the sample space $\mathcal{K}$ the quadratic subspace $\mathcal{F}$ with the usual trace inner product defined by $[A,B] = \operatorname{tr} A B$ for $A, B \in \mathcal{F}$. Since $\Delta = 1$ by assumption, it follows that there exists a vector $B \in \mathcal{F}$ such that span $\{B\} = \mathcal{E}^{\perp}$. Moreover, a vector $A \in \mathcal{F}$ is best locally at Γ_δ iff

$$[B, \Gamma_\delta A] = 0 \tag{5.1}$$

or, equivalently, iff

$$(B1, \Gamma_\delta(A)\underline{1}) = 0,$$

where $(.,.)$ stands for the standard inner product in $\mathcal{R}^m$, while $\underline{1} = (1,\ldots,1)' \in \mathcal{R}^m$. Using the fact that $\Gamma_\delta(A) = A\Gamma_\delta(I)$ we easily see that

$$\Gamma_\delta(A)1 = A\,K\delta,$$

where

$$K = (W_1\underline{1},\ldots,W_k\underline{1},W_1W_2\underline{1},\ldots,W_{k-1}W_k\underline{1})$$

is an $m \times k(k+1)/2$ matrix, while, as usual,

$$\delta = (\delta_1,\ldots,\delta_k,\delta_{12},\ldots,\delta_{k-1\;k})' \in H(S).$$

Thus

$$[B, \Gamma_\delta(A)] = (B\underline{1}, AK\delta) = (K'A\,B\,\underline{1}, \delta).$$

This shows that A is best locally at Γ_δ iff $K'A\,B\,\underline{1}$ is orthogonal to δ. Hence, to prove the assertion of the theorem, it is sufficient to show that there exists a nonzero vector $\xi \in S$ which is orthogonal to $K'AB\,\underline{1}$ when $K'A\,B\,\underline{1}$ is orthogonal to some vector $\delta \in H(S)$. As known any element in $H(S)$ may be presented as a convex linear combination of elements belonging to S, e.i. there exist vectors $\xi_1,\ldots,\xi_u \in S$ and coefficients $t_1,\ldots,t_u \in \mathcal{R}^+$, $\sum t_i = 1$, such that $t_1\xi_1 + \ldots + t_u\xi_u = \delta$. Since

$$t_1(K'A\,B\,\underline{1},\xi_1) + \ldots + t_u(K'AB\,\underline{1},\xi_u) = (K'AB\,\underline{1},\delta) = 0$$

and since S is a compact connected set we conclude that there exists at least one element $\xi \in S$ such that $\Gamma_\xi A \in \mathcal{E}$. Otherwise, we would arrive at a contradiction with (5.1). Thus the proof is terminated.

Remark. The assertion of Theorem 5.1 for $k = 2$ is due to Olsen at al. [8], but the proof given here is slightly different from that one given by these authors.

The following two theorems show that no reduction is possible when Δ is too large. The first one is reproduced from [8].

Theorem 5.2. Let $k = 2$ and suppose that $\Delta = 2$. If $\xi, \delta \in H(S)$ are such that $\Gamma_\xi \neq \Gamma_\delta$ and such that $\Gamma_\delta A \in \mathcal{E}$ and $\Gamma_\delta B \in \mathcal{E}$ for some $A, B \in \mathcal{X}$ then $A \neq B$.

P r o o f: First take $\xi, \delta \in H(S)$ such that both Γ_ξ and Γ_δ are invertible. We may assume that $\xi_2 > 0$, where $\xi = (\xi_1, \xi_2, \xi_{12})'$. As noted by Olsen et al. [8] it is sufficient to show that

$$\Gamma_\xi^{-1}(\mathcal{E}) \cap \Gamma_\delta^{-1}(\mathcal{E}) = \{0\}. \tag{5.2}$$

Because Γ_ξ and Γ_δ commute, this condition is equivalent to

$$\Gamma_\delta(\mathcal{E}) \cap \Gamma_\delta(\mathcal{E}) = \{0\}.$$

Letting

$$H = (\Gamma_\xi(I)\underline{1}, \Gamma_\xi(W_2)\underline{1}, \Gamma_\delta(I)\underline{1}, \Gamma_\delta(W_2)\underline{1})$$

we note that (5.2) is met if $\operatorname{rank}(H) = 4$. Now observe that we may write $H = L\Xi'$, where

$$L = (\underline{1}, W_2\underline{1}, W_2^2\underline{1}, W_2^3\underline{1}).$$

while

$$\Xi = \begin{pmatrix} \xi_1 & \xi_{12} & \xi_2 & 0 \\ 0 & \xi_1 & \xi_{12} & \xi_2 \\ \delta_1 & \delta_{12} & \delta_2 & 0 \\ 0 & \delta_1 & \delta_{12} & \delta_2 \end{pmatrix}$$

Writing the determinant of Ξ in the form

$$\xi_1^2\xi_2^2 \begin{vmatrix} 1 & 0 & 1 & 0 \\ 0 & 1 & 0 & 1 \\ \delta_1/\xi_1 & (\delta_{12}-\xi_{12}\delta_1/\xi_1)/\xi_1 & \delta_2/\xi_2 & 0 \\ 0 & \delta_1/\xi_1 & (\delta_{12}-\xi_{12}\delta_2/\xi_2)/\xi_2 & \delta_2/\xi_2 \end{vmatrix}$$

it is easily checked that it is different from zero when $\Gamma_\xi \neq \Gamma_\delta$. Thus rank $(\Xi) = 4$. Moreover, because I, W, W^2 and W^3 are linearly independent when $\Delta = 2$, we conclude that rank $(H) = 4$. Now suppose that Γ_ξ, say, is not invertible, e.i. that $\xi = (0,1,0)'$. If there would exist and estimator $A \in \mathcal{B}$ such that $\Gamma_\xi A \in \mathcal{E}$ and $\Gamma_\eta A \in \mathcal{E}$ for some invertible Γ_η, $\eta \in H(S)$, then A would be best locally at each $\alpha\Gamma_\xi + (1-\alpha)\Gamma_\eta$, $0 \leq \alpha < 1$, which are invertible. This would be a contradiction with the conclusion reached above.

Now we prove an analogous theorem for $k = 3$ under the assumption that Γ_ξ and Γ_δ are invertible.

<u>Theorem 5.3.</u> Let $k = 3$ and suppose that rank$(H) = 6$, where $H = L\Xi'$, while

$$L = (\underline{1}, W_2\underline{1}, W_2^2\underline{1}, W_2^3\underline{1}, W_3\underline{1}, W_2W_3\underline{1}, W_2^2W_3\underline{1}, W_3^2\underline{1}, W_2W_3^2\underline{1}, W_3^3\underline{1})$$

and

$$\Xi = \begin{pmatrix} \xi_1 & \xi_{12} & \xi_{12} & \xi_2 & 0 & \xi_3 & 0 & 0 & \xi_{23} \\ 0 & \xi_1 & 0 & \xi_{12} & \xi_2 & 0 & \xi_3 & 0 & \xi_{13} \\ 0 & 0 & \xi_1 & 0 & 0 & \xi_{13} & \xi_{23} & \xi_3 & \xi_{12} \\ \delta_1 & \delta_{12} & \delta_{13} & \delta_2 & 0 & \delta_3 & 0 & 0 & \delta_{23} \\ 0 & \delta_1 & 0 & \delta_{12} & \delta_2 & 0 & \delta_3 & 0 & \delta_{13} \\ 0 & 0 & \delta_1 & 0 & 0 & \delta_{13} & \delta_{23} & \delta_3 & \delta_{12} \end{pmatrix}.$$

If $\xi, \delta \in H(S)$ are such that Γ_ξ and Γ_δ are invertible, $\Gamma_\xi \neq \Gamma_\delta$, $\Gamma_\xi A \in \mathcal{E}$ and $\Gamma_\delta B \in \mathcal{E}$ for some $A, B \in \mathcal{B}$, then $A \neq B$.

P r o o f: As for $k = 2$ it is sufficient to show that under the specified assumptions

$$\Gamma_\xi(\varepsilon) \cap \Gamma_\delta(\varepsilon) = \{0\}.$$

This condition is met if the rank of the matrix given by

$$(\Gamma_\xi(I)\underline{1}, \Gamma_\xi(W_2)\underline{1}, \Gamma_\xi(W_3)\underline{1}, \Gamma_\delta(I)\underline{1}, \Gamma_\delta(W_2)\underline{1}, \Gamma_\delta(W_3)\underline{1}) \qquad (5.3)$$

is equal to 6, because I, W_2 and W_3 are assumed to be linearly independent. This terminates the proof of Theorem 5.3, because (5.3) coincides with H.

Remark. One can show along similar lines as for $k = 2$ that rank $(\Xi) = 6$, when $\Gamma_\xi \neq \Gamma_\delta$. Consequently, rank(H) = 6, when

$$I, W_2^2, W_2^3, W_3, W_2W_3, W_2^2W_3, W_3^2, W_2W_3^2, W_3^3$$

are linearly independent. Also note that rank(H) = 6 when $\Delta \geq 7$. It should be clear that for $k = 3$ it may be possible to reduce the description of the minimal complete class to some subsets of H(S) when rank(H) < 6. However a full description of the situation seems to be rather complex and therefore it is not considered in the paper.

6. Limits of admissible estimators

In general a limit of a sequence of admissible estimators must not be admissible. It turns out however that within the framework of a mixed linear model as defined in Section 4 the limit of admissible estimators is admissible. For $k = 2$ this fact has been first noted by Olsen et al. [8].

Theorem 6.1. The limit of a sequence of estimators $\{A_n\}$ admissible within set $\mathcal{B}$ is admissible within $\mathcal{B}$.

P r o o f: We establish Theorem 6.1 for $k = 2$ and $k = 3$ only. For $k > 3$ the proof is omitted, because it is lengthy and uninformative.

Let $\{A_n\}$ be a sequence of estimators admissible within $\mathcal{B}$ and suppose that it converges to A_o. By Theorem 2.2 each A_n is best locally at some $\Gamma_n = \Gamma_{\xi_n}$, say, where $\xi_n \in H(S)$. Because $H(S)$ is a compact set, there exists a subsequence $\{\xi_{n_i}\}$ such that $\Gamma_{n_i} \to \Gamma_o$, where $\Gamma_o = \Gamma_{\xi_o}$, while $\xi_o \in H(S)$. Because $\Gamma_{n_i} A_{n_i} \to \Gamma_o A_o$, it follows that A is best locally at Γ_o. Now if Γ_o is invertible, A_o is admissible by Theorem 2.3. To show that A_o is admissible when Γ_o is not invertible we introduce the following notation. Let $A_n = \text{diag}(a_1^n,\ldots,a_m^n)$, $A_o = \text{diag}(a_1^o,\ldots,a_m^o)$, $W_i = \text{diag}(w_{i1},\ldots,w_{im})$, $i = 1,2,\ldots,k$, $\xi_n = (\xi_1^n,\ldots,\xi_k^n, \xi_{12}^n,\ldots,\xi_{k-1,k}^n)'$ and $\xi_o = (\xi_1^o,\ldots,\xi_{k-1,k}^o)'$. For the purpose of simplicity of notation let $\xi_n \to \xi_o$. Also without loss of generality we may assume that the Γ_n's are invertible.

Because $\Gamma_n A_n \in \mathcal{E}$, there exist coefficients $x_n, y_n, \ldots, z_n \in R$ such that

$$A_n = \Gamma_n^{-1}(I)(x_n I + y_n W_2 + \ldots + z_n W_k). \tag{6.1}$$

It is clear that $x_n, y_n, \ldots, z_n$ are determined for $n = 1,2,\ldots$ uniquely by a system of linear equations

$$\begin{aligned} \text{tr } A_n &= c_1 , \\ \text{tr } A_n W_2 &= c_2 , \\ &\cdots\cdots\cdots\cdots \\ \text{tr } A_n W_k &= c_k , \end{aligned} \tag{6.2}$$

which insures that $[A_n, EZ] = g$, so that $A_n \in \mathcal{B}$.

(i) For $k = 2$ notice that $\Gamma_o = W_2 \otimes W_2$. Thus in this particular case $\xi_1^n \to 0$ and $\xi_{12}^n \to 0$ as $n \to \infty$. Moreover from the adopted assumptions it follows that $x_n \to 0$ and $y_n \to y_o$, say. Rewriting (6.1) in the form of

$$a_i^n = \frac{x_n + y_n w_{2i}}{\xi_1^n + \xi_{12}^n w_{2i} + \xi_2^n w_{2i}^2}$$

it is easily checked that $a_i^n \to y_o/w_{2i}$, when $w_{2i} > 0$ and that $a_i^n \to x_o$ when $w_{2i} = 0$. Thus

$$A_o = \rho_1 W_2^+ + \rho_2(I - W_2W_2^+).$$

This shows that A_o has the form of (4.2), and, therefore, it is an admissible estimator within $\mathcal{B}$.

(ii) For $k = 3$ we shall prove Theorem 6.1 in case when $\mathcal{R}(W_2) = \neq \mathcal{R}(W_3)$ and $\mathcal{R}(W_2 + W_3) \neq \mathcal{R}^m$. Under these assumptions

$$\Gamma_o = \xi_2^o W_2 \otimes W_2 + \xi_3^o W_3 \otimes W_3 + \xi_{23}^o W_2 \otimes W_3$$

so that $\xi_1^n \to 0$, $\xi_{12}^n \to 0$, $\xi_{13}^n \to 0$ whereas $\xi_2^n \to \xi_2^o$, $\xi_3^n \to \xi_3^o$, $\xi_{23}^n \to \to \xi_{23}^o$. Also notice that $x_n \to 0$, $y_n \to y_o$, $z_n \to z_o$, say. For the remaining part of the proof it will be convenient to write (6.1) in the form of

$$a_i^n = \frac{x_n + y_n w_{2i} + z_n w_{3i}}{\xi_1^n + \xi_2^n w_{2i}^2 + \xi_3^n w_{3i}^2 + \xi_{12}^n w_{2i} + \xi_{13}^n w_{3i} + \xi_{23}^n w_{2i} w_{3i}}.$$

The proof of Theorem 6.1 for $k = 3$ will be accomplished by considering a number of particular cases. Firstly we consider the case when $\xi_2^o > 0$ and $\xi_3^o > 0$. Under these assumptions it follows straightforward from (6.3) that

$$a_i^o = \frac{y_o w_{2i} + z_o w_{3i}}{\xi_2^o w_{2i}^2 + \xi_3^o w_{3i}^2 + \xi_{23}^o w_{2i} w_{3i}}$$

when $w_{2i} + w_{3i} > 0$, and that $a_i^o = u_o$ otherwise. This means that A_o may be written as

$$A_o = \rho_1\Gamma_o^+(I)W_2 + \rho_2\Gamma_o^+(I)W_3 + \rho_3\left[I - (W_2 + W_3)(W_2 + W_3)^+\right]$$

and, by virtue of (4.4), the assertion follows.

Next let us consider the case when $\xi_3^o = 0$. Then, evidently, $\Gamma_o = W_2 \otimes W_2$. Thus all coordinates of ξ_n except ξ_2^n converge to zero. Also $z_o = 0$, so that $x_n \to 0$, $y_n \to y_o$ and $z_n \to 0$. Moreover, it follows from (6.3) that $a_i^n \to y_o/w_{2i}$ when $w_{2i} > 0$ and that $a_i^n \to u_o$ when $w_{2i} + w_{3i} = 0$. When $w_{2i} = 0$ and $w_{3i} > 0$ formula (6.3) reduces to

$$a_i^n = \frac{x_n + z_n w_{3i}}{\xi_1^n + \xi_3^n w_{3i}^2 + \xi_{13}^n w_{3i}} \; . \tag{6.4}$$

To investigate the limit of (6.4) we may assume without loss of generality that one of the following situations must hold:

(i) $\xi_3^n/\xi_1^n \to 0$,

(ii) $\xi_1^n/\xi_3^n \to 0$,

(iii) $\xi_3^n/\xi_1^n \to a$, $\xi_{13}^n/\xi_1^n \to b$, $0 < a < \infty$, $0 \leq b < \infty$.

It is easily seen from (6.4) that in case (i) the sequence z_n/ξ_1^n must converge to a finite limit v_o, say, as $n \to \infty$, because $x_n/\xi_1^n \to u_o$ as $n \to \infty$. Thus $a_i^n \to u_o + v_o w_{3i}$. Under (ii) we have $x_n/\xi_3^n \to 0$ what follows by noting that $x_n/\xi_1^n \to u_o$ and that $x_n/\xi_3^n = (x_n/\xi_1^n)(\xi_1^n/\xi_3^n)$. Also z_n/ξ_3^n must converge to a finite number w_o, say. Thus $a_i^n \to w_o/w_{3i}$. Finally, in case (iii) formula (6.4) implies immediately that $a_i^n \to (u_o + v_o w_{3i})/(1 + a w_{3i}^2 + b w_{3i})$, where $u_o = \lim x_n/\xi_1^n$, while $v_o = \lim z_n/\xi_1^n$. Combining all the above obtained results we easily notice that in cases (i) - (iii) vector A_o may be presented in the form of

$$A_o = \rho_1 W_2^+ + \rho_2\Gamma_\delta^+(I)W_3(I - W_2W_2^+) + \rho_3\Gamma_\delta^+(I)(I - W_2W_2^+),$$

where $\delta = (\delta_1, 0, \delta_3, \delta_{13}, 0, 0)' \in H(S)$, while $\delta_1 > 0$. And this is an admissible estimator (compare formula (4.6)). Clearly in the remaining case when $\xi_2^o = 0$ and $\xi_3^o > 0$ the assertion follows by reasons of symmetry.

When $\mathcal{R}(W_2) = \mathcal{R}(W_3)$ and/or $\mathcal{R}(W_2 + W_3) = \mathcal{R}^m$ Theorem 6.1 may be proved exactly along the same lines, and the proof is omitted here.

Acknowledgements. I would like to thank Dr. S. Gnot and Dr. R. Zmyślony for very helpful discussions.

References

[1] F e r g u s o n, T. S., Mathematical statistics, a decision theoretic approach, Academic Press, New York and London 1967.

[2] G n o t, S., K l o n e c k i, W., Z m y ś l o n y, R., Best umbiased estimation, a coordinate free approach, Preprint 124, Institute of Mathemetics, Polish Academy of Sciences, Warszawa 1978.

[3] H a r v i l l e, D. A., Quadratic unbiased estimation of variance components for one-way classification, Biometrika, 56, 313--326 (1969).

[4] H a r v i l l e, D. A., Maximum likelihood approaches to variance component estimation and related problems, JASA, 72, 320-338 (1977).

[5] L a M o t t e, L. R., Invariant quadratic estimators in the random, one-way ANOVA model, Biometrics, 32, 793-804 (1976).

[6] L a M o t t e, L. R., A canonical form for the general linear model, The Annals of Statistics, 5, 787-789 (1977).

[7] L a M o t t e, L. R., On admissibility and completeness of linear ubiased estimators in a general linear model, JASA, 72, 438-441 (1977).

[8] O l s e n, A., S e e l y, J., B i r k e s, D., Invariant quadratic unbiased estimation for two variance components. The Annals of Statistics, 4, 878-890 (1976).

[9] W i s t u b a, I., Ph. D. Thesis (1978).

[10] Z m y ś l o n y, R., Kwadratowe dopuszczalne estymatory komponentów wariancyjnych w modelach losowych, Matematyka Stosowana, 117-122 (1976).

MIXTURES OF INFINITELY DIVISIBLE DISTRIBUTIONS AS LIMIT LAWS FOR SUMS OF DEPENDENT RANDOM VARIABLES

by

Andrzej Kłopotowski
Nicolaus Copernicus University, Toruń

1. Introduction

Let there be given a double sequence X of random variables(rv's)

$$\begin{array}{l} x_{11}, x_{12}, \ldots, x_{1,k_1}, \\ x_{21}, x_{22}, \ldots, x_{2,k_2}, \\ \cdots\cdots\cdots\cdots \\ x_{n1}, x_{n2}, \ldots, x_{n,k_n}, \\ \cdots\cdots\cdots\cdots \end{array}$$

and a sequence of its row sums

$$S_n := \sum_{k=1}^{k_n} x_{nk}, \quad n \in N$$

(where all these rv's are defined on a common probability space $(\Omega, \mathcal{F}, P)$).

The problem of asymptotic behaviour of probability distributions of $S_n, n \in N$, is mainly contained in two questions:

1° Which measures can appear as weak limit laws?

2° Which properties of X imply the weak convergence of $S_n, n \in N$, to the specified probability measure?

Bawly's idea of accompanying laws allows us to find the complete solution of this problem for row-wise independent arrays. It is based on the fact that the characteristic function (chf) of the sum S_n can be approximated by a suitably constructed infinitely divisible chf. If we assume some "smallness" conditions on X then these two sequences of chf's have the same common limit. Of course, in this case the limit law of $S_n, n \in N$, must be infinitely divisible. Moreover, the method of the above construction gives necessary and sufficient conditions for the weak convergence of the sum distributions given in terms of X. If we omit the postulate of independence of rv's in the same rows, then the situation is more complicated e.g. every probability measure can appear as a limit law. One way to look for limit theorems for dependent rv's is to generalize the classical situation i.e. to give such conditions which imply the weak convergence and which turn out to be known from the classical theory if we apply them to independent rv's. An essential step in this direction was made by Brown and Eagleson ([1]) by putting to a good use the idea of accompanying laws. Observe that with X we can associate(non-uniquely) a double array $\underline{\mathbf{F}}$ of σ-fields:

$$\mathcal{F}_{1,0} \subset \mathcal{F}_{1,1} \subset \ldots \subset \mathcal{F}_{1,k_1} \subset \mathcal{F},$$
$$\mathcal{F}_{2,0} \subset \mathcal{F}_{2,1} \subset \ldots \subset \mathcal{F}_{2,k_2} \subset \mathcal{F},$$
$$\ldots\ldots\ldots\ldots\ldots\ldots\ldots\ldots\ldots$$
$$\mathcal{F}_{n,0} \subset \mathcal{F}_{n,1} \subset \ldots \subset \mathcal{F}_{n,k_n} \subset \mathcal{F},$$
$$\ldots\ldots\ldots\ldots\ldots\ldots\ldots\ldots\ldots$$

such that every rv X_{nk} is $\mathcal{F}_{nk}$-measurable. The pair $(X,\underline{\mathbf{F}})$ will be called a system. Thanks to the certain approximation lemma (see [1] and lemma 3.2 of [3]) we can imitate the classical case by the construction of some "conditional chf's", which are defined in terms of conditional quantities of rv's from X with respect to σ-fields from $\underline{\mathbf{F}}$ and approximate chf's of $S_n, n \in N$. If in known necessary and sufficient conditions all mean values are replaced by conditional mean values with respect to σ-fields from $\underline{\mathbf{F}}$ and such obtained sequences of rv's are convergent in probability, then these new conditions guarantee the convergence of "conditional chf's". The possibility of such approximation is given by the analogous conditional "smallness" properties. The above procedure was applied in the first place for infinitely divisible laws with finite variance by Brown and Eagleson in [1]; then it was extended by Kłopotowski for an arbitrary infinitely divisible law in R^d [3]. Now we extend the class of possible weak limits taking into account mixtures of infinitely divisible laws. This extension is maximal; every probability distribution in R^1 induced by some rv on Ω can be trivially decomposed as the mixture of infinitely divisible laws. Eagleson in [2] has proved the limit theorem for martingale difference sequences with finite variances, which generalizes the preceeding situation giving sufficient conditions for weak convergence to mixtures of laws with finite variance. The proof of this theorem is based on a very artificial construction involving some regular conditional probabilities on R^∞. A purpose of this note is to show that in the case of mixtures the idea of the accompanying laws can also be applied. Proofs thereof will be given in the most general case without any assumptions about the existence of moments of rv's and mixed limit laws.

2. Mixtures

For every $t \in R^1$ let us define a function $g_t \colon R^1 \to \underline{C}$ as follows

$$g_t(x) := \begin{cases} \left(e^{itx} - 1 - \dfrac{itx}{1+x^2}\right) \dfrac{1+x^2}{x^2} & \text{for } x \neq 0, \\ -\dfrac{t^2}{2} & \text{for } x = 0. \end{cases}$$

The function g_t is continuous and bounded on R^1, i.e. there exists a constant $M = M_t > 0$ such that $|g_t(x)| \leq M$ for $x \in R^1$. For the distribution function K of some finite mesure on R^1 and for some $a \in R^1$ the function

$$\varphi(t) := \exp\left\{ita + \int_{-\infty}^{+\infty} g_t(x)\,dK(x)\right\}, \quad t \in R^1, \tag{1}$$

is the chf of some infinitely divisible law on R^1. Conversely, every chf of the infinitely divisible law on R^1 can be uniquely decomposed in the form (1).

Now let us assume that both parameters a and K in (1) are random i.e. $a(\cdot)$ and $K(x,\cdot)$, $x \in R^1$, are rv's and for a.e. $\omega\in\Omega$ $K(\cdot,\omega)$ is bounded, nondecreasing, left continous, $\lim_{x\to-\infty} K(x,\omega) = 0$. Thus we have the family of the chf's

$$\varphi(t,\omega) = \exp\left\{ita(\omega) + \int_{-\infty}^{+\infty} g_t(x)\,dK(x,\omega)\right\}, \quad t\in R^1, \tag{2}$$

defined for a.e. $\omega\in\Omega$. Integrating both sides of (2) with respect to P we obtain the chf

$$\psi(t) := \int_{\Omega} \varphi(t,\omega)\,dP(\omega), \quad t \in R^1. \tag{3}$$

Its corresponding probability measure is called the mixture of laws given by (2) and will be denoted by Mix(a,K).

If, instead of g_t, we use the functions h_t, $t \in R^1$, defined by

$$h_t(x) := \begin{cases} \dfrac{e^{itx} - 1 - itx}{x^2} & \text{for } x \neq 0, \\[2ex] -\dfrac{t^2}{2} & \text{for } x = 0, \end{cases}$$

then

$$\rho(t) := \int_{\Omega} \exp\left\{ita(\omega) + \int_{-\infty}^{+\infty} h_t(x)\,dK(x,\omega)\right\} dP(\omega), \quad t \in R^1,$$

is the chf of the mixture of some infinitely divisible laws with finite variances; it will be denoted by mix(a,K).

3. Accompanying conditional laws

For given $\mathbf{F}$ let us define the σ-field $\mathcal{F}_0 := \bigcap_{n=1}^{\infty} \mathcal{F}_{n,0}$. Our fundamental assumption about Mix (a,K) and $\mathbf{F}$ is:

$$\begin{array}{ll} 1^\circ & a(\cdot) \text{ is } \mathcal{F}_0\text{-measurable,} \\ 2^\circ & \text{for every fixed } x \in R^1 \ K(x,\cdot) \text{ is } \mathcal{F}_0\text{-measurable,} \\ 3^\circ & K(+\infty,\cdot) := \lim_{x \to +\infty} K(x,\cdot) \text{ is a.e. finite.} \end{array} \tag{C.0}$$

Because of 3° all $K(x,\cdot)$, $-\infty \leqslant x \leqslant +\infty$, are finite a.e. (Of course $K(-\infty,\cdot) := \lim_{x \to -\infty} K(x,\cdot) = 0$), 2° implies $\mathcal{F}_0$-measurability of $K(+\infty,\cdot)$. In this part we shall consider only systems $(X,\mathbf{F})$ having the following properties:

$$\sum_{k=1}^{k_n} \left\{ A_{nk} + E_{n,k-1}\left(\frac{Y_{nk}}{1+Y_{nk}^2}\right)\right\} \xrightarrow{\text{a.e.}} \alpha(\cdot); \tag{C.1}$$

$$\sum_{k=1}^{k_n} E_{n,k-1}\left(\frac{Y_{nk}^2}{1+Y_{nk}^2}\, I\,(Y_{nk} < x)\right) \xrightarrow{\text{a.e}} K(x,\cdot) \tag{C.2}$$

for every x belonging to some countable dense set $D \subset R^1$;

$$\sum_{k=1}^{k_n} E_{n,k-1}\left(\frac{Y_{nk}^2}{1+Y_{nk}^2}\right) \xrightarrow{\text{a.e.}} K(+\infty,\cdot), \tag{C.3}$$

where

$$A_{nk} := E_{n,k-1}\,(X_{nk} I(|X_{nk}| \leqslant \tau)), \quad 1 \leqslant k \leqslant k_n, n \in N,$$

$$Y_{nk} := X_{nk} - A_{nk}, \quad 1 \leqslant k \leqslant k_n, \ n \in N,$$

for arbitrarily fixed $\tau > 0$.

(Here and in the sequel we use the notation

$$E_{n,k-1}\,(x) := E(X \mid \mathcal{F}_{n,k-1}),$$

$$P_{n,k-1}(A) := P(A \mid \mathcal{F}_{n,k-1}).$$

All equalities and inequalities between rv's are considered in the sense "with probability one", $\xrightarrow{a.e.}$ denotes the convergence almost sure, $\xrightarrow{p}$ denotes the convergence in probability).

Observe that (C.2) implies

$$\sum_{k=1}^{k_n} E_{n,k-1}\left(\frac{Y_{nk}^2}{1+Y_{nk}^2}\, I(a\leq Y_{nk}<b)\right) \xrightarrow{a.e.} K(b,\cdot)-K(a,\cdot) \tag{4}$$

for every $a,b\in D$.

We are going to prove that if, in addition, the system $(X,\underline{\mathbf{F}})$ satisfies

$$\sum_{k=1}^{k_n} E_{n,k-1}\left(\frac{Y_{nk}^2}{1+Y_{nk}^2}\right)\leq C,\quad n\in N, \tag{C.4}$$

for some $\mathcal{F}_0$-measurable rv $0\leq C<+\infty$, then for every fixed $t\in R^1$ we have

$$\sum_{k=1}^{k_n} E_{n,k-1}\left(g_t(Y_{nk})\,\frac{Y_{nk}^2}{1+Y_{nk}^2}\right)\xrightarrow{a.e.}\int_{-\infty}^{+\infty} g_t(x)\,dK(x,\cdot). \tag{5}$$

Let us fix a st[illegible]y decreasing to zero sequence of real numbers $\varepsilon_s, s\in N$. For every $s\in N$ we choose sufficiently large number of points on the real line

$$x_0^{(1)}< x_1^{(s)}<\dots<x_{m_s}^{(s)}$$

from the dense set D such that

(a) $$x_0^{(s)}\searrow-\infty,\quad x_{m_s}^{(s)}\nearrow+\infty,$$

(b) $$\max_{1\leq j\leq m_s}\left|x_j^{(s)}-x_{j-1}^{(s)}\right|\searrow 0,\quad s\to+\infty$$

(c) $$\max_{1\leq j\leq m_s}\left|g_t(x_j^{(s)})-g_t\left(x_{j-1}^{(s)}\right)\right|<\varepsilon_s,\quad s\in N.$$

Since the improper Lebesgue-Stieltjes (stochastic) integral $\int_{-\infty}^{+\infty} g_t(x)\,dK(x,\cdot)$ is well defined for a.e. $\omega\in\Omega$, then by continuity and boundedness of g_t this integral is equel to the improper Riemann-Stieltjes one. Therefore (for fixed t)

$$\sum_{j=1}^{s} g_t\left(x_{j-1}^{(s)}\right)\left[K\left(x_{j,}^{(1)}\cdot\right) - K\left(x_{j-1}^{(s)},\cdot\right)\right] \xrightarrow[s\to\infty]{a.e.} \int_{-\infty}^{+\infty} g_t(x)\,dK(x,\omega) \qquad (6)$$

and the integral is $\mathcal{F}_0$-measurable, a.e. finite rv.

For a moment let us fix $s \in N$. From (4)

$$\sum_{k=1}^{k_n}\sum_{j=1}^{m_s} E_{n,k-1}\left(g_t\left(x_{j-1}^{(s)}\right)\frac{Y_{nk}^2}{1+Y_{nk}^2} I\left(x_{j-1}^{(s)} \leqslant Y_{nk} < x_j^{(s)}\right)\right) \xrightarrow[n\to\infty]{a.e.} \qquad (7)$$

$$\longrightarrow \sum_{j=1}^{m_s} g_t\left(x_{j-1}^{(s)}\right)\left[K\left(x_j^{(s)},\cdot\right) - K\left(x_{j-1}^{(s)},\cdot\right)\right].$$

Next, observe that for this fixed s and every $n \in N$

$$\Bigg|\sum_{k=1}^{k_n}\sum_{j=1}^{m_s} E_{n,k-1}\left(g_t\left(x_{j-1}^{(s)}\right)\frac{Y_{nk}^2}{1+Y_{nk}^2} I\left(x_{j-1}^{(s)} \leqslant Y_{nk} < x_j^{(s)}\right)\right) -$$

$$- \sum_{k=1}^{k_n}\sum_{j=1}^{m_s} E_{n,k-1}\left(g_t(Y_{nk})\frac{Y_{nk}^2}{1+Y_{nk}^2} I\left(x_{j-1}^{(s)} \leqslant Y_{nk} < x_j^{(s)}\right)\right) \leqslant \varepsilon_s \cdot C\,. \qquad (8)$$

Finally

$$\sum_{k=1}^{k_n} E_{n,k-1}\left(g_t(Y_{nk})\frac{Y_{nk}^2}{1+Y_{nk}^2}\right) - \sum_{k=1}^{k_n} E_{n,k-1}\left(g_t(Y_{nk})\frac{Y_{nk}^2}{1+Y_{nk}^2} I\left(x_0^{(s)} \leqslant Y_{nk} <\right.\right.$$

$$\left.\left.< x_{m_s}^{(s)}\right)\right) \xrightarrow{a.e} K(+\infty,\cdot) - K\left(x_{m_s}^{(s)},\cdot\right) + K\left(x_0^{(s)},\cdot\right). \qquad (9)$$

For every natural number s each of conditions (7) - (9) determines the P-null set on which this condition is not fulfilled; the same concerns (6) and (C,4). The sum N_0 of these sets is P-null set. Moreover, we can assume that rv C is finite on $\Omega\setminus N_0$. On the set $\Omega\setminus N_0$ all sequences of rv's in (6) - (9) are pointwise convergent. It remains to prove that it is also true for (5) i.e. for every $\omega\in\Omega\setminus N_0$

$$\sum_{k=1}^{k_n} E_{n,k-1}\left(g_t(Y_{nk})\frac{Y_{nk}^2}{1+Y_{nk}^2}\right)(\omega)\xrightarrow[n\to\infty]{}\int_{-\infty}^{+\infty} g_t(x)\,dK(x,w). \tag{10}$$

Let us fix $\varepsilon > 0$. Choose s so large that

1° $\varepsilon_s \cdot c(\omega) < \frac{\varepsilon}{8}$,

2° $K(+\infty,\omega) - K\left(x_{m_s}^{(s)},\omega\right) + K\left(x_0^{(s)},\omega\right) < \frac{\varepsilon}{8}$,

3° the absolute value of the difference between both sides in (6) at ω is less than $\frac{\varepsilon}{2}$.

For this s there exists n_ε so large that for $n > n_\varepsilon$ the absolute value of the difference between both sides in (7) and (9) at ω is less than $\frac{\varepsilon}{8}$. This gives (10). From (C.1) and (5) by the continuity of exp we obtain the final conclusion of this part

$$\exp\left\{it\sum_{k=1}^{k_n}\left[A_{nk}+E_{n,k-1}\left(\frac{Y_{nk}}{1+Y_{nk}^2}\right)\right]+\sum_{k=1}^{k_n} E_{n,k-1}\left(g_t(Y_{nk})\frac{Y_{nk}^2}{1+Y_{nk}^2}\right)\right\}$$
$$\xrightarrow{a.e.} \exp\left\{ita(\cdot)+\int_{-\infty}^{+\infty} g_t(x)\,dK(x,\cdot)\right\}. \tag{11}$$

The idea of the accompanying laws is contained in this fact. To see this it suffices to take into account a row-wise independent system $(X,\underline{F})$ and nonrandom a,K. For technical reasons we shall use a property which is very similar to (11) and follows from identical argumentation.

<u>Lemma 1.</u> Assumptions (C.0) - (C.4) imply

$$\exp\left\{-it\sum_{k=1}^{k_n}\left[A_{nk}+E_{n,k-1}\left(\frac{Y_{nk}}{1+Y_{nk}^2}\right)\right]-\sum_{k=1}^{k_n} E_{n,k-1}\left(g_t(Y_{nk})\frac{Y_{nk}^2}{1+Y_{nk}^2}\right)\right\} \tag{12}$$
$$\xrightarrow{a.e.} \exp\left\{-ita(\cdot)-\int_{-\infty}^{+\infty} g_t(x)\,dK(x,\cdot)\right\}.$$

4. Comments about (C.4)

Putting

$$V_{nk} := \sum_{j=1}^{k} E_{n,j-1}\left(\frac{Y_{nj}^2}{1+Y_{nj}^2}\right), \quad 1 \leq k \leq k_n, \quad n \in N,$$

$$C := K(+\infty, \cdot) + 1.$$

We obtain from (C.3) that

$$P\left(\lim_{n\to\infty} \sup \left[V_{n,k_n} > C\right]\right) = 0. \tag{13}$$

If we truncate the rv's X_{nk} in the following manner

$$X_{nk}^{*} := X_{nk}\, I\, (V_{nk} \leq C), \quad 1 \leq k \leq k_n, \quad n \in N,$$

then by $\mathcal{F}_{n,k-1}$-measurability of $V_{nk}-C$ and $I(V_{nk} \leq C)$ we have

$$A_{nk}^{*} := E_{n,k-1}(X_{nk}^{*}\, I\left(|X_{nk}^{*}| \leq \tau\right)) = A_{nk} I(V_{nk} \leq C),$$

$$Y_{nk}^{*} := X_{nk}^{*} - A_{nk}^{*} = Y_{nk} I\, (V_{nk} \leq C),$$

$$V_{nk}^{*} := \sum_{j=1}^{k} E_{n,j-1}\left(\frac{Y_{nj}^{*2}}{1+Y_{nj}^{*2}}\right) = \sum_{j=1}^{k} E_{n,j-1}\left(\frac{Y_{nj}^2}{1+Y_{nj}^2}\right) I\left(V_{nj} \leq C\right),$$

for $1 \leq k \leq k_n, n \in N$. Thus the last equality gives the property (C.4) for $(\mathcal{X}^{*}, \underline{F})$. Using (13) one can easily prove that $(\mathcal{X}^{*}, \underline{F})$ satisfies (C.1) - (C.3) and the limit distributions of $S_n, n \in N$, $S_n^{*} := \sum_{k=1}^{k_n} X_{nk}^{*}$, $n \in N$, are equal. Moreover

$$e^{itS_n} - e^{itS_n^{*}} \xrightarrow{a.e.} 0$$

and then by a bounded convergence theorem for conditional expectation we have

$$E(e^{itS_n} \mid \mathcal{F}_0) - E\left(e^{itS_n^{*}} \mid \mathcal{F}_0\right) \xrightarrow{a.e.} 0. \tag{14}$$

5. Approximation

The following version of Brown-Eagleson lemma gives us the possibility to approximate the conditional chf's of S_n by the previously constructed conditional chf's of accompanying laws.

Lemma 2. Let there be given a σ-field $\mathcal{F}_0 \subset \mathcal{F}$ and a function $f : R^1 \times \Omega \longrightarrow \underline{C}$ such that $f(t,\cdot)$ is F_0-measurable for every $t \in R^1$, $|f(t,\omega)| \leq 1$, $f(t,\omega) \neq 0$ for all $t \in R^1$ and almost all $\omega \in \Omega$. For fixed $t \in R^1$ let U_n, W_n be sequences of rv's such that

1° for some $\mathcal{F}_0$-measurable a.e. finite rv C, $|W_n^{-1}| \leq C$,

2° $f(t,\cdot) \cdot W_n^{-1} \xrightarrow{\text{a.e.}} 1$.

Then

3° $E\left(e^{itU_n} \mid \mathcal{F}_0\right) \xrightarrow{\text{a.e.}} f(t,\cdot)$

if and only if

4° $E\left(W_n^{-1} \; e^{itU_n} \mid \mathcal{F}_0\right) \xrightarrow{\text{a.e.}} 1$.

Applying lemmas 1 and 2 to

$$U_n := S_n, n \in N,$$

$$W_n := \exp\left\{it \sum_{k=1}^{k_n}\left[A_{nk} + E_{n,k-1}\left(\frac{Y_{nk}}{1+Y_{nk}^2}\right)\right] + \sum_{k=1}^{k_n} E_{n,k-1}\left(g_t(Y_{nk}) \frac{Y_{nk}^2}{1+Y_{nk}^2}\right)\right\}$$

$$n \in N,$$

we obtain:

Theorem 1. If the system $(X, \underline{F})$ satisfies (C.0) - (C.4), then for every $t \in R^1$

$$E\left(e^{itS_n} \mid \mathcal{F}_0\right) \xrightarrow{\text{a.e.}} \exp\left[ita(\cdot) + \int_{-\infty}^{+\infty} g_t(x)\, dK(x,\cdot)\right] \tag{15}$$

if and only if

$$E\left(\exp\left[it\sum_{k=1}^{k_n} Y_{nk} - \sum_{k=1}^{k_n} E_{n,k-1}\left(e^{itY_{nk}}-1\right)\right]\Big|\mathcal{F}_o\right)\xrightarrow{a.e.} 1. \quad (16)$$

Corollary 1. If the system $(X,\underline{F})$ satisfies (C.0) - (C.3) then it fulfils (15) if and only if each of the equivalent conditions is satisfied:

1° (16) holds for every X^* given (in the manner described at section 4) by $\mathcal{F}_o$-measurable rv C, $K(+\infty) < C < +\infty$;

2° (16) holds for X^* given by some $_o$-measurable rv $C, K(+\infty)<C<+\infty$.

Observe that if $(X,\underline{F})$ satisfies (C.0) - (C.4), then the sequence integrated in (16) is bounded and therefore the property

$$\exp\left[it\sum_{k=1}^{k_n} Y_{nk} - \sum_{k=1}^{k_n} E_{n,k-1}\left(e^{itY_{nk}}-1\right)\right]\xrightarrow{a.e.} 1 \quad (17)$$

is sufficient for (15) to be satisfied. Then using the $*$-procedure one can prove:

Corollary 2. If the system $(X,\underline{F})$ satisfies (C.0)-(C.3) and (17), then it fulfils (15).

It seems that conditions (16) and (17) have no intuitive meaning (thanks to their generality). Now we are going to show that they are implied by more restrictive properties of rv's to be uniformly asymptotically negligible in some sense. In the first place such property is contained in the following

Theorem 2. If the system $(X, \underline{F})$ satisfies (C.0)-(C.3) and

$$\sum_{k=1}^{k_n} \left| E_{n,k-1}\left(e^{itY_{nk}}-1\right)\right|^2 \xrightarrow{a.e.} 0, \quad t\in R^1, \quad (C.5)$$

then it satisfies (15).

P r o o f: Applying the above definition of X^* we obtain

$$\left|E_{n,k-1}\left(e^{itY^*_{nk}}-1\right)\right|^2 = I\ (V_{nk}\leq C)\left|E_{n,k-1}\left(e^{itY_{nk}}-1\right)\right|^2 \leq$$

$$\leq \left|E_{n,k-1}\left(e^{itY_{nk}}-1\right)\right|^2$$

and thus we can assume (C.4) for $(X,\underline{F})$. Repeating the arguments used in the proof of Theorem 3.4 of [3] we can prove that

(a) the sequence in (C.5) is bounded from above by $2(M^2 + t)\cdot C$ and then we have

$$\sum_{k=1}^{k_n} E\left(\left|E_{n,k-1}\left(e^{itY_{nk}} - 1\right)\right|^2 \middle| \mathcal{F}_0\right) \xrightarrow{a.e.} 0, \tag{18}$$

(b) the following inequality holds

$$\left|E\left(\exp\left[it\sum_{k=1}^{k_n} Y_{nk} - \sum_{k=1}^{k_n} E_{n,k-1}\left(e^{itY_{nk}}-1\right)\right]-1 \middle| \mathcal{F}_0\right)\right| \leq$$

$$\leq \frac{1}{2}\exp\ (2 + M\cdot C)\sum_{k=1}^{k} E\left(\left|E_{n,k-1}\left(e^{itY_{nk}}-1\right)\right|^2 \middle| \mathcal{F}_0\right), \quad n\in N.$$

Then (18) implies the required result.

Now let us assume that $(X,\underline{F})$ is strongly conditionally infinitesimal i.e. for every $\varepsilon>0$

$$\max_{1\leq k\leq k_n} P_{n,k-1}\left(|X_{nk}|>\varepsilon\right) \xrightarrow{a.e.} 0. \tag{C.6}$$

Observe that if (C.6) is satisfied for every element of some strictly decreasing to zero sequence $\varepsilon_s, s\in N$, then $(X,\underline{F})$ is strongly conditionally infinitesimal. This condition implies

$$\max_{1\leq k\leq k_n} |A_{nk}| \xrightarrow{a.e.} 0,\ \tau>0;$$

$$\max_{1\leq k\leq k_n} P_{n,k-1}\left(|Y_{nk}|>\varepsilon\right) \xrightarrow{a.e.} 0,\ \varepsilon>0. \tag{19}$$

The last one is equivalent to

$$\max_{1\leq k\leq k_n} E_{n,k-1}\left(\frac{Y_{nk}^2}{1+Y_{nk}^2}\right) \xrightarrow{a.e.} 0,$$

which with (C.3) gives

$$\sum_{k=1}^{k_n}\left[E_{n,k-1}\left(\frac{Y_{nk}^2}{1+Y_{nk}^2}\right)\right]^2 \xrightarrow{a.e.} 0.$$

In [3] we have proved that if

$$\max_{1\leq k\leq k_n} \left|A_{nk}\right| \leq \frac{1}{2}, \quad n\in N, \tag{20}$$

then for every $t \in R^1$ there exists a constatnt $\eta(t) > 0$, independent of k and n, such that

$$\left|E_{n,k-1}\left(e^{itY_{nk}} - 1\right)\right| < \eta(t)\ E_{n,k-1}\left(\frac{Y_{nk}^2}{1+Y_{nk}^2}\right).$$

Because of (19) we can assume (20), eventually applying a construction similar to the $*$-procedure. Therefore we have obtained:

Corollary 3. If the system $(X,\underline{F})$ satisfies (C.0)-(C.3) and (C.6), then (15) holds.

Integrating both sides in (15) we obtain the final result of this part:

Theorem 3. If the system $(X,\underline{F})$ satisfies (C.0)-(C.3) and at least one of conditions (17), (C.5) or (C.6), then its row sums $S_n, n \in N$, converage in distribution to Mix(a,K).

6. Weaker assumptions

It would be very sophisticated if proving limit theorems for mixtures we had to assume almost sure convergence while for infinitely divisible laws we have criteria with weaker convergence in probability of relative sequences. Eagleson in [2] has shown how we can weaken the assumptions of Theorem 3. Using this method we can prove

Theorem 4. If the system $(X,\underline{F})$ satisfies (C.0) and

$$\max_{1\leq k\leq k_n} P_{n,k-1}\left(|X_{nk}|>\varepsilon\right)\xrightarrow{p} 0,\quad \varepsilon>0, \tag{C.7}$$

$$\sum_{k=1}^{k_n}\left\{A_{nk}+E_{n,k-1}\left(\frac{Y_{nk}}{1+Y_{nk}^2}\right)\right\}\xrightarrow{p} a(\cdot), \tag{C.8}$$

$$\sum_{k=1}^{k_n} E_{n,k-1}\left(\frac{Y_{nk}^2}{1+Y_{nk}^2}\, I\,(Y_{nk}<x)\right)\xrightarrow{p} K(x,\cdot) \tag{C.9}$$

for every x belonging to some countable dense set $D\subset R^1$;

$$\sum_{k=1}^{k_n} E_{n,k-1}\left(\frac{Y_{nk}^2}{1+Y_{nk}^2}\right)\xrightarrow{p} K(+\infty,\cdot), \tag{C.10}$$

then its row sums $S_n, n\in N$, converge in law to Mix(a,K).

P r o o f: It suffices to show that every subsequence $\{S_{n_k}\}\subset\{S_n\}$ contains further sequence $\{S_{n_{k_l}}\}$ which is convergent in law to Mix (a,K). Of course we can consider only the case of $\{S_{n_k}\}=\{S_n\}$. Observe that in (C.8)-(C.10) we have only countably many conditions; the same is true for (C.7) as we have remarked above. Then using the diagonal method we can choose subsequence $\{S_{n_k}\}\subset\{S_n\}$ such that all these conditions are fulfilled with almost sure convergence. Theorem 3 implies the conclusion.

7. Case of finite variances

In this part we shall consider only systems $(X,\underline{F})$ with all rv's having finite variance.

Theorem 5. If the system $(X,\underline{F})$ satisfies (C.0) and

$$\sum_{k=1}^{k_n} B_{nk} \xrightarrow{\text{a.e.}} a(\cdot), \tag{C.11}$$

$$\sum_{k=1}^{k_n} E_{n,k-1}\left(Z_{nk}^2 \, I\,(Z_{nk} < x)\right) \xrightarrow{\text{a.e.}} K(x,\cdot) \tag{C.12}$$

for every x belonging to some countable dense subset $D \subset R^1$, there exists $\mathcal{F}_0$-measurable rv $0 < C < +\infty$ such that

$$P\left(\limsup_{n\to\infty}\left[\sum_{k=1}^{k_n} E_{n,k-1}\,(Z_{nk}^2) > C\right]\right) = 0, \tag{C.13}$$

where

$$B_{nk} := E_{n,k-1}\,(X_{nk}), \quad 1 \leqslant k \leqslant k_n, \quad n \in N,$$

$$Z_{nk} := X_{nk} - B_{nk}, \quad 1 \leqslant k \leqslant k_n, \quad n \in N,$$

then each of the conditions

$$\exp\left[it\sum_{k=1}^{k_n} Z_{nk} - \sum_{k=1}^{k_n} E_{n,k-1}\left(e^{itZ_{nk}} - 1\right)\right] \xrightarrow{\text{a.e.}} 1, \quad t \in R^1, \tag{C.14}$$

$$\sum_{k=1}^{k_n} \left| E_{n,k-1}\left(e^{itZ_{nk}} - 1\right)\right|^2 \xrightarrow{\text{a.e.}} 0, \quad t \in R^1, \tag{C.15}$$

$$\max_{1\leqslant k\leqslant k_n} P_{n,k-1}\,(|Z_{nk}| > \varepsilon) \xrightarrow{\text{a.e.}} 0, \ \varepsilon > 0, \tag{C.16}$$

$$\max_{1\leqslant k\leqslant k_n} E_{n,k-1}\left(Z_{nk}^2\right) \xrightarrow{\text{a.e.}} 0, \tag{C.17}$$

is sufficient for

$$E\left(e^{itS_n} \mid \mathcal{F}_0\right) \xrightarrow{\text{a.e.}} \exp\left[ita(\cdot) + \int_{-\infty}^{+\infty} h_t(x)\,dK(x,\cdot)\right], \quad t \in R^1,$$

and hence for the convergence in law of $S_n, n \in N$, to mix(a,K).

Theorem 6. If the system $(X, \underline{F})$ satisfies (C.0) and

$$\sum_{k=1}^{k_n} B_{nk} \xrightarrow{p} a(\cdot), \tag{C.18}$$

$$\sum_{k=1}^{k_n} E_{n,k-1}\left(Z_{nk}^2\, I\,(Z_{nk} < x)\right) \xrightarrow{p} K(x,\cdot), \tag{C.19}$$

for every x belonging to some countable dense set $D \subset R^1$,

there exists $\mathcal{F}_0$-measurable a.e. finite rv $C > 0$ such that

$$\lim_{n\to\infty} P\left[\sum_{k=1}^{k_n} E_{n,k-1}\left(Z_{nk}^2\right) > C\right] = 0, \tag{C.20}$$

then each of the conditions

$$\max_{1\leq k\leq k_n} P_{n,k-1}\left(|Z_{nk}| > \varepsilon\right) \xrightarrow{p} 0,\ \ \varepsilon > 0, \qquad \text{(C.21)}$$

$$\max_{1\leq k\leq k_n} E_{n,k-1}\left(Z_{nk}^2\right) \xrightarrow{p} 0, \qquad \text{(C.21)}$$

is sufficient for the convergence in law of $S_n, n\in N$, to mix (a,K).

We omit the proofe of these theorems because they are very similar to the previous one.

7. Normal case

Let us assume that for almost every $\omega\in\Omega$

$$K(t,\omega) = \begin{cases} 0 & \text{for } t\leq 0, \\ \eta(\omega)\geq 0 & \text{for } t>0. \end{cases}$$

Then $N(a,\eta) :=$ mix(a,K) = Mix(a,K) is a mixture of normal distributions on R^1 with the chf

$$\varphi(t) = E\exp\left[ita(\cdot) - \tfrac{1}{2}t^2\eta(\cdot)\right],\quad t\in R^1.$$

Conditions (C.1)-(C.3) in this case are equivalent to

$$\sum_{k=1}^{k_n} \left\{ A_{nk} + E_{n,k-1}\left(\frac{Y_{nk}}{1+Y_{nk}^2}\right)\right\} \xrightarrow{\text{a.e.}} a(\cdot), \tag{CN.1}$$

$$\sum_{k=1}^{k_n} E_{n,k-1}\left(\frac{Y_{nk}^2}{1+Y_{nk}^2} I\left(|Y_{nk}|>\varepsilon\right)\right) \xrightarrow{\text{a.e.}} 0,\ \varepsilon>0, \tag{CN.2}$$

$$\sum_{k=1}^{k_n} E_{n,k-1}\left(\frac{Y_{nk}^2}{1+Y_{nk}^2}\right) \xrightarrow{\text{a.e.}} \eta(\cdot). \tag{CN.3}$$

Condition (CN.2) is equivalent to

$$\sum_{k=1}^{k_n} P_{n,k-1}\left(|Y_{nk}|>\varepsilon\right) \xrightarrow{\text{a.e.}} 0,\ \varepsilon>0. \tag{CN.4}$$

One can easily prove that

Theorem 7. If the system $(X, \underline{\mathbf{F}})$ satisfies (C.0) and

$$\sum_{k=1}^{k_n} P_{n,k-1}\left(|X_{nk}|>\varepsilon\right) \xrightarrow{\text{a.e.}} 0,\ \varepsilon>0, \tag{CN.5}$$

$$\sum_{k=1}^{k_n} E_{n,k-1}\left(X_{nk}\, I\left(|X_{nk}|\leqslant\varepsilon\right)\right) \xrightarrow{\text{a.e.}} a(\cdot),\ \varepsilon>0, \tag{CN.6}$$

$$\sum_{k=1}^{k_n} E_{n,k-1}\left(X_{nk}^2\, I\left(|X_{nk}|\leqslant\varepsilon\right)\right) - \sum_{k=1}^{k_n}\left[E_{n,k-1}\left(X_{nk}\, I\left(|X_{nk}|\leqslant\varepsilon\right)\right)\right]^2 \xrightarrow{\text{a.e.}} \eta(\cdot),\ \varepsilon>0 \tag{CN.7}$$

then the sums $S_n, n\in N$, converge in law to $N(a, \eta)$. ■

We can prove more, namely, that under the assumption (C.6) conditions (CN.5)-(CN.7) are equivalent to (CN.1)-(CN.3). If we put the convergence in probability into (CN.5)-(CN.7), Then the conclusion also holds.

If all rv's of the system $(\mathcal{X},\underline{F})$ have finite variances then the straight reformulation of Theorems 5 and 6 gives sufficient conditions, which improve the results obtained in [2].

8. Convergence of randomly weighted sums of independent random variables

Let there be given a sequence $X_n, n \in N$, of independent, identically distributed (i.i.d.) rv's.

Jamison, Orey and Pruitt [4] have considered the problem of the convergence in probability of the following weighted sums

$$S_n := \sum_{k=1}^{n} \alpha_{nk} X_k, \quad n \in N, \tag{21}$$

where

$$\alpha_{nk} := \frac{\omega_k}{\sum_{i=1}^{n} \omega_i}, \quad 1 \leqslant k \leqslant n,$$

and $\omega_k, k \in N$, is a sequence of positive real numbers such that

$$\max_{1 \leqslant k \leqslant n} \alpha_{nk} \xrightarrow[n \to \infty]{} 0. \tag{22}$$

They have obtained the following result:

Theorem 8 [4]. The weighted sums (21) converge in probability to the finite constant m for all sequences $\omega_k, k \in N$, satisfying (22) if and only if

$$\lim_{T \to +\infty} T\, P\left[|X_1| \geqslant T \right] = 0 \tag{23}$$

and

$$\lim_{T \to +\infty} E\left(X_1\, I\left(|X_1| \leqslant T \right) \right) = m. \tag{24}$$

(Observe that the conditions (23) and (24) are equivalent to the existence of the derivative of the characteristic function of X_1 and this property is weaker than $E\,|X_1| < +\infty$).

Now one can consider a more general problem of the asymptotic behaviour of sums (21) in which constant weights are replaced by positive rv's. In this situation a method of characteristic function is completely useless. The first result in this area was obtained by Brown, Eagleson and Fisher [5]. They have proved the following generalisation of Theorem 8:

Theorem 9 [5] . Let us assume (23) and (24). If for each $n \in N$ the nonnegative rv's $\alpha_{n1}, \alpha_{n2}, \ldots, \alpha_{nn}$ are independent of the sequence $X_m, m \in N$, (but possibly dependent among themselves) and they satisfy the following conditions:

$$\max_{1 \leq k \leq n} \alpha_{nk} \xrightarrow{p} 0, \tag{25}$$

$$\lim_{n \to \infty} P\left[\sum_{k=1}^{n} \alpha_{nk} > C\right] = 0 \tag{26}$$

for some constant $0 < C < +\infty$, then

$$\sum_{k=1}^{n} \alpha_{nk} (X_k - m) \xrightarrow{p} 0,$$

In particular, if

$$\sum_{k=1}^{n} \alpha_{nk} \xrightarrow{p} \alpha = \text{const}, \tag{27}$$

then

$$\sum_{k=1}^{n} \alpha_{nk} X_k \xrightarrow{p} \alpha m. \blacksquare$$

If we omit condition (27), then the conclusion fails. Brown, Eagleson and Fisher [5] have shown that if, instead of (27), we assume another natural condition, then for a special class of $X_m, m \in N$, we obtain asymptotical normality of randomly weighted sums (21). More precisely:

<u>Theorem 10</u> [5]. Let $X_m, m \in N$, be a sequence of i.i.d. rv's with zero mean and unit variance. Assume that for each $n \in N$ the nonnegative rv's $\alpha_{nk}, 1 \leq k \leq n$, are independent of $X_m, m \in N$ (or, more generally, for each $1 \leq k \leq n$ the rv's $\alpha_{n1}, \dots, \alpha_{nk}$ are independent of X_k, $X_{k+1}, \dots$). If these random weights satisfy (25) or, equivalently

$$\max_{1 \leq k \leq n} \alpha_{nk}^2 \xrightarrow{p} 0, \text{ and } \sum_{k=1}^{n} \alpha_{nk}^2 \xrightarrow{p} \sigma^2,$$

then the weighted sums (21) converge in distribution to the normal law $\mathcal{N}(0, \sigma^2)$.

(Observe that for such sequences $X_m, m \in N$, Theorem 9 is a consequence of Theorem 10, because

$$\sum_{k=1}^{n} \alpha_{nk}^2 \leq (\max_{1 \leq k \leq n} \alpha_{nk}) \sum_{k=1}^{n} \alpha_{nk} \xrightarrow{p} 0).$$

This last theorem is an example of the application of the general limit theorems for martingale difference arrays proved by Brown and Eagleson in [1]. Now we can give similar results applying limit theorems for mixtures obtained in the present paper.

Let us take into account the conditions which appear in Th.6.

If $K(\cdot, \omega) \equiv 0$ for almost all $\omega \in \Omega$, then mix $(a, 0)$ is a mixture of degenerate distributions on R^1 with the characteristic function

$$\varphi(t) = E\left(e^{ita(\cdot)}\right), \quad t \in R^1,$$

and in this way we can obtain all possible distributions on R^1 induced by random variables defined on Ω. Therefore, if (C.18) and

$$\sum_{k=1}^{n} E_{n,k-1}\left(Z_{nk}^2\right) \xrightarrow{p} 0 \tag{28}$$

are satisfied, then the sums $\sum_{k=1}^{n} X_{nk}$, $n \in N$, converge in law to $a(\cdot)$.

Moreover, by corollary 6.5 of [3], condition (28) implies

$$\sum_{k=1}^{n} Z_{nk} \xrightarrow{p} 0$$

and then, by (C.18), we obtain

$$\sum_{k=1}^{n} X_{nk} \xrightarrow{p} a(\cdot).$$

Now we apply these facts for randomly weighted sums. Let us define

$$\mathcal{F}_{n0} := \mathcal{B}(a, \alpha_{n1}, \ldots, \alpha_{nn}),$$

$$\mathcal{F}_{n,k} := \mathcal{B}(a, \alpha_{n1}, \ldots, \alpha_{nn}, X_1, \ldots, X_k),$$

$$X_{nk} := \alpha_{nk} X_k,$$

for $1 \leq k \leq n$, $n \in \mathbb{N}$. Of course $a(\cdot)$ is independent of X's and then

$$B_{nk} = E\left(X_{nk} \mid \mathcal{F}_{n,k-1}\right) = \alpha_{nk}\, EX_k,$$

$$E\left(Z_{nk}^2 \mid \mathcal{F}_{n,k-1}\right) = \alpha_{nk}^2 \ \mathrm{Var} X_k.$$

We have to consider two possibilities:

1° $$E\, X_k = 0, \ k \in \mathbb{N}.$$

In this case condition (C.18) can be satisfies only for $a(\cdot) = 0$: (28) is fulfilled, if

$$\sum_{k=1}^{n} \alpha_{nk}^2 \xrightarrow{p} 0, \tag{29}$$

and then we are in the situation described in Theorem 10 for degenerate normal law.

2° $EX_k \neq 0$, $k \in \mathbb{N}$. We can assume, for simplicity, that

$$E\, X_k = \mathrm{Var}\, X_k = 1, \ k \in \mathbb{N}.$$

In this case

$$\sum_{k=1}^{n} \alpha_{nk} \xrightarrow{p} a(\cdot) \tag{30}$$

implies (C.18), and similarly we obtain (28) from (29), which under the assumption (30) is equivalent to (25). In this way we have obtained the following result:

<u>Theorem 11.</u> Let there be given a double sequence of (possibly dependent) nonnegative rv's $\alpha_{nk}, 1 \leq k \leq n, n \in N$, which satisfies (25) and (30) for some rv $a(\cdot)$. Moreover, let there be given a sequence $X_m, m \in N$, of independent rv's such that $\alpha_{nk}, 1 \leq k \leq n$ are independent of $X_m, m \in N$, for all $n \in N$.

If $EX_m = VarX_m = 1$, $m \in N$, then the weighted sums (21) converge in probability to $a(\cdot)$. If $EX_m = 0$, $m \in N$, then these sums converge in probability to zero.

Unfortunately, in all above theorems the normal law is prefered or we obtain more stringent convergence in probability. It is an open question to give sufficient conditions for the weak convergence of weighted sums to an arbitrary infinitely divisible distribution. Recently Beśka has proved in [6] the following result:

<u>Theorem 12.</u> Let there be given a double sequence X_{nk}, $1 \leq k \leq n$, $n \in N$, of row-wise independent nonnegative rv's such that

$$E\,X_{nk} = \lambda > 0, \quad 1 \leq k \leq n, \quad n \in N.$$

Moreover, let us take a double sequence α_{nk}, $1 \leq k \leq n, n \in N$, of nonnegative rv's such that for each $n \in N$ its n-th row is independent of $X_{n1}, \dots, X_{nn}$. If conditions (25), (27) for $\alpha = 1$ and

$$\max_{1 \leq k \leq n} E(X_{nk} I\,(\,|\alpha_{nk}(\omega) X_{nk} - 1| > \varepsilon)) \xrightarrow{p} 0 \tag{31}$$

are satisfied, then the sums $\sum_{k=1}^{n} \alpha_{nk} X_{nk}$, $n \in N$, converge in law to the Poisson distribution with the parameter λ.

Observe that (25) and (27) imply (29). This is the reason the a consideration of the double sequence of X's and together with (31) it shows some difficulties for other sensible results.

References

[1] B r o w n, B. M., E a g l e s o n, G. K., Martingale convergence to infinitely divisible laws with finite variances, Trans. Amer. Math. Soc., 162, 449-453 (1971).

[2] E a g l e s o n, G. K., Martingale convergence to mixtures of infinitely divisible laws. Ann. of Prob., 3, 557-562 (1975).

[3] Kłopotowski, A., Limit theorem for sums of dependent random vectors in R^d. Diss. Math. CLI 1-58 (1977).

[4] J a m i s o n, B., O r e y S., P r u i t t, W., Convergence of weighted averages of independent random variables. Z. Wahrscheinlichkeitstheorie verw. Gebiete, 4, 40-44 (1965).

[5] B r o w n, B. M., E a g l e s o n, G. K., F i s h e r, N. I. On the central limit problem for sums with random coefficients, Z. Wahrscheinlichkeitstheorie und verw. Gebiete 27, 171-178 (1973).

[6] B e ś k a, M., Poisson distribution as a limit law for sums of dependent random variables, Preprint 6/78, Inst. of Math., N. Copernicus Univ. Toruń.

CONDITIONAL EXPECTATIONS OF SELECTORS AND JENSEN'S INEQUALITY

by

Andrzej Kozek and Zdzisław Suchanecki

Polish Academy of Sciences

and

Technical University of Wrocław

1. Summary and introduction

Let $(T, \mathcal{A}, P)$ be a P-complete probability space, $\mathcal{B}$ a P-complete σ-subfield of $\mathcal{A}$ and X a separable locally convex Fréchet space. In the paper we prove that if a strongly $\mathcal{A}$-measurable function $x(\cdot)$ with values in X is a selector of a random closed convex set $C(t)$ which varies in a $\mathcal{B}$-measurable manner, then the same holds true for any version of its conditional expectation $E^{\mathcal{B}}x(\cdot)$ (Theorem 1(i), Section 3). A version of conditional expectation can run, on some $\mathcal{B}$-measurable subset of T, along extremal points of $C(t)$ only if it coincides on this subset with the original function (Theorem 1 (ii), Section 3).

These results are used to obtain an extended version of Jensen's inequality for a random convex function $f(t, \cdot)$ defined on a random convex subset $S(t)$ of a separable locally convex Fréchet space V, where the convexity of f is with respect to a random convex cone $K(t)$ in a separable locally convex Fréchet space U. The only essential assumptions needed here are $\mathcal{B}\otimes\mathcal{B}(V \times U)$-measurability of the epigraph of f and the closedness of its t-sections. If $f(t,\cdot)$ is strictly convex, then the obtained Jensen's inequality is also strict (Theorem 2, Section 4).

If the original function $x(\cdot)$ is a selector of the random convex set on a subset A of T, where $A \in \mathcal{A}$, then any version of $E^{\mathcal{B}}(1_A x(\cdot)) / E^{\mathcal{B}} 1_A$ is a selector of $C(\cdot)$ on the smallest $\mathcal{B}$-measurable set containing A. This function can run, on an $\mathcal{B}$-measurable subset of A, along extremal points of C(t) only if it coincides on this subset of A with the original function $x(\cdot)$ (Theorem 1', Section 3).

The obtained results described above are analogous to those obtained by Pfanzagl [8] and Daures [2]. Pfanzagl considered the particular case when $X = R^n$ and $C(t) = C$ is a constant Borel subset of R^n. Daures proved the corresponding results assuming that $C(\cdot)$ is a $\mathcal{B}$-measurable multifunction such that C(t) does not contain lines and is a weakly locally compact and convex subset of a locally convex separable Fréchet space X. In the case when X is a Banach space this result can also be derived from a result of Hiai and Umegaki [5; Theorem 5.3.1°] provided C(t) is contained in a ball with a varying and integrable radius, or, implicitly, if $C(\cdot)$ contains $\mathcal{B}$-measurable Bochner integrable selectors.

We are interested here in the case when X is an infinite dimensional separable locally convex Fréchet space and $C(\cdot)$ is a convex closed $\mathcal{B}$-measurable multifunction. The aim of this paper is to prove that the Daures's theorems remain valid without the restrictive and inconvenient assumption that C(t) is weakly locally compact and does not contain lines. Our proof is, however, completely different from that of Daures since the method he used is based on a Klee-Olech characterization of convex sets which is useless in the present framework.

2. Preliminaries and notation

Let $(T, \mathcal{A}, P)$ be a probability space with a complete probability measure P and $\mathcal{B}$ a σ-subfield of $\mathcal{A}$ containing all P-null sets. Since we consider separable locally convex metrizable complete vector spaces only, there is no danger of confusion to call them, simply, Fréchet spaces. Denote by X a Fréchet space and by $\mathcal{B}(X)$ the σ-field of its Borel subsets (we shall use analogous notations for σ-fields of Borel subsets of other Fréchet spaces). Let C be a $\mathcal{B} \otimes \mathcal{B}(X)$-measurable subset of $T \times X$ such that $C(t) = \{x \in X : (t,x) \in C\}$ is for P-a.e. $t \in T$ a closed non-empty subset of X. Then, multifunction $C(\cdot) : t \to C(t)$ from T into the space of closed convex subsets of X has its graph equal to C, i.e., $\mathrm{Gr}\, C(\cdot) = \{(t,x) \in T \times X : x \in C(t)\} = C$. An $\mathcal{A}$-measurable function $x(\cdot) : T \to X$ is called a selector of $C(\cdot)$ on set A, $A \in \mathcal{A}$, whenever $x(t) \in C(t)$ for P-a.e. $t \in A$. In some papers (see e.g. [10]) such a $x(\cdot)$ is called a P-a.e. selector. Since we shall deal with P-a.e. selectors only, for simplicity of notation we omit the term "P-a.e.".

For $A \in \mathcal{A}$ we denote by $c(A,\cdot)$ any fixed version of conditional probability of A, i.e., $c(A,\cdot) \in E^{\mathcal{B}} 1_A(\cdot)$, and by $c^-(A,\cdot)$ a function given by

$$c^-(A,t) = \begin{cases} 1/c(A,t) & \text{if } c(A,t) > 0, \\ 0 & \text{elsewhere.} \end{cases} \tag{1}$$

Let $x(\cdot) : T \to X$ be a weakly measurable function. It is known that $x(\cdot)$ is $(\mathcal{A}, \mathcal{B}(X))$-measurable. A measurable function $x(\cdot)$ is called strongly integrable whenever

$$\int_T p(x(t))\, P(dt) < \infty \tag{2}$$

holds for each continuous seminorm $p(\cdot)$ on X. Note that condition (2) is equivalent to the following one

$$\sup \left\{ \langle x(t), x' \rangle : x' \in V_i^o \right\} \leq h_i(t), \quad i=1,2,\ldots, \tag{3}$$

where $h_i(\cdot)$ are P-integrable functions, x' stands for a continuous linear functional on X, V_i is the i-th element of a base of neighbourhoods of zero in X consisting of balanced convex sets such that $V_{i+1} \subset V_i$ and, finally, V_i^o denotes the polar set of V_i. The Banach-Grothendieck theorem ([3], Corollary 8.5.2) implies that if $x(\cdot)$ is strongly integrable, then there exists $x_0 \in X$ such that for every continuous linear functional $x' \in X'$

$$\int_T \langle x(t), x' \rangle \, P(dt) = \langle x_0, x' \rangle$$

holds ([1] and [11], Theorem 3 (1)). Vector x_0 is called a strong integral of $x(\cdot)$ and it is denoted by $\int_T x(t) \, P(dt)$.

It is known that for any σ-subfield $\mathcal{B}$ of $\mathcal{A}$ and for any strongly integrable function $x(\cdot)$ there exists a $\mathcal{B}$-measurable strongly integrable function $y(\cdot)$ such that

$$\int_B x(t) \, P(dt) = \int_B y(t) \, P(dt) \tag{4}$$

holds for each $B \in \mathcal{B}$ (cf. [1]). The function $y(\cdot)$ is called a version of the conditional expectation of $x(\cdot)$ with respect to $\mathcal{B}$. The equivalence class of functions equal P-a.e. to $y(\cdot)$ will be denoted by $E^{\mathcal{B}} x(\cdot)$.

Since X is a Polish space and P is complete on $\mathcal{B}$, a closed multifunction $C(\cdot)$ is $\mathcal{B}$-measurable if and only if its graph Gr $C(\cdot)$ belongs to $\mathcal{B} \otimes \mathcal{B}(X)$ ([10], Theorem 4.2 (f)). So, for the purposes of this paper, we can adopt this equivalence for the definition of $\mathcal{B}$-measurability of a multifunction with values in the space of closed subsets of X.

For closed convex multifunction $C(\cdot)$, and $t \in T$, we denote by $C_e(t)$ the set of extremal points of $C(t)$.

3. Conditional expectations of selectors

Theorem 1. Let $x(\cdot)$ be an $\mathcal{A}$-measurable, strongly integrable function from T into X which is a selector on the whole T of a $\mathcal{B}$-measurable closed convex multifunction $C(\cdot)$. If $y(\cdot)$ is a version of $E^{\mathcal{B}}x(\cdot)$, then

(i) $y(\cdot)$ is a selector of $C(\cdot)$ on T,

(ii) $x(t) = y(t)$ P-a.e. on set

$A_e = \{t \in T : y(t) \in C_e(t)\}$.

For $A \in \mathcal{A}$, let

$$y_A(t) = c^-(A,t)y'_A(t), \tag{5}$$

where $y'_A(\cdot) \in E^{\mathcal{B}}(1_A(\cdot)x(\cdot))$ and $c^-(A,\cdot)$ is given by (1).

Theorem 1'. Let $x(\cdot)$ be an $\mathcal{A}$-measurable, strongly integrable function from T into X which is a selector on A of a $\mathcal{B}$-measurable closed convex multifunction $C(\cdot)$, where $A \in \mathcal{A}$ and $P(A) > 0$. Then

(i) $y_A(\cdot)$ is a selector of $C(\cdot)$ on set $\{t \in T : c(A,t) > 0\}$ and hence on A,

(ii) $x(t) = y_A(t)$ P-a.e. on set $A \cap \{t \in T : y_A(t) \in C_e(t)\}$.

If A appearing in Theorem 1' is equal to T, then Theorem 1' reduces to Theorem 1. Moreover, it is easy to verify that $y_A(\cdot)$ is a version of $E^{\mathcal{B}}_{P'}x(\cdot)$, where P' is a probability measure on $\mathcal{A}$ given by $P'(D) = P(D \cap A)/P(A)$, $D \in \mathcal{A}$.

Consequently, Theorem 1 implies immediately Theorem 1' (ii) and a "weak" version of Theorem 1' (i) concerning the P-a.e. selectors on set A only. However, since $y_A(\cdot)$ is $\mathcal{B}$-measurable and $C \in \mathcal{B} \otimes \mathcal{B}(X)$ we infer that $\{t \in T : y(t) \in C(t)\} \in \mathcal{B}$ ([10] Theorem 4.2 (g) (i), (v), (vii)). Because $\{t \in T : c(A,t) > 0\}$ is the smallest (exact to sets of P-measure zero) $\mathcal{B}$-measurable set containing A ([4] Lemma p. 443) we obtain then that $y_A(\cdot)$ is a selector of $C(\cdot)$ on set $\{t : c(A,t) > 0\}$. Thus we get the following.

Proposition. Theorem 1 (i) implies Theorem 1′(i) and, conversely, Theorem 1 (i) follows from Theorem 1′ (i). Moreover, Theorem 1 (ii) implies Theorem 1′ (ii) and the converse implication is valid, too.

Our interest in Theorem 1′ is motivated by its use in the proof of Theorem 1 (ii). The differences between Theorem 1 and Theorem 1′ will be illuminated by Example 3 in Section 5.

Proof of Theorem 1, part (i). Firstly suppose that $C(\cdot)$ admits a selector $x(\cdot)$ on a set A, $A \in \mathcal{A}$, which is a step function, i.e. $x(t) = \sum_{i=1}^{k} 1_{A_i}(t)\, x_i$, where $A_i \in \mathcal{A}$, $A_i \cap A_j = \emptyset$ when $i \neq j$, $\bigcup_{i=1}^{k} A_i = A$, $P(A_i) > 0$ and $x(t) \in C(t)$ for P-a.e. $t \in A$. Let

$$y_A(t) = \sum_{i=1}^{k} x_i f_i(t), \tag{6}$$

where

$$f_i(t) = c^{-}(A,t) \cdot c(A_i,t).$$

Denote by T_0 the set of all $t \in T$ satisfying the following conditions

1) $f_i(t) \in [0,1]$ for $i = 1,\dots,k$,
2) if $f_i(t) > 0$, then $t \in C^{x_i}$,
3) $\sum_{i=1}^{k} f_i(t) = 1$,

where $C^x = \{t \in T : (t,x) \in C\}$. Since $\{t \in T : c(A_i,t) > 0\}$ is the smallest $\mathcal{B}$-measurable set containing A_i we have

$$C^{x_i} \supset \{t \in T : f_i(t) > 0\} \supset A_i.$$

Hence $T_0 = \{t \in T : c(A,t) > 0\}$ P-a.s. (we write "$A_1 \supset A_2$ P-a.s." if $P(A_2 \setminus A_1) = 0$ and "$A_1 = A_2$ P-a.s." if $A_1 \supset A_2$ P-a.s. and $A_2 \supset A_1$ P-a.s.). Thus, if $f_i(t) > 0$ and $t \in T_0$, then $x_i \in C(t)$. Hence convexity of $C(t)$ yields $\sum_{i=1}^{k} f_i(t)x_i \in C(t)$. Therefore, $y_A(\cdot)$ given by (6) is a selector of $C(\cdot)$ on T_o. Hence it is a selector on A, too. Thus, we have proved

Lemma 1. If $x(\cdot)$ is a step function which is a selector of $C(\cdot)$ on A, $A \in \mathcal{G}$, and $x(t) = 0$ for $t \in T \setminus A$, then $y_A(\cdot)$ given by (6) is a selector of $C(\cdot)$ on T_o and hence on A.

It is clear that for step functions Lemma 1 coincides with Theorem 1$'$ (i).

The remaining part of the proof is divided into several steps.

1. Construction of a decreasing sequence of multifunctions $C^{(m)}(\cdot)$. Denote by $C^{(m)}$ a subset of $T \times X$ given by $C^{(m)} = \{(t,x) \in T \times X : x \in cl(C(t) + V_m)\}$, where V_m is the m-th element of the countable base of balanced convex neighbourhoods of zero in X such that $V_{m+1} \subset V_m$. Then $C^{(m)} \in \mathcal{B} \otimes \mathcal{B}(X)$ because $C^{(m)}(\cdot)$ admits a Castaing representation([10] Theorem 4.2 (g), (v), (ix)), int $C^{(m)}(t) \supset C(t)$ P-a.s. and $x(t) \in$ int $C^{(m)}(t)$ for P-a.e. $t \in T$. For further considerations let m be fixed.

2. Construction of an increasing sequence of sets T_n such that $P\left(\bigcup_{n=1}^{\infty} T_n\right) = 1$. Let x_i, $i = 1,2,\ldots$ be a dense subset of X. Let

$$T_{n,1} = T'_{n,1}, \quad T_{n,k} = T'_{n,k} \setminus \bigcup_{i=1}^{k-1} T_{n,i} \qquad \begin{matrix} n = 1,2,\ldots \\ k = 2,3,\ldots \end{matrix}$$

and

$$T'_{n,i} = \left\{t \in T : x(t) \in x_i + V_n\right\} \qquad n,i = 1,2,\ldots$$

The $T_{n,i}$'s are $\mathcal{G}$-measurable, pairwise disjoint and $\bigcup_{i=1}^{\infty} T_{n,i} = T$ for every n. Let

$$N_n = \inf\left\{k : P\left(\bigcup_{i=1}^{k} T_{n,i}\right) \geqslant 1 - 1/2^n\right\}$$

and put

$$T_n = \bigcap_{l \geqslant n} \bigcup_{i=1}^{N_l} T_{l,i} .$$

By the Borel-Cantelli lemma we have $P\left(\bigcup_{n=1}^{\infty} T_n\right) = 1$.

3. Construction of countable partitions of T. For each j=1,2,... we define a partition of T onto a countable collection of disjoint measurable sets. In the j-th partition we consider sets $T_j, T_{j+1}, \ldots$ Set T_j is partitioned by sets $T_{j,i}$ $(i = 1,\ldots,N_j)$, set $T_{j+1} \setminus T_j$ is partitioned by sets $T_{j+1,i}$ $(i = 1,\ldots,N_{j+1})$ and finally sets $T_{j+k+1} \setminus \setminus T_{j+k}$ $(k \geqslant 1)$ are partitioned by sets $T_{j+k+1,i}$ $(i = 1,\ldots, N_{j+k+1})$. In this way the j-th partition of T consists of sets $\left\{T^k_{j,i} : k=0,1,\ldots, i = 1,2,\ldots,N_k\right\}$, where

$$T^0_{j,i} = T_{j,i} \cap T_j, \qquad i = 1,2,\ldots,N_j,$$

$$T^k_{j,i} = T_{j+k,i} \cap (T_{j+k} \setminus T_{j+k-1}), \qquad \begin{array}{l} k = 1,2,\ldots, \\ i = 1,2,\ldots,N_{j+k}. \end{array}$$

Note that in the j-th partition the T_n's are divided onto a finite number of disjoint sets. Moreover, if $k > 0$, then the j-th and the (j+k)-th partitions of set $T \setminus T_{n+k}$ coincide.

4. Countable valued functions $x_j(\cdot)$ approximating $x(\cdot)$. Given the j-th partition we put $x_j(t) = x_i$ whenever $t \in \bigcup_{k=0}^{\infty} T^k_{j,i}$, $i = 1,2,\ldots$

From the definition of sets $T_{n,i}$ we have for each j $x_j(t) \in x(t) + V_j$.

Hence $x_j(t)$ converges to $x(t)$ for P-a.e. $t \in T$. Moreover, if $t \in T_{j+k}$, then $x_j(t) \in x(t) + V_{j+k}$. This implies that the functions $x_j(\cdot)$ are strongly integrable. Indeed, given V_n we have for each $t \in T_{j \vee n}$

$$\sup_{z \in V_n^0} | \langle x_j(t), z \rangle | \leq \text{const} \tag{7}$$

because $x_j(T_{j \vee n})$ is a finite set. If $t \notin T_{j \vee n}$, then $x_j(t)-x(t) \in V_n$ and hence

$$\sup_{z \in V_n^0} | \langle x_j(t), z \rangle | \leq h_n(t) + 1. \tag{8}$$

Thus, (7) and (8) imply that $x(\cdot)$ is strongly integrable.

5. Step functions $x_{j,k}(\cdot)$ approximating $x_j(\cdot)$. Functions $x_j(\cdot)$ can be approximated by step functions of the form

$$x_{j,1}(t) = x_j(t) \cdot 1_{T_{j+1}}(t).$$

Note that if $s > r$, then

$$x_{j,r}(t) - x_{j,s}(t) = x_j(t) 1_{T_{j+s} \setminus T_{j+r}}(t) \in \text{cl } V_{j+r}$$

holds for P-a.e. $t \in T$. Hence $x_{j,r}(\cdot) - x_{j,s}(\cdot)$ is a selector of cl V_{j+r} on the whole T.

6. Functions $y_{j,1}(\cdot)$. Let us put

$$y_{j,1}(t) = \sum_{k=0}^{1} \sum_{i=1}^{N_{j+k}} x_i c(T_{j,i}^k, t),$$

where $c(A, \cdot) \in E^{\mathcal{B}} 1_A(\cdot)$. In view of step 5 and Lemma 1 we obtain that $y_{j,r}(\cdot) - y_{j,s}(\cdot)$ is a selector of cl V_{j+r} on T. Hence $y_{j,1}(t)$ is a Cauchy sequence for P-a.e. $t \in T$.

7. Functions $y_j(t) = \lim_{l\to\infty} y_{j,l}(t)$ are selectors of $C^{(m)}(\cdot)$, $j \geq m$. By step 6 functions $y_j(\cdot)$ given by

$$y_j(t) = \sum_{k=0}^{\infty} \sum_{i=1}^{N_{j+k}} x_i \cdot c(T^k_{j,i}, t) = \lim_{l\to\infty} y_{j,l}(t)$$

are well defined. If $j \geq m$, then the step functions are selectors of $C^{(m)}(\cdot)$ on T_{j+l} (compare steps 4 and 5). Consequently, Lemma 1 implies that functions $y'_{j,l}(\cdot)$ are selectors of $C^{(m)}(\cdot)$ on T_{j+l}, where

$$y'_{j,l}(\cdot) = c^-(T_{j+l}, \cdot) y_{j,l}(\cdot)$$

and $c^-(.,.)$ is given by (1). Since $\lim_{l\to\infty} c^-(T_{j+l}, t) = 1$ and $\lim_{l\to\infty} y_{j,l}(t) = y_j(t)$ hold P-a.e. we obtain

$$y_j(t) = \lim_{l\to\infty} y'_{j,l}(t).$$

Hence $y_j(\cdot)$ is a limit of functions which are selectors of $C^{(m)}(\cdot)$ on an increasing sequence of sets. Hence $y_j(\cdot)$ is a selector of $C^{(m)}(\cdot)$ on T.

8. Function $y(t) = \lim_{j\to\infty} y_j(t)$ is a well defined selector of $C(\cdot)$. Functions $x_j(\cdot) - x_{j+k}(\cdot)$ are step functions (cf. step 3) which take values in $2V_j$ (cf. step 4). Therefore, $y_j(\cdot) - y_{j+k}(\cdot)$ take values in $2\mathrm{cl}\, V_j$ by Lemma 1. This shows that $y_j(\cdot)$ is a Cauchy sequence. We put

$$y(t) = \lim_{j\to\infty} y_j(t). \tag{9}$$

It is clear that $y(\cdot)$ is $\mathcal{B}$-measurable. Since $y_j(\cdot)$ is a selector of the closed multifunction $C^{(m)}(\cdot)$ for $j \geq m$ we obtain that $y(\cdot)$ is a selector of $C^{(m)}(\cdot)$ too. Thus $y(t) \in \bigcap_{m=1}^{\infty} C^{(m)}(t) = C(t)$ for P-a.e. $t \in T$.

9. $y(\cdot) \in E^{\mathcal{B}}x(\cdot)$. The definition of $y(\cdot)$ given in step 8 implies that $y_j(\cdot) - y(\cdot)$ is a selector of $2\,\mathrm{cl}\,V_j$. For a given n we take $j > n$ such that $2\,\mathrm{cl}V_j \subset V_n$. If $z \in V_n^0$, then we have

$$<x_j(\cdot), z> = <x(\cdot), z> + <x_j(\cdot)-x(\cdot), z> \leqslant h_n(\cdot)+1,$$

$$E^{\mathcal{B}}<x_j(\cdot), z> = <y_j(\cdot), z> \leqslant E^{\mathcal{B}}h_n(\cdot) + 1,$$

$$<y(\cdot), z> = <y_j(\cdot), z> + <y(\cdot) - y_j(\cdot), z> \leqslant E^{\mathcal{B}}h_n(\cdot) + 2.$$

This proves that $y(\cdot)$ is strongly integrable. Now we note that functions $y_{j,1}(\cdot)$ and $x_{j,1}(\cdot)$ satisfy condition (4). In view of the Dominated Convergence Theorem we then obtain that (4) is also valid for $y(\cdot)$ and $x(\cdot)$.

So we have proved that $y(\cdot)$ is a version of $E^{\mathcal{B}}x(\cdot)$ and that it is a selector of $C(\cdot)$.

Proof of part (ii) of Theorem 1. Firstly we shall show that set

$$A_e = \left\{ t \in T : y(t) \in C_e(t) \right\}$$

belongs to $\mathcal{B}$. Denote by C^2 and D subsets of $T \times X \times X$ defined by

$$C^2 = \left\{ (t, x, y,) \in T \times X \times X : (x, y) \in C(t) \times C(t) \right\}$$

and

$$D = \left\{ (t, x, x) \in T \times X \times X : t \in T, \quad x \in X \right\},$$

respectively.

Moreover, we define a multifunction $y^2(\cdot)$ from T into the space of closed subsets of $X \times X$ by

$$y^2(t) = \left\{ (x, y) \in X \times X : \tfrac{1}{2}(x+y) = y(t) \right\}.$$

From Theorem 4.2 (g) in [10] it easily follows that $C^2 \setminus D \in \mathcal{B} \otimes \mathcal{B}(X \times X)$ and that $y^2(\cdot)$ is a $\mathcal{B}$-measurable multifunction. Since

$$A_e^c = \left\{ t \in T : y(t) \notin C_e(t) \right\} = \left\{ t \in T : y^2(t) \cap (C^2(t) \setminus D(t) \neq \emptyset \right\}$$

we obtain that $A_e^c \in \mathcal{B}$ (see [10]; Theorem 4.2 (g) (iii), (vii)). Hence, A_e, the complement of A_e^c, belongs to $\mathcal{B}$ too.

Now we return to the proof of part (ii) of Theorem 1. We shall prove that $1_{A_e}(t)y(t) = 1_{A_e}(t)x(t)$ for P-a.e. $t \in T$.

It is enough to show that

$$\int_{A_e \cap A} y(t)P(dt) = \int_{A_e \cap A} x(t)P(dt)$$

holds for every $A \in \mathcal{A}$. Since $A_e \in \mathcal{B}$ this is obviously true when $P(A) = 0$ or $P(A) = 1$. If $P(A) \in (0,1)$, then the following argumentation similar to that of Pfanzagl ([8] p. 492) may be applied (see also [2]).

Let $P(A) \in (0,1)$. Then $x(\cdot)$ is a selector of $C(\cdot)$ both on A and on its complement A^c. By Theorem 1$'$(i) we obtain that functions $y_A(\cdot)$ and $y_{A^c}(\cdot)$ given by (5) are selectors of $C(\cdot)$ on sets $\{t \in T : c(A,t) > 0\}$ and $\{t \in T : c(A^c,t) > 0\}$, respectively. There is no loss of generality in assuming that $c(A,\cdot)$ and $c(A^c,\cdot)$ are such versions of $P(A|\mathcal{B})$ and $P(A^c|\mathcal{B})$, respectively, that $c(A,t) \in [0,1]$, $c(A^c,t) \in [0,1]$ and $c(A,t) + c(A^c,t) = 1$ for every $t \in T$. Let

$$y'(t) = c(A,t)y_A(t) + c(A^c,t)y_{A^c}(t).$$

Then, by (1) and (5) we obtain that $y'(\cdot)$ is a version of $E^{\mathcal{B}}x(\cdot)$. Consequently, there exists $A_e^* \in \mathcal{B}$ such that $A_e^* \subset A_e$, $P(A_e \setminus A_e^*) = 0$ and $y'(t) \in C_e(t)$ for every $t \in A_e^*$. Let $D_0 = \{t \in T : c(A,t) = 0\}$, $D_1 = \{t \in T : c(A^c,t) = 0\}$ and $D_2 = \{t \in T: 0 < c(A,t) < 1\}$. By part (i) of Theorem 1 functions $y_A(\cdot)$ and $y_{A^c}(\cdot)$ are selectors of $C(\cdot)$ on D_2. Thus, if $t \in A_e^* \cap D_2$, we have $y'(t) \in C_e(t)$ and hence $y_A(t) = y_{A^c}(t)$. Thus, we have

$$\int_{A_e \cap A} y(t)\, P(dt) = \int_{A_e^* \cap A} y'(t)\, P(dt)$$

$$= \int_{A_e^* \cap A} c(A,t) y_A(t)\, P(dt) + \int_{A_e^* \cap A} c(A^c,t) y_{A^c}(t)\, P(dt)$$

$$= \int_{A_e^* \cap A \cap D_1} y_A(t)\, P(dt) + \int_{A_e^* \cap A \cap D_2} c(A,t) y_A(t)\, P(dt) +$$

$$+ \int_{A_e^* \cap A \cap D_0} y_A(t)\, P(dt) + \int_{A_e^* \cap A \cap D_2} c(A^c,t) y_{A^c}(t)\, P(dt)$$

$$= \int_{A_e^* \cap A} y_A(t)\, P(dt) = \int_{A_e^* \cap A} x(t)\, P(dt).$$

The last equality holds because $y_A(\cdot)$ is a version of $E_{P'}^{\mathcal{B}} x(\cdot)$, where $P'(A_1) = P(A_1 \cap A)/P(A)$, $A_1 \in \mathcal{A}$. This proves Theorem 1.

4. Jensen's inequality

Let U and V be separable Fréchet spaces and let $\{K(t),\ t \in T\}$ be a class of closed convex cones in U. Let S be a $\mathcal{B} \otimes \mathcal{B}(V)$-measurable subset of $T \times V$ and f a function from S into U. Moreover, we assume that the following conditions are fulfiled:

C1. for every $t \in T$ set $S(t) = \{v \in V : (t,v) \in S\}$ is convex and non-empty,

C2. for every $t \in T$ function $f(t,\cdot)$ is convex on $S(t)$ with respect to the partial ordering induced in U by the cone $K(t)$, i.e.

$$f(t, \alpha v_1 + (1-\alpha) v_2) \leqslant_{K(t)} \alpha f(t,v_1) + (1-\alpha) f(t,v_2) \tag{10}$$

holds for every $\alpha \in [0,1]$ and $v_1, v_2 \in S(t)$, where $u_1 \leqslant_{K(t)} u_2$ means $u_2 - u_1 \in K(t)$,

C3. f is $(\mathcal{B} \otimes \mathcal{B}(V), \mathcal{B}(U))$-measurable,

C4. multifunction epi f $(\cdot)$ takes values in the space of closed subsets of $V \times U$, where

$$\text{epi } f(t) = \left\{ (v,u) \in V \times U : v \in S(t),\ f(t,v) \leq_{K(t)} u \right\},$$

C5. the graph of epi $f(\cdot)$ is $\mathcal{B} \otimes \mathcal{B}(V \times U)$-measurable, i.e.

$$\text{Gr epi } f = \left\{ (t,v,u) \in T \times V \times U : v \in S(t),\ f(t,v) \leq_{K(t)} u \right\}$$

is an element of $\mathcal{B} \otimes \mathcal{B}(V \times U)$.

Conditions C1, C2, and C4 imply that for each $t \in T$ epi $f(t)$ is a convex closed and non-empty subset of $V \times U$. Conditions C1 and C2 admit the following intuitive interpretation: $f(t,\cdot)$ is a variable convex function defined on a variable convex set $S(t)$ and the convexity of $f(t,\cdot)$ is with respect to a variable partial ordering given by a variable convex cone $K(t)$.

Condition C3 implies the measurability of $f(\cdot,v(\cdot))$ provided $v(\cdot)$ is an $\mathcal{A}$-measurable function from T into V and $v(t) \in S(t)$ for every $t \in T$. Conditions C4 and C5 guarantee that epi $f(\cdot)$, the epigraph of f, is a convex closed multifunction satisfying the same measurability conditions as $C(\cdot)$ in the preceding section. This is a sufficient condition for Jensen's inequality to hold in the case of infinite dimensional spaces (see Example 2 in Section 5, where this inequality is not valid in the case of a Borel convex function on l^2).

Theorem 2 (Jensen's inequality). Assume that $U,V,S,K(\cdot)$ and f are defined as at the begining of this section and that conditions C1-C5 are fulfiled. Let $v(\cdot)$ be a strongly integrable selector of $S(\cdot)$ such that $f(\cdot,v(\cdot))$ is strongly integrable. If $w(\cdot)$ is a version of $E^{\mathcal{B}}v(\cdot)$ and $g(\cdot)$ is a version of $E^{\mathcal{B}}f(\cdot,v(\cdot))$, then

(i) $f(t,w(t)) \leq_{K(t)} g(t)$ for P-a.e. $t \in T$,

(ii) if, moreover, $f(t,\cdot)$ is strictly convex for $t \in A, A \in \mathcal{A}, P(A) > 0$ and $K(t) \cap (-K(t)) = \{0\}$ when $t \in A$, then $v(t) = w(t)$ P-a.e. on set

$$A \cap \left\{ t \in T : f(t,w(t)) = g(t) \right\}.$$

P r o o f: (i). Clearly, $(v(t), f(t,v(t))) \in \operatorname{epi} f(t)$ for every $t \in T$. Since values of epi $f(\cdot)$ are convex closed sets and epi $f(\cdot)$ is $\mathcal{B}$-measurable, we infer from Theorem 1 (i) that $(w(t), g(t)) \in \operatorname{epi} f(t)$ P-a.e. This is equivalent to the inequality (i).

(ii). The strict convexity of $f(t,\cdot)$ implies that if $\alpha \in (0,1)$, v_1, $v_2 \in S(t)$, $v_1 \neq v_2$, then $\alpha f(t,v_1) + (1-\alpha) f(t,v_2) + f(t, \alpha v_1 + (1-\alpha) v_2) \neq 0$. Hence, if $t \in A$, then every point of the graph of $f(t,\cdot)$ is extremal in epi $f(t)$. Thus, if $t \in A$ and $f(t,w(t)) = g(t)$, then $(w(t), f(t,w(t)))$ is extremal in epi $f(t)$. In view of Theorem 1 (ii) we conclude that $(w(t), f(t,w(t))) = (v(t), f(t,v(t)))$ and hence $w(t) = v(t)$ P-a.e. on $A \cap \{t \in T : f(t,w(t)) = g(t)\}$.

5. Remarks

1. Theorems 1 and 2 extend the corresponding theorems proved by Pfanzagl in [8]. He considered the case where $X = R^n$ and where C is a constant Borel subset of R^n. However, for infinite dimensional spaces expectations of selectors of constant convex Borel subsets may lie outside of this convex set. The following example shows that this is the case even for a separable Hilbert space.

Example 1. Let $T = \underline{N}$, $\mathcal{A} = 2^T$, $\mathcal{B} = \{\emptyset, T\}$, $P(t) = 1/2^t$, $X = l^2$, $x(t) = e_t$, where e_t is from l^2 and its t-th component is equal to 1 and the other ones equal to 0, $C = \operatorname{conv}\{e_t, t = 1,2,\ldots\}$. Then C is a countable union of compact sets $\operatorname{conv}\{e_1,\ldots,e_n\}$, $n = 1,2,\ldots$, $x(t) \in C$ for every $t \in T$ and $Ex(\cdot) \notin C$.

Consequently, Borel convex functions defined on Borel convex subsets of infinite dimensional spaces need not to satisfy Jensen's integral inequality. We give an example.

Example 2. Let $T, \mathcal{A}, \mathcal{B}, P, X$ and $x(\cdot)$ be defined as in Example 1. Now let

$$C = \operatorname{conv}\left\{ Ex(\cdot), e_1 e_2, \ldots \right\}.$$

Clearly, C is a countable union of compact sets and $Ex(\cdot)$ is an extremal point of C (but not an extremal point of cl C). Let $f: C \to \underline{R}$ be given by

$$f(x) = \begin{cases} 0 & \text{if } x \in C \setminus Ex(\cdot), \\ 1 & \text{if } x = Ex(\cdot). \end{cases}$$

Then f is Borel and convex on C and $0 = Ef(x(\cdot)) < f(Ex(\cdot)) = 1$. However this inequality is opposite to the Jensen's one.

On the other hand it is not difficult to find multifunctions, even not measurable which can be used in Theorem 1 instead of $C(\cdot)$ with values in the space of closed convex subsets of a Frechet space. For example, if we remove from each (closed) $C(t)$ any subset of its extremal points, Theorem 1 remains still valid. Indeed, if $y(\cdot) \in E^{\mathcal{B}} x(\cdot)$, then $y(t) \in \text{cl } C(t)$ by Theorem 1 (i), and in view of Theorem 1 (ii) function $y(\cdot)$ cannot take values in the removed extremal points.

The assumption on the closedness of the multifunction can also be dropped in the following case. Let

$$C(t) = A(t, G(t)) \quad \text{and} \quad x(t) = A(t, z(t)),$$

where 1. $A(t, \cdot)$ is for every $t \in T$ an affine continuous operator from a Fréchet space Z into another Fréchet space X; 2. $A(\cdot, z)$ is strongly $\mathcal{B}$-measurable for every $z \in Z$; 3. $G(\cdot)$ is a $\mathcal{B}$-measurable multifunction from T into the space of closed convex subsets of Z and 4. $A(\cdot, 0)$, $x(\cdot)$ and $z(\cdot)$ are strongly integrable.

If the foregoing conditions are fulfiled, then the assertion of Theorem 1 (i) remains valid. This easily follows from Theorem 1 and from

$$E^{\mathcal{B}} A(\cdot, z(\cdot))(t) = A(t, E^{\mathcal{B}} z(\cdot)(t)) \quad P\text{-a.e.}$$

The last equality follows from Jensen's inequality (Theorem 2) if we put $\underline{A} = f$ and $K(t) = \{0\}$ for every $t \in T$. If, additionally, we assume that $(T, \mathcal{B})$ is a Blackwell space and that the map$(t,z) \longrightarrow (t, \underline{A}(t,z))$ has a measurable graph, then, applying Theorem II.4.1.1 in [6] and repeating the proof of part (ii) of Theorem 1, we obtain that an assertion analogous to that of Theorem 1 (ii) is also true.

2. In order to illustrate the difference between Theorem 1 and Theorem 1′ consider the following example.

<u>Example 3.</u> Let $T = \left[-\frac{1}{2},\frac{1}{2}\right]\left[-\frac{1}{2},\frac{1}{2}\right]$, let $\mathcal{A}$ be the set of Lebesgue measurable subsets of T. Moreover, let $\underline{P}$ be the Lebesgue measure restricted to $\mathcal{A}$ and let $\mathcal{B}$ be the P-completed σ-algebra of subsets of T generated by the map $(t_1,t_2) \longrightarrow t_2$. Finally, let $A = \operatorname{conv}\left\{(-\frac{1}{2}, \frac{1}{2}), (0,0), (\frac{1}{2},\frac{1}{2})\right\} \cup \operatorname{conv}\left\{(-\frac{1}{2},\frac{1}{2}), (0,0), (\frac{1}{2},-\frac{1}{2})\right\}$, $x(t_1,t_2) = 1_A(t_1,t_2) \cdot -\frac{1}{t_2}$ (= 0 if $t_2 = 0$) and

$$C(t_1,t_2) = \begin{cases} \left[-\frac{1}{t_2}, +\infty\right), & \text{if } t_2 > 0, \\ 0, & \text{if } t_2 = 0, \\ \left(-\infty,-\frac{1}{t_2}\right], & \text{if } t_2 < 0. \end{cases}$$

Then $x(t_1,t_2) \in C(t_1,t_2)$ for $(t_1,t_2) \in A$, $x(.,.)$ is P-integrable, $E^{\mathcal{B}} 1_A x(.,.) = 1$ P-a.e., $c(A;t_1,t_2) = t_2$ and $y_A(t_1,t_2) = -\frac{1}{t_2}$(cf. (5)). Clearly, $y_A(t_1,t_2) \in C(t_1,t_2)$ P-a.e. and this is covered by Theorem 1′ (i) because $c(A;t_1,t_2) > 0$ P-a.e. On the other hand the assertion of Theorem 1 (i) implies in the case of measure P', $P'(A_1)=P(A_1 \cap A)/P(A)$, $A_1 \in \mathcal{A}$, that $y_A(t_1,t_2) \in C(t_1,t_2)$ for $(t_1,t_2) \in A$ only. Notice that selector $y_A(\cdot,\cdot)$ is not P-summable, however it is P'-integrable. Moreover, for P-a.e. $(t_1,t_2) \in T$ point $y_A(t_1,t_2)$ is extremal in $C(t_1,t_2)$.

However, $y_A(t_1, t_2) = x(t_1, t_2)$ P-a.e. on set A, (only), but not on the whole T.

3. In Section 4 the Jensen's inequality (Theorem 2 (i))has been derived from a theorem on conditional expectations of selectors of a closed convex multifunction $C(\cdot)$ (Theorem 1 (i).). A similar method has been previousely used by Pfanzagl [8] and Daures [2]. It is interesting to note that the converse implication is also valid. Indeed, let $p_i(\cdot)$ be the Minkowski functional (gauge) of the i-th element V_i of the base of balanced and convex neighbourhoods of zero in X. Let

$$d_i(x, C(t)) = \inf\left\{p_i(x-y) : y \in C(t)\right\}.$$

Since $d_i(\cdot, C(\cdot))$ is convex and continuous with respect to the first argument and $\mathcal{B} \otimes \mathcal{B}(X)$-measurable, Theorem 2 (i) applies to d_i. So, if $x(\cdot)$ is a selector of $C(\cdot)$, then $d_i(x(t), C(t)) = 0$ for each $t \in T$. By the Jensen's inequality $d_i(y(t), C(t)) \leqslant E^{\mathcal{B}} d_i(x(\cdot), C(\cdot))(t) = 0$. Thus, $d_i(y(t), C(t)) = 0$ P-a.e. for any d_i and $y(\cdot) \in E^{\mathcal{B}} x(\cdot)$. Hence $y(t) \in C(t)$ P-a.e., i.e., $y(\cdot)$ is a selector of $C(\cdot)$.

Finally, we note that if X is a Banach space, then, in view of the implication given above, Theorem 1 (i) and Theorem 2 (i) follow from a theorem of To Ting On and Wing Yip Kai given in [9], where a Jensen's inequality for continuous convex random functions was proved.

Acknowledgments. The authors wish to express their gratitude to Professor C. Castaing for the informations about the results of Daures and for a discussion on an earlier version of the paper. In particular, part 3 of Section 5 provides an answer to a question which has been rised by him.

References

[1] D a u r è s, J.-P., Version multivoque du théorème de Doob, Ann. Inst. Henri Poincaré, 9, 167-176 (1973).

[2] D a u r è s, J.-P., Quelques nouvelles proprietes de l'esperance conditionnelle d'une multi-application. Seminaire d'Analyse Convexe. Montpellier: Exposé No 11 (1975).

[3] E d w a r d s, R. E., Functional analysis. Holt, Rinehart and Winston. New York 1965.

[4] H a n e n, A., N e v e u, J.,Atomes conditionnels d'une espace de probabilité. Acta Math. Acad. Sci. Hungarica, 17, 443-449 (1966).

[5] H i a i, F., U m e g a k i, H., Integrals,conditional expectations and martingales of multivalued functions. J.Multivariate Analysis, 7, 149-182 (1977).

[6] H o f f m a n n-J ø r g e n s e n, J., The theory of analytic spaces. Various Publication Series No 10. Mathematisk Institut, Aarhus Universitet 1970.

[7] P e r l m a n, M. D., Jensen's inequality for a convex vector-valued function on an infinite-dimensional space.J.Multivariate Analysis, 4, 52-65 (1974).

[8] P f a n z a g l, J., Convexity and conditional expectations. Ann. Probability, 2, 490-494 (1974).

[9] T o T i n g O n, W i n g Y i p K a i, A generalized Jensen's inequality. Pacific J. of Math., 58, 255-259 (1975).

[10] W a g n e r, D. H., Survey of measurable selection theorems. SIAM J. Control and Optimisation, 15, 859-903 (1977).

[11] V a k h a n i y a, N. N., T a r i e l a d z e, V. I.,Covariance operators of probability measures in locally convex spaces. Theory Prob. Appl., 23, 1, 3-26 (1978).

SOME RESULTS ON BIASED LINEAR ESTIMATION APPLIED TO VARIANCE COMPONENT ESTIMATION

by

Lynn Roy LaMotte

University of Houston

1. Introduction

Shortcomings of commonly used estimators of variance components have been noted often in previous work. Hodges and Lehmann [2] noted that the sample variance is dominated by a simple multiple of itself. Klotz, Milton and Zacks [3] demonstrated that the customarily-used estimator of the among-groups variance component in the balanced, random, one-way analysis of variance (ANOVA) model may be dominated by various estimators. Harville [1] established similar results for the two-way mixed ANOVA model. Olsen, Seely and Birkes [6] established conditions for admissibility in the class of unbiased invariant quadratic estimators in models with two variance components and noted that some standard estimators are not admissible in this class (see also Seely [7]). LaMotte [4] identified biased invariant quadratic estimators which dominate the ANOVA estimator of the among groups variance component in the unbalanced one-way random ANOVA model.

Both Pukelsheim [8] and Olsen et al. [6] note that invariant quadratics in normal random variables follow linear models in the variance components. A distinguishing characteristic of linear models for variance components is that the mean vector and the variance-covariance matrix are functionally related in that elements of both are (linear and quadratic) functions of the variance components. This relationship is used below to establish that no linear combination of invariant quadratics in normal random variables is admissible among invariant quadratic estimators of its expectation. This result is in contrast to the situation in linear models in which the mean vector and variance-covariance matrix are functionally independent: in such cases it can be demonstrated that, under fairly general conditions, best (in the sense of Olsen et al. [6] and as defined below) unbiased linear estimators are admissible among linear estimators.

In the following section two results are established for linear estimation in the general linear model. In the third section these results are applied to variance component estimation.

2. Results on linear estimation

Let Y denote a random n-variate with mean vector $E(Y) = \mu$ and variance-covariance matrix $\mathrm{Var}(Y) = V$. The parameters for this model are (V, μ): denote the parameter space by $\mathcal{P}$, which is a subset of the cartesian product of the set of $n \times n$ nonnegative definite symmetric matrices and R^n.

For a matrix A, the linear subspace spanned by the columns of A will be denoted by $R(A)$, the transpose of A by A', the trace of A (if A is square) by $\mathrm{tr}(A)$, and the linear subspace $\{x: Ax = 0\}$ (the null space of A) by $N(A)$. For matrices M and H for which the matrix MH' is defined, the inner product $\mathrm{tr}(MH')$ will be used.

Let C be an $n \times p$ matrix of constants. A linear estimator of $C'\mu$ is a function $L'Y$, with L an $n \times p$ matrix of constants. In comparing linear estimators, total mean squared error will be used:

$$\begin{aligned} \mathrm{TMSE}_L(V, \mu\mu') &= E[(L'Y - C'\mu)'(L'Y - C'\mu)] \\ &= \mathrm{tr}[L'VL + (L - C)'\mu\mu'(L - C)]. \end{aligned} \tag{2.1}$$

Notation and definitions here usually follow Olsen et al. (1976). Thus $L_1'Y$ is <u>as good as</u> $L_2'Y$ if $TMSE_1(V, \mu\mu') \leq TMSE_2(V, \mu\mu')$ throughout $\mathcal{P}$; and $L_1'Y$ is <u>better than</u> $L_2'Y$ if, in addition, strict inequality holds for some $(V, \mu) \in \mathcal{P}$. The linear estimator $L'Y$ is <u>admissible</u> (among linear estimators) if no other linear estimator is better than $L'Y$. Definitions of <u>complete</u> and <u>essentially complete</u> classes of estimators are standard, as given in Olsen et al. [6]

Note that $\mathcal{T} = \{(V, \mu\mu') : (V, \mu) \in \mathcal{P}\}$ is a subset of the linear subspace $\mathcal{S} = \{(A, B) : A \text{ and } B \text{ are } n \times n \text{ symmetric matrices}\}$. Following Olsen et al. (1976), for any subset $\mathcal{U}$ of $\mathcal{S}$ let $[\mathcal{U}]$ denote the minimal closed convex cone in $\mathcal{S}$ containing $\mathcal{U}$. It will be said that $\mathcal{U}$ generates $\mathcal{T}$ if $[\mathcal{U}] = [\mathcal{T}]$. It may be seen that relations (as good as, better than, admissible) among linear estimators are unchanged if $\mathcal{T}$ is replaced by $[\mathcal{T}]$. The elements of $[\mathcal{T}]$ are of the form (V, Φ) with V and Φ $n \times n$ nonnegative definite matrices.

With minor and fairly obvious modifications, the arguments used by Olsen et al. [6] to show the minimal completeness of the class of admissible linear unbiased estimators may be adapted to show that the admissible linear estimators of $C'\mu$ form a minimal complete class. Their argument is based on finding an essentially complete class by eliminating parts of coefficient vectors which lie in the null space of every V. A similar reduction may be accomplished for linear estimators as follows. Let F be a matrix whose q columns form a basis for the linear subspace orthogonal to the intersection $\mathcal{N}$ of all $N(V + \Phi)$, $(V, \Phi) \in [\mathcal{T}]$. Then each $n \times p$ matrix L has a unique representation $L = FA + B$, with columns of B in $\mathcal{N}$, from which it may be seen that $A'F'Y$ is as good as $L'Y$. Results analogous to Lemma 3.1, Lemma 3.2 and Proposition 3.3 of Olsen et al. [6] may be established, with the conclusion that the class of admissible linear estimators is minimal complete among linear estimators.

Following LaMotte [5] it may be seen that $\{F'(V + \Phi)F : (V,\Phi) \in [\mathcal{T}]\}$ contains only positive definite matrices in its relative interior. Thus for purposes of linear estimation there is no loss of generality in assuming that $[\mathcal{T}]$ is a closed convex cone such that for all (V, Φ) in the relative interior of $[\mathcal{T}]$, $V + \Phi$ is positive definite.

A linear estimator L'Y will be called a <u>best linear estimator</u>, or <u>best at</u> (V_0, Φ_0) if there exists a point $(V_0, \Phi_0) \in [\mathcal{T}]$ such that for any $n \times p$ matrix M, $\mathrm{TMSE}_L(V_0, \Phi_0) - \mathrm{TMSE}_M(V_0, \Phi_0) \leq 0$. The following theorem may be proved easily.

<u>Theorem 2.1.</u> Let V and Φ be $n \times n$ symmetric nonnegative matrices.

(i) There exists an $n \times p$ matrix H such that $(V + \Phi)H = \Phi C$.

(ii) In order that L'Y be best at (V, Φ) it is necessary and sufficient that $(V + \Phi)L = \Phi C$.

(iii) For any $n \times p$ matrix L,

$$\mathrm{TMSE}_L(V, \Phi) \geq \mathrm{tr}\left[C'V C - H'(V + \Phi)H\right] \geq 0,$$

where H is as in (i), and the right-hand side is invariant to the choice of H.

(iv) There exists exactly one linear estimator best at (V, Φ) if and only if $V + \Phi$ is positive definite.

If $V + \Phi$ is positive definite and L'Y is best at $(V,\Phi) \in [\mathcal{T}]$ then L'Y is admissible because the existence of a better linear estimator would contradict the fact that only L'Y is best at (V, Φ). On the other hand, it is easy to construct examples of linear estimators, best at (V, Φ) with $V + \Phi$ singular, which are not admissible.

Theorem 2.2 is an obvious adaption of Proposition 3.6 in Olsen et al. [6] ; with some modifications, their proof may be used. It may be shown that there exists a compact convex subset of C of $[\mathcal{T}]$, not containing (0,0), which generates $[\mathcal{T}]$.

Theorem 2.2. In order that $L'Y$ be admissible among linear estimators of $C'\mu$ it is necessary and sufficient that there exist a nonzero $(V,\Phi)\in[\mathcal{T}]$ such that $L'Y$ is best at (V,Φ) and is admissible among linear estimators of $C'\mu$ best at (V,Φ).

The following theorem establishes conditions under which no unbiased linear estimator of $C'\mu$ is admissible.

Theorem 2.3. If there exists a number $m > 0$ such that $V - m\mu\mu' \geq 0$ (i.e., is nonnegative definite) for all $(V,\mu)\in\mathcal{P}$, and if $L'Y$ is an admissible linear estimator of $L'\mu$, then $L = 0$.

P r o o f: Note that $V \geq m\mu\mu'$ in $\mathcal{P}$ implies that, for $(V,\Phi)\in[\mathcal{T}]$, $V=0$ implies that $\Phi = 0$. If $L'Y$ is an admissible linear estimator of $L'\mu$, then by Theorem 2.2 there exists a nonzero $(V_0,\Phi_0)\in[\mathcal{T}]$ such that

$$(V_0 + \Phi_0)L = \Phi_0 L,$$

which implies that $V_0 L = 0$ ($V_0 \neq 0$ for otherwise $\Phi_0 = 0$). Let the s columns of the matrix N_0 form a basis for $N(V_0)$. Then $L = N_0 Z_1$. Because $L'Y$ is admissible, it is admissible among $\{Z'N_0'Y\colon Z \text{ sxp}\}$. Let

$$\mathcal{T}_0 = \{(N_0'VN_0, N_0'\Phi N_0) : (V,\Phi)\in[\mathcal{T}]\}.$$

Note that for any $(A,B)\in[\mathcal{T}_0]$, A and B are nonnegative definite and there exist nonnegative definite $n \times n$ matrices V_1 and Φ_1 (perhaps not in $[\mathcal{T}]$) such that $A = N_0'V_1N_0$ and $B = N_0'\Phi_1N_0$. By Theorem 2.2, there exists a nonzero $(N_0'V_1N_0, N_0'\Phi_1N_0)$ in $[\mathcal{T}_0]$ such that

$$N_0'(V_1 + \Phi_1)\, N_0Z_1 = N_0'\Phi_1N_0Z_1,$$

so that $V_0L = 0$ and $N_0'V_1L = 0$.

Because $N_0'V_1 \neq 0$, the rank of $(V_0, V_1N_0)'$ is greater than the rank of V_0. With columns of N_1 forming a basis for $N((V_0,\ V_1N_0)')$, $L = N_1Z_2$ and, proceeding as before, there exists a nonzero $N_1'V_2$ such that $N_1'V_2L = 0$, and the rank of $(V_0,\ V_1N_0,\ V_2N_1)'$ is greater than the rank of (V_0, V_1N_0). Continuing in this way, increasing the rank by at least one at each step, it is seen that $L = 0$. Q.E.D.

Corollary 2.4. Under the conditions of Theorem 2.3, there exists a number c, $0 \leq c < 1$, such that $cL'Y$ is a better linear estimator of $L'\mu$ than $L'Y$.

P r o o f: Let c be a number such that $\max\{0, (1-m)/(1+m)\} < c < 1$. Then $L'VL \geq mL'\Phi L$ and $m \geq (1-c)/(1+c)$, so

$$L'VL \geq \frac{1-c}{1+c} L'\Phi L.$$

Then it follows that

$$c^2 tr(L'VL) + (1-c)^2 tr(L'\Phi L) \leq L'VL$$

for all $(V, \Phi) \in [\mathcal{J}]$, so that $cL'Y$ is better than $L'Y$. Note that if $m \geq 1$, 0 is as good as $L'Y$. Q.E.D.

3. Application to invariant quadratic estimation

Let A be an $n \times n$ symmetric matrix. The quadratic $Y'AY$ in the n-vector Y is said to be invariant if $(Y-b)'A(Y-b) = Y'AY$ for all n-vectors Y and all n-vectors b in the minimal linear subspace $\mathcal{L}$ containing $\{\mu: (V, \mu) \in \mathcal{P}\}$. Let the columns of H form a basis for the subspace of vectors orthogonal to $\mathcal{L}$. Then if $Y'AY$ is invariant, $A =$ $= HBH'$ for some symmetric matrix B. Thus when restricting attention to invariant quadratics we need examine only quadratics in $Z = H'Y$. Note that $E(Z) = 0$ and $Var(Z) = H'VH$.

Let $Q_i = Y'A_iY$, $i = 1, \ldots, k$, where $Y \sim N_n(0, \Sigma), \Sigma \in \Omega$, each A_i is an $n \times n$ symmetric matrix, and Ω is a subset of $n \times n$ symmetric non-negative definite matrices. The k-vector $Q = (Q_1, \ldots, Q_k)'$ will now take the role of Y in the previous section. By the comments in the previous paragraph, there is no loss of generality in assuming $E(Y) = 0$ for purposes of invariant quadratic estimation. Let

$$\mathcal{T} = \{(\mathrm{Var}(Q),\ E(Q)E(Q)'): \Sigma \in \Omega\}$$

where the (i, j) - components of Var(Q) and $E(Q)E(Q)'$ are, respectively, $2\,\mathrm{tr}(A_i \Sigma A_j \Sigma)$ and $\mathrm{tr}(A_i \Sigma)\,\mathrm{tr}(A_j \Sigma)$.

<u>Lemma 3.1.</u> For any $\Sigma \in \Omega$, $\mathrm{Var}(Q) \geq \frac{2}{n} E(Q)E(Q)'$.

P r o o f: Let t be a k-vector. Then

$$t'E(Q)E(Q)'t = \left[\mathrm{tr}(A_t \Sigma)\right]^2,$$

where $A_t = \Sigma_i t_i A_i$. Let B be a symmetric nonnegative definite matrix such that $\Sigma = B^2$. Then, using the Cauchy-Schwartz inequality in the form

$$\mathrm{tr}(MN')^2 \leq \mathrm{tr}(MM')\mathrm{tr}(NN'),$$

note that

$$\left[\mathrm{tr}(A_t \Sigma)\right]^2 = \left[\mathrm{tr}(BA_t BI_n)\right]^2 \leq \mathrm{tr}(BA_t BBA_t B)\mathrm{tr}(I^2) = n\,\mathrm{tr}(A_t \Sigma A_t \Sigma)$$

$$= \frac{n}{2} t'\mathrm{Var}(Q)t. \quad \text{Q.E.D.}$$

From Lemma 3.1 and Theorem 2.3, it follows immediately that no invariant quadratic in an n-variate normal random variable is an admissible estimator of its expected value. In particular, unbiased invariant quadratic estimators of variance components are not admissible. For any unbiased invariant quadratic estimator of a variance component, there exists a better invariant quadratic estimator. The gain in efficiency obtained by using a biased invariant quadratic estimator may be considerable, as shown by examples in LaMotte [4].

In the conventional formulation of variance components models, $E(Y) = X\beta$ and $\mathrm{Var}(Y) = \Sigma_i \gamma_i V_i$, where X is an n × p matrix of constants, each V_i is a known symmetric matrix of constants, β is a p-vector of unknown parameters and $\gamma = (\gamma_1, \ldots, \gamma_k)'$ is a k-vector of unknown parameters called variance components(but perhaps including covariances as well). Assume that the rank of X is p. Let H be an $n \times (n-p)$ matrix whose columns form an orthonormal basis for $N(X')$. Then invariant quadratics in Y are of the form $Z'AZ$ with $Z = H'Y$. If the $(n-p)$-vector Z is normally distributed then there exists another invariant quadratic which is better than $Z'AZ$ in estimating $E(Z'AZ) = \Sigma_i \gamma_i \operatorname{tr}(AH'V_iH)$.

By Corollary 2.4 it may be seen that for any positive number c such that $(n-p-2)/(n-p+2) < c < 1$, $cZ'AZ$ is better than $Z'AZ$, but $cZ'AZ$ may not be admissible among invariant quadratics.For instance, if $Z'AZ \sim \sigma^2 \chi^2_\nu$, then $Z'AZ/\nu$ is an unbiased estimator of σ^2, $(\nu-2)/(\nu+2)\ Z'AZ/\nu$ is better, but $\nu/(\nu+2)\ Z'AZ/\nu$ (the Hodges-Lehmann [2] estimator) is better than any positive multiple of $Z'AZ/\nu$ The Hodges-Lehmann fixup yields a better estimator, but it in turn may not be admissible among invariant quadratics.

It should be noted that unbiased invariant quadratic estimators form a proper subset of the class of unbiased quadratic estimators. Quadratics which are not invariant do not apparently satisfy the conditions of Theorem 2.3, so whether unbiased quadratics are inadmissible among quadratics has not been established.

4. Acknowledgements

Most of the work in this paper was done while I was supported by the Department of Statistics at Iowa State University as a visiting faculty member in the summer of 1978.I am grateful to the Department of Statistics, and particularly to David A. Harville. I am grateful to the Institute of Mathematics, Polish Academy of Sciences, for the privelege of presenting this paper at the Sixth International Conference on Mathematical Statistics in Wisła, Poland.

References

[1] H a r v i l l e, D a v i d A. (1978). Alternative formulations and procedures for the two-way mixed model. Biometrics (1978).

[2] H o d g e s, J. L. and L e h m a n n, E. L., Some applications of the Cramer-Rao inequality. Proceedings of the Second Berkeley Symposium on Mathematical Statistics and Probability, 1 (1951), 13-22. University of California Press, Berkeley.

[3] K l o t z, J. H., M i l t o n, R. C., and Z a c k s, S., Mean Square Efficiency of Estimators of Variance Components. Amer. Statist. Assoc., 64, 1383-1402 (1969).

[4] L a M o t t e, L. R., Invariant quadratic estimators in the random, one-way ANOVA model. Biometrics, 32, 793-804 (1976).

[5] L a M o t t e, L. R., A canonical form for the general linear model. Ann. Statist., 5, 787-789 (1977).

[6] O l s e n, A., S e e l y, J., and B i r k e s, D., Invariant quadratic unbiased estimation for two variance components. Ann. Statist., 4, 878-890 (1976).

[7] S e e l y, J., An example of an inquadmissible analysis of variance estimator for a variance component. Biometrika, 62, 689--690 (1976).

[8] P u k e l s h e i m, F., Estimating variance components in linear models. Journal of Multivariate Analysis, 6, 626-629 (1976).

ESTIMATION PROBLEM FOR THE EXPONENTIAL CLASS OF DISTRIBUTIONS FROM DELAYED OBSERVATIONS

by

Ryszard Magiera
Wrocław Technical University, Poland

1. Introduction and summary

Let $X_1,\ldots,X_n$ be independent random variables with the same probability distribution depending on an unknown parameter ϑ. Suppose that X_i, $i=1,\ldots,n$ is observed at time t_i, where $0 \leqslant t_1 \leqslant t_2 \leqslant \ldots \leqslant t_n$, and $t_1,\ldots,t_n$ are independent of $X_1,\ldots,X_n$. We will, in fact, suppose that $t_1,\ldots,t_n$ are the order statistics of positive exchangeable random variables $U_1,\ldots,U_n$ which are independent of $X_1,\ldots,X_n$. We shall be interested in the problem of estimating the parameter ϑ when an information as described above is accessible at random moments of time. We assume that the loss incurred by the statistician in estimating ϑ is not only due to the error of estimation but also to the cost of observation.

The decision of the statistician is determined by a Markov stopping time τ which denotes the moment when the statistician decides to stop the observation and an estimator f of ϑ chosen by him when he does stop. The estimator f is assumed to be a function of observations and of the number of observations up to time τ. At least one observation is assumed to be taken. The problem is to find sequential decisions which minimize the expected value of the over-all loss due to estimation error and due to observation costs.

We will adopt a Bayesian approach by placing a prior distribution over ϑ and we shall find a class of optimal decisions for the statistician under the assumption that the loss due to estimation error is the squared error loss and that the observation cost equals c units for a unit time. For independent normally distributed random variables $X_1,\dots,X_n$ with unknown mean and known variance the underlying problem was considered by Starr, Wardrop and Woodroofe in [4]. In the present paper we consider a family $\mathcal{E}(\vartheta,\alpha)$ of exponential distributions. It is shown that for a prior distribution Φ from a family $\mathcal{E}_0(\alpha_0,\gamma)$, which is conjugate with $\mathcal{E}(\vartheta,\alpha)$, the Bayes estimator of ϑ has the form

$$\hat{f}_\tau = \frac{\gamma + \sum_{i=1}^{k(\tau)} X_i}{\alpha_0 + \beta + \alpha k(\tau)},$$

where $k(\tau)$ denotes the the number of observations made up to time τ, while β is a constant indicating certain subclass of $\mathcal{E}(\vartheta,\alpha)$. Further, we establish that the posterior risk corresponding to $\hat{f}_\tau$ equals $[\alpha_0+\beta+\alpha k(\tau)]^{-1}$. Therefore the sequential estimation problem could be reduced to an optimal stopping problem. Under given assumptions on $\mathcal{E}(\vartheta,\alpha)$ and on the common distribution function with a failure rate ρ of $U_1,\dots,U_n$ it is shown that for a weighted quadratic loss function the sequential plan $\delta^o = (\tau_o, \hat{f}^o_{\tau_o})$ with

$$\tau_o = \inf\left\{ t \geqslant 0 : [n-k(t)]\,\rho(t) \leqslant c\alpha^{-1}\{\beta+\alpha[k(t)+1]\}[\beta+\alpha k(t)]\right\}$$

and

$$\hat{f}^o_{\tau_o} = \frac{\sum_{i=1}^{k(\tau_o)} X_i}{\beta + \alpha k(\tau_o)}$$

is minimax.

The considered problem arises when data, no matter how we choose to manipulate our environment, is forthcoming at random moments of time only. Following [4] let us quote a few examples. In studying the effectiveness of safety devices in mobile objects the relevant data can be obtained only as the results of a failure or an accident. For instance, medical data on effectiveness of a medicine can be obtained at random times when patients seek help and are examined. Further examples of the described situation supply mail service and archeological discoveries.

2. Preliminaries

Let $(\Omega,\mathcal{F},P)$ be a probability space. Denote by $(\mathcal{X},\mathcal{B})$ a measurable space, where $\mathcal{X}\subseteq R$ (R denotes the real line) and $\mathcal{B}$ is the σ-algebra of Borel subsets of $\mathcal{X}$. Consider random variables $X_1,\ldots,X_n$ defined on $(\Omega,\mathcal{F},P)$, with values in $(\mathcal{X},\mathcal{B})$ and having the same distribution $\mathcal{P}_\vartheta$ defined on $(\mathcal{X},\mathcal{B})$ and depending on $\vartheta\in D$. We assume that D is an open interval (possibly infinite or semi-infinite) of the real line. Further, we suppose that the $\mathcal{P}_\vartheta's$ are absolutely continuous with respect to a σ-finite measure ν on $(\mathcal{X},\mathcal{B})$. The probability measure P may be interpreted as an element of the family of probability measures P_ϑ, $\vartheta\in D$, on $(\Omega,\mathcal{F})$. By $E_\vartheta(\cdot)$ and $D_\vartheta(\cdot)$ we denote, respectively, the expected value and the variance evaluated with respect to measure P_ϑ. We assume that $E_\vartheta(X_i^2)<\infty$ for all $\vartheta\in D$.

We suppose that the distributions $\mathcal{P}_\vartheta$, $\vartheta\in D$, belong to a family $\mathcal{E}(\vartheta,\alpha)$ of exponential distributions which is defined as follows.

Definition. Let $\mathcal{E}(\vartheta,\alpha)$ be a family of distributions $\mathcal{P}_\vartheta$, $\vartheta\in D$, having densities with respect to measure ν of the form

$$\frac{d\mathcal{P}_\vartheta}{d\nu}(x) = p(x;\vartheta,\alpha) = s(x,\alpha)\exp\left[\alpha w_1(\vartheta) + x\, w_2(\vartheta)\right], \tag{1}$$

where (a) α is a positive constant,

(b) $s(x,\alpha)$ is a (non-negative) $\mathcal{B}$- measurable function independent of ϑ and

(c) $w_1(\vartheta)$ and $w_2(\vartheta)$ are functions defined on D_x, twice continuously differentiable in D, and with the derivatives $w_1'(\vartheta)$ and $w_2'(\vartheta)$ satisfying the condition: $w_2'(\vartheta) > 0$ and $w_1'(\vartheta)/w_2'(\vartheta)$ being strictly decreasing in the whole interval D.

The expected value and variance of a random variable X with distribution in $\mathcal{E}(\vartheta,\alpha)$ is given by

$$E_\vartheta(X) = -\alpha \frac{w_1'(\vartheta)}{w_2'(\vartheta)} \tag{2}$$

and

$$D_\vartheta(X) = -\frac{\alpha}{w_2'(\vartheta)} \frac{d}{d\vartheta} \left[\frac{w_1'(\vartheta)}{w_2'(\vartheta)}\right], \tag{3}$$

respectively.

Let us remark that the normal $\mathcal{N}(\alpha\vartheta,\alpha)$ distribution with $\vartheta\in(-\infty,\infty)$, the gamma $\mathcal{G}(\vartheta^{-1},\alpha)$ distribution, the Poisson $\mathcal{P}(\alpha\vartheta)$ distribution and the negative-binomial $n\mathcal{B}(\vartheta(1+\vartheta)^{-1},\alpha)$ distribution with $\vartheta\in(0,\infty)$ belong to $\mathcal{E}(\vartheta,\alpha)$.

Now let $X_1,\dots,X_n$ be independent random variables with the same distribution $\mathcal{P}_\vartheta$ belonging to $\mathcal{E}(\vartheta,\alpha)$ with an unknown ϑ and a known α. We consider the problem of estimating ϑ when the observations become available at random times. Suppose that X_i is observed at time t_i, $i=1,\dots,n$, where $t_1,\dots,t_n$ are the order statistics of positive exchangeable random variables $U_1,\dots,U_n$. We assume that $U_1,\dots,U_n$ are independent of $X_1,\dots,X_n$. Let

$$k(t) = \sum_{i=1}^{n} I_{[0,t]}(U_i)$$

be the number of observations made during time $t \geq 0$, and let

$$F_t = \sigma\left\{k(s),\ s \leq t,\ X_1,\dots,X_{k(t)}\right\},$$

which is the information available to the statistician at time t.

An F_t-measurable random variable f will be called an estimator of ϑ. We suppose that the loss due to estimation error is determined by a weighted quadratic loss function $L(\vartheta, f)$ and that the cost of observing the process for unit time is constant $c > 0$. Thus, if the statistician decides to stop at time t, then the loss incurred by him when ϑ is the true value of the parameter and f is the chosen estimator is determined by

$$L_t(\vartheta, f) = L(\vartheta, f) + ct.$$

By a *stopping time* we mean an extended random variable τ for which $P_\vartheta(0 \leq \tau < \infty) = 1$ for all $\vartheta \in D$ and $\left\{\tau > t\right\} \in F_t$ for every $t \geq 0$, and by a *sequential plan* we understand any pair $\delta = (\tau, f)$.

The statistician decides when to stop to observe the process and what estimator to take when he does stop. He is interested to choose τ and f so that the expected value of the over-all loss function $L_\tau(\vartheta, f) = L(\vartheta, f) + c\tau$ be small. The *risk function* is defined by

$$R(\vartheta, \delta) = E_\vartheta\left[L_\tau(\vartheta, f)\right],$$

where $\delta = (\tau, f)$ is the chosen sequential plan and ϑ is the true value of the parameter.

We restrict attention to sequential plans δ such that $R(\vartheta, \delta) < \infty$ for all $\vartheta \in D$.

We shall use a Bayesian approach introducing a prior probability distribution of the parameter ϑ. Let us formulate this more formally. We invent a random variable Θ with values ϑ on D. Let $\mathcal{M}$

be the σ-algebra of Borel subsets of D, and let Φ be a prior probability distribution of Θ on $(D,\mathcal{M})$. We suppose that the random variables $X_1,\dots,X_n$, given $\Theta=\vartheta$, are independent and have a common distribution $\mathcal{P}_\vartheta\in\mathcal{E}(\vartheta,\alpha)$ and that $\Theta, X_1,\dots,X_n$ are independent of $U_1,\dots,U_n$. The unconditional probability measure is denoted by P_Φ. Let $E_\Phi(\Theta^2)<\infty$. For a given sequential plan δ the e x p e c t e d r i s k with respect to Φ is defined by

$$r(\Phi,\delta) = E_\Phi(L_\tau) = \int_D R(\vartheta,\delta)\,\Phi(d\vartheta).$$

Suppose that for the prior distribution Φ, the posterior distribution $\Phi^{\mathcal{F}_t}$, given $\mathcal{F}_t$, is well defined. Then the conditional expected loss, given $\mathcal{F}_t$, corresponding to Φ and an estimator f is defined by

$$r^{\mathcal{F}_t}(\Phi, f) = \int_D L(\vartheta, f)\,\Phi^{\mathcal{F}_t}(d\vartheta).$$

$r^{\mathcal{F}_t}(\Phi,f)$ will be called the p o s t e r i o r r i s k corresponding to Φ and f.

It is clear that for any stopping time τ the fuctional $r^{\mathcal{F}_t}(\Phi,f)$ is minimized by $f=\hat{f}$ being a Bayes estimator with respect to Φ. Thus the problem of finding Bayes sequential plans may be reduced to an optimal stopping problem.

In sequential estimation problems without delaying of observations it turns out that for some processes with a proper loss function the only minimax sequential plans are the fixed-time ones. For example, it is known [1, 3] that for processes with independent increments, most frequently involved in mathematical statistics, and for a weighted quadratic loss function, the minimax(sequential) plan reduces to a fixed-time plan. The solution of the problem to be considered in this paper leads to the plans which are essentially sequential.

3. Bayes and minimax sequential plans

Throughout this section let $D = (a,b)$ be an open interval. We establish a class of optimal sequential plans $\delta = (\tau, f)$ for $\vartheta \in D$ based on a sequence $X_1,\dots,X_n$ of independent random variables with a common distribution $\mathcal{P}_\vartheta \in \mathcal{E}(\vartheta,\alpha)$ satisfying the following conditions:

(i) for each $\vartheta \in D$

$$\vartheta = -\frac{w_1'(\vartheta)}{w_2'(\vartheta)}\ , \tag{4}$$

(ii) there exists a constant $\beta \geq 0$ such that the relation

$$\int_D exp\left[\alpha w_1(\vartheta) + x w_2(\vartheta)\right] d\vartheta = \frac{1}{(\alpha-\beta)s(x,\alpha)} \tag{5}$$

is valid for all $\alpha > \beta$ and $x \in \mathcal{X}$ such that $s(x,\alpha) > 0$,

(iii) $\lim\limits_{\vartheta\to\alpha^+} \exp\left[\alpha w_1(\vartheta) + x w_2(\vartheta)\right] = \lim\limits_{\vartheta\to b^-} \exp\left[\alpha w_1(\vartheta) + x w_2(\vartheta)\right]$

and

$$\lim_{\vartheta\to\alpha^+} \vartheta\exp\left[\alpha w_1(\vartheta) + x w_2(\vartheta)\right] = \lim_{\vartheta\to b^-} \vartheta\exp\left[\alpha w_1(\vartheta) + x w_2(\vartheta)\right]$$

for every $\alpha > \beta$ and each $x \in \mathcal{X}$ except perhaps $x = \inf \mathcal{X}$.

It is easily verified that (i), (ii) and (iii) are fulfilled for all the above mentioned distributions, i.e. for $\mathcal{N}(\alpha\vartheta,\alpha)$ with $a = -\infty$ $b = \infty$, $\beta = 0$, for $\mathcal{G}(\vartheta^{-1},\alpha)$ and $n\mathcal{B}(\vartheta(1+\vartheta)^{-1},\alpha)$ with $a = 0$, $b = \infty$, $\beta = 1$, and for $\mathcal{P}(\alpha\vartheta)$ with $a=0, b=\infty$, $\beta = 0$, which belong to $\mathcal{E}(\vartheta,\alpha)$.

Let $\varphi(\vartheta)$ be the density (with respect to the Lebesgue measure) of the probability distribution Φ on $(D,\mathcal{M})$, and let $\varphi^{\mathcal{F}_t}(\vartheta)$ be the density of the conditional probability distribution $\Phi^{\mathcal{F}_t}$.

Assume that $\varphi(\vartheta)$ is of the form

$$\varphi(\vartheta) = \alpha_o p(\gamma;\vartheta,\alpha_o+\beta) = \alpha_o s(\gamma,\alpha_o+\beta)\, exp\left[(\alpha_o+\beta)\, w_1(\vartheta) + \gamma w_2(\vartheta)\right] \tag{6}$$

where α_0 and γ are constants such that $\alpha_0 > 0$ and function s is positive. Note that $\varphi(\vartheta)$ is a density of a probability distribution on D, since

$$\int_D \varphi(\vartheta)\, d\vartheta = 1 .$$

This follows from (5), because in view of (6)

$$\int_D \varphi(\vartheta)\, d\vartheta = \alpha_0 s(\gamma, \alpha_0 + \beta) \int_D \exp\left[(\alpha_0+\beta) w_1(\vartheta) + \gamma w_2(\vartheta)\right] d\vartheta .$$

Let $\mathcal{E}_0(\alpha_0, \gamma)$ denote the family of all probability distributions on D with densities defined by (6). The following result may be established easily.

<u>Lemma 1.</u> Suppose that $\mathcal{P}_\vartheta \in \mathcal{E}(\vartheta, \alpha)$ and that (5) is met. If $\Phi \in \mathcal{E}_0(\alpha_0, \gamma)$, then $\Phi^{\mathcal{F}_t} \in \mathcal{E}_0\left(\alpha_0 + \alpha k(t), \gamma + \sum_{i=1}^{k(t)} X_i\right)$.

P r o o f. Since the considered random variables are independent it follows from Bayes theorem that

$$\varphi^{\mathcal{F}_t}(\vartheta) = \frac{\varphi(\vartheta) \prod_{i=1}^{k(t)} p(X_i; \vartheta, \alpha)}{\int_D \varphi(\vartheta) \prod_{i=1}^{k(t)} p(X_i; \vartheta, \alpha)\, d\vartheta} .$$

Making use of (6) and (1) we obtain

$$\varphi^{\mathcal{F}_t}(\vartheta) = \frac{\exp\left\{\left[\alpha_0 + \beta + \alpha k(t)\right] w_1(\vartheta) + \left(\gamma + \sum_{i=1}^{k(t)} X_i\right) w_2(\vartheta)\right\}}{\int_D \exp\left\{\left[\alpha_0 + \beta + \alpha k(t)\right] w_1(\vartheta) + \left(\gamma + \sum_{i=1}^{k(t)} X_i\right) w_2(\vartheta)\right\} d\vartheta} .$$

Now, taking into account (5) we have

$$\varphi^{\mathcal{F}_t}(\vartheta) =$$

$$= \left[\alpha_0 + \alpha k(t)\right] s\left(\gamma + \sum_{i=1}^{k(t)} X_i, \alpha_0 + \beta + \alpha k(t)\right) \exp\left\{\left[\alpha_0 + \beta + \alpha k(t)\right] w_1(\vartheta) + \left(\gamma + \sum_{i=1}^{k(t)} X_i\right) w_2(\vartheta)\right\} ,$$

This proves the lemma.

The above lemma shows that the family $\mathcal{E}_0(\alpha_0,\gamma)$ of prior distributions Φ is conjugate with the family $\mathcal{E}(\vartheta,\alpha)$ of distributions $\mathcal{P}_\vartheta$ satisfying (5).

From Lemma 1 and from strong Markov property the following result is obtained.

Corollary. For any stopping time τ

$$\Phi^{\mathcal{F}_t} \in \mathcal{E}_0\left(\alpha_0 + \alpha k(\tau),\ \gamma + \sum_{i=1}^{k(\tau)} X_i\right). \tag{7}$$

Conditions (i) and (iii) imply the following relations which are useful in our further considerations.

$$\alpha \int_D \vartheta w_2'(\vartheta) \exp\left[\alpha w_1(\vartheta) + x w_2(\vartheta)\right] d\vartheta = x \int_D w_2'(\vartheta) \exp\left[\alpha w_1(\vartheta) + x w_2(\vartheta)\right] d\vartheta, \tag{8}$$

$$\int_D \vartheta(x-\alpha\vartheta) w_2'(\vartheta) \exp\left[\alpha w_1(\vartheta) + x w_2(\vartheta)\right] d\vartheta = -\int_D \exp\left[\alpha w_1(\vartheta) + x w_2(\vartheta)\right] d\vartheta. \tag{9}$$

By using (5), (8) and (9) we have

$$\int_D (x-\alpha\vartheta)^2 w_2'(\vartheta) \exp\left[\alpha w_1(\vartheta) + x w_2(\vartheta)\right] d\vartheta = \frac{\alpha}{(\alpha-\beta)s(x,\alpha)}. \tag{10}$$

In view of (2), (3) and (4) the expected value and the variance of X_i, i=1,...,n, is given by

$$E_\vartheta(X_i) = \alpha\vartheta \tag{11}$$

and

$$D_\vartheta(X_i) = \frac{\alpha}{w_2'(\vartheta)}, \tag{12}$$

respectively.

Taking as the loss function the squared error measured in terms of the variance

$$L(\vartheta,f) = w_2'(\vartheta)(f-\vartheta)^2, \tag{13}$$

we establish the following result.

Lemma 2. Suppose that $\mathcal{P}_\vartheta \in \mathcal{E}(\vartheta,\alpha)$ and that (i)-(iii) are satisfied. Then for loss function (13) and any stopping time τ the Bayes estimator of ϑ with respect to $\Phi \in \mathcal{E}_0(\alpha_0, \gamma)$ has the form

$$\hat{f}_\tau = \frac{\gamma + \sum_{i=1}^{k(\tau)} X_i}{\alpha_0 + \beta + \alpha k(\tau)} . \tag{14}$$

P r o o f: In view of (13) the posterior risk $r^{\mathcal{F}_\tau}(\Phi, f)$ takes the form

$$r^{\mathcal{F}_\tau}(\Phi, f) = \int_D w_2'(\vartheta)(f-\vartheta)^2 \varphi^{\mathcal{F}_\tau}(\vartheta)\,d\vartheta . \tag{15}$$

This risk is minimized by taking

$$f = \hat{f}_\tau = \frac{\int_D \vartheta w_2'(\vartheta)\varphi^{\mathcal{F}_\tau}(\vartheta)\,d\vartheta}{\int_D w_2'(\vartheta)\varphi^{\mathcal{F}_\tau}(\vartheta)\,d\vartheta} .$$

Taking into account (7) we obtain

$$\hat{f}_\tau = \frac{\int_D \vartheta w_2'(\vartheta) \exp\left\{\left[\alpha_0+\beta+\alpha k(\tau)\right] w_1(\vartheta) + \left(\gamma + \sum_{i=1}^{k(\tau)} X_i\right) w_2(\vartheta)\right\} d\vartheta}{\int_D w_2'(\vartheta) \exp\left\{\left[\alpha_0+\beta+\alpha k(\tau)\right] w_1(\vartheta) + \left(\gamma + \sum_{i=1}^{k(\tau)} X_i\right) w_2(\vartheta)\right\} d\vartheta} .$$

Then, using (8) we get (14).

Let G be the common distribution function of the independent random variables $U_1, \dots, U_n$. We suppose that $G(0) = 0$, $G(t) > 0$ for $t > 0$, where G is absolutely continuous with density g, while g is the right hand derivative of G on $(0,\infty)$. The class of distribution functions fulfilling these conditions is denoted by $\mathcal{G}$. Let $\zeta = \sup\{t: G(t) < 1\}$. Moreover, let $\rho(z) = g(z)\left[1-G(z)\right]^{-1}$, $0 \leq z < \zeta$, denote the failure rate.

We have the following theorem.

Theorem 1. Suppose that $G \in \mathcal{G}$ has an non-increasing failure rate ρ. If $\Phi \in \mathcal{E}_0(\alpha_0, \gamma)$, $0 < \alpha_0 < \infty$, then the Bayes sequential plan is $\delta_{\alpha_0} = (\tau_{\alpha_0}, \hat{f}_{\tau_{\alpha_0}})$, where

$$\tau_{\alpha_0} = inf\left\{t \geq 0 : \left[n-k(t)\right]\rho(t) \leq c\alpha^{-1}\left\{\alpha_0+\beta+\alpha\left[k(t)+1\right]\right\}\left[\alpha_0+\beta+\alpha k(t)\right]\right\}. \quad (16)$$

P r o o f: First we evaluate the posterior risk $r^{\mathcal{F}_\tau}(\Phi, f)$ corresponding to (14). Substituing (14) into (15) and taking ino account (7) we get

$$\left(r^{\mathcal{F}_\tau}(\Phi, \hat{f}_\tau)\right.$$

$$= \left[\alpha_0 + \alpha k(\tau)\right] s\left(\gamma + \sum_{i=1}^{k(\tau)} X_i, \alpha_0 + \beta + \alpha k(\tau)\right) \times$$

$$\times \int_D w_2'(\vartheta)\left(\frac{\gamma + \sum_{i=1}^{k(\tau)} X_i}{\alpha_0 + \beta + \alpha k(\tau)} - \vartheta\right)^2 exp\left\{\left[\alpha_0+\beta+\alpha k(\tau)\right] w_1(\vartheta) + \left(\gamma + \sum_{i=1}^{k(\tau)} X_i\right) w_2(\vartheta)\right\} d\vartheta$$

$$= \frac{\left[\alpha_0 + \alpha k(\tau)\right] s\left(\gamma + \sum_{i=1}^{k(\tau)} X_i, \alpha_0 + \beta + \alpha k(\tau)\right)}{\left[\alpha_0 + \beta + \alpha k(\tau)\right]^2} \times$$

$$\times \int_D w_2'(\vartheta)\left\{\gamma + \sum_{i=1}^{k(\tau)} X_i - \vartheta\left[\alpha_0+\beta+\alpha k(\tau)\right]\right\}^2 exp\left\{\left[\alpha_0+\beta+\alpha k(\tau)\right] w_1(\vartheta) + \left(\gamma + \sum_{i=1}^{k(\tau)} X_i\right) w_2(\vartheta)\right\} d\vartheta.$$

By (10) we have

$$r^{\mathcal{F}_\tau}(\Phi, \hat{f}_\tau) = \frac{1}{\alpha_0 + \beta + \alpha k(\tau)} \, .$$

Thus, if $\Phi \in \mathcal{E}_0(\alpha_0, \gamma)$ then the problem of finding Bayes sequential plans reduces to the problem of minimizing

$$V_\Phi(\alpha_0, \tau) = E_\Phi\left\{\left[\alpha_0+\beta+\alpha k(\tau)\right]^{-1} + c\tau\right\}$$

with respect to τ .

If $G \in \mathcal{G}$ has a non-increasing failure rate, then it follows from Theorem 2.1 in [4] that for any prior distribution Φ (and not just for $\Phi \in \mathcal{E}_0(\alpha_0, \gamma)$) $V_\Phi(\alpha_0, \tau)$ is minimized by (16). Thus Lemma 2 yields the theorem.

It also follows from Theorem 2.1 in [4] that $V_\Phi(\alpha_0, \tau_{\alpha_0})$ is independent of Φ.

The minimax sequential plans are given in the following theorem.

Theorem 2. Suppose $X_1, \ldots, X_n$ are independent random variables with a common distribution $\mathcal{P}_\vartheta \in \mathcal{E}(\vartheta, \alpha)$ satisfying conditions (i) - - (iii), and suppose that for $\beta > 0$

(iv) $$\sup_{\vartheta \in D} \vartheta^2 w_2'(\vartheta) = \beta^{-1}.$$

If $E(t_1) < \infty$, then for loss function given by (13), the sequential plan $\delta^o = (\tau_0, \hat{f}^o_{\tau_0})$, where

$$\hat{f}^o_{\tau_0} = \frac{\sum_{i=1}^{k(\tau_0)} X_i}{\alpha k(\tau_0) + \beta}, \tag{17}$$

is minimax.

P r o o f. We write V instead of V_Φ when the distribution Φ is degenerated at $\alpha_0 = 0$. Then, as remarked above, $r(\Phi, \delta_{\alpha_0}) = V(\alpha_0, \tau_{\alpha_0})$. Now we take into consideration estimator (17). After simple computation and making use of (iv) we establish that

$$\sup_{\vartheta \in D} R(\vartheta, \delta^o) = V(0, \tau_0)$$

is finite. According to the well-known method of finding minimax rules in decision theory (e.g. [2] page 90), it suffices to show that $V(\alpha_0, \tau_{\alpha_0}) \to V(0, \tau_0)$ as $\alpha_0 \to 0$. To show this one can use an argument used in [4] in the proof of Theorem 4.2.

Note that condition (iv) is fulfilled, for example, for the negative-binomial and gamma distributions ($\beta = 1$).

Adopting the methods used in [4] one may consider a certain modification of the model discussed by assuming that the common distribution of the random variables $U_1,\ldots,U_n$ belongs to an exponential family of distributions with an unknown parameter and taking a conjugate prior distribution of this parameter.

In an analogous way as in [4] one may propose an adaptive plan which requires knowledge of neither n nor G and performs nearly as well as is possible when n is large for a wide class of G.

References

[1] Dvoretzky, A., Kiefer, J. and Wolfowitz, J., Sequential decision problems for processes with continuous time parameter. Problems of estimation, Ann. Math. Statist. 24, 403--415 (1953).

[2] Ferguson, T., Mathematical Statistics, A Decision Theorem Approach, New York, Academic Press 1967.

[3] Magiera, R., On sequential minimax estimation for the exponential class of processes, Zastosow. Matem. XV, 4, 445-454 (1977).

[4] Starr, N., Wardrop, R. and Woodroofe, M., Estimating a mean from delayed observations, Z. Wahrscheinlichkeitstheorie verw. Gebiete 35, 2, 103-113 (1976).

STATISTICAL ANALYSIS OF NONESTIMABLE FUNCTIONALS

Part I: Estimation

by

Dibyen Majumdar and Sujit Kumar Mitra
Indian Statistical Institute, New Delhi

1. Introduction

In the Gauss-Markov model $(Y, X\beta, \sigma^2\Lambda)$ a linear parametric functional $p'\beta$ is said to be estimable if it has a linear unbiased estimator. Experimenters using a fractional replicate of a factorial design are required to estimate important factorial effects which are nonestimable on account of incomplete replication. Very few theoretical investigations on Gauss-Markov models, on the other hand, are concerned with nonestimable linear parametric functionals - though such investigations would clearly be valuable from the point of view of practical applications.

Let $\mathcal{P}_{\beta,\sigma^2}$ denote the probability distribution of Y. The parametric functional $p'\beta$ is said to be identifiable by distribution if for distinct parameter points (β, σ^2) and (β_0, σ_0^2)

$$\mathcal{P}_{\beta,\sigma^2} = \mathcal{P}_{\beta_0,\sigma_0^2} \Rightarrow p'\beta = p'\beta_0. \tag{1.1}$$

Consider now the case where $\mathcal{P}_{\beta,\sigma^2}$ depends on β only through $X\beta$. This would be the situation for example when Y is distributed as n-variate normal with mean vector $X\beta$ and dispersion matrix $\sigma^2\Lambda$. It was shown by Bunke and Bunke [3] that for such families the following statements are equivalent:

(1) $p'\beta$ is linearly estimable,

(2) $p'\beta$ is identifiable by distribution,

(3) $p'\beta$ has an unbiased estimator not necessarily linear.

For nonestimable parametric functionals $q'\beta$, nonidentifiability by distribution could have disastrous consequences. The worst to happen would be when the probability distribution of the estimator we propose for $q'\beta$ is insensitive to value assumed by $q'\beta$, and this could be the case unless one has some prior information on the parameters to supplement the information contained in the observations Y. This prior information could either be in the form of an assumed prior distribution for the parameters or at least as bounds on values the parameters could assume. The latter assumption is sometimes quite realistic and indeed in several applications the experimenter on the basis of past experience would be able to place appropriate bounds on the parameters that are universally acceptable.

In section 3 of this paper we shall be concerned with the situation where the parameter space Ω_β of β is an ellipsoid and consists of all such β in R^m such that

$$\beta' H \beta \leqslant \delta^2. \tag{1.2}$$

Unless explicitly stated otherwise the parameter space Ω_σ of σ^2 will be assumed to be the whole of the positive half of the real line. The parameter space Ω for (β, σ^2) is the cartesian product of Ω_β and Ω_σ.

Let us examine the nonidentifiability of a nonestimable $q'\beta$ in this new context. To keep our present discussions simple we restrict ourselves to the case where the matrix H in (1.2) is p.d. Writing $q = X'b+s$ we observe that the part $p'\beta = b'X\beta$ of $q'\beta$ being an estimable part is identifiable by distribution while $s'\beta$ can assume any value between $-\delta\sqrt{s'H^{-1}s}$ and $\delta\sqrt{s'H^{-1}s}$ on Ω_β. This interval of uncertainty of $s'\beta$ is shortest when one chooses $p=P_{X'}q = p_0$ (say) and $P_{X'}$ is the orthogonal projector onto $\mathcal{M}(X')$ under the inner product induced by H^{-1}. Accordingly one could propose $\delta\sqrt{s_0'H^{-1}s_0}$ as a measure of nonidentifiability of a nonestimable $q'\beta$, where $s_0=(I-P_{X'})q$. In fact one could do even better noting that a measure of nonidentifiability should indicate the extent to which $q'\beta$ is indeterminable given $\mathcal{P}_{\beta,\sigma^2}$ (that is given the value assumed by $X\beta$). Given the parameter space Ω_β as in (1.2), given that

$$X\beta = \frac{\delta}{\sqrt{p'H^{-1}p}} XH^{-1}p,$$

where p is a specified vector in $\mathcal{M}(X')$, it is seen that $q'\beta$ is uniquely determined as

$$\frac{\delta}{\sqrt{p_0'H^{-1}p}} q'H^{-1}p.$$

Thus here $q'\beta$ is as identifiable as any estimable parametric functional. The situation is more revealing when one considers the testing problem. When $\Omega_\beta = R^m$, it is not difficult to see that a hypothesis H_0 such as $q'\beta = c$ on a nonestimable parametric functional $q'\beta$ is essentially nontestible in the sense that for any test of size α such a hypothesis H_0 has a constant power also equal to α. Thus here the hypothesis H_0 and alternative $q'\beta = d \neq c$ are indistinguishable on the basis of the observations on Y. On the other hand, when Ω_β is bounded as in (1.2), using an approximation to $q'\beta$ by an estimable parametric functional $p'\beta$ of the type we have just now discussed it is possible to construct a test function φ such that

$$\sup_{\beta:q'\beta=c} E\varphi = \alpha \tag{1.3}$$

and for $d \neq c$,

$$\sup_{\beta:q'\beta=d} E\varphi > \alpha \quad . \tag{1.4}$$

In fact, for d sufficiently removed from c, even

$$\inf_{\beta:q'\beta=d} E\varphi > \alpha \quad . \tag{1.5}$$

For completeness, in section 2, we have reviewed some methods of estimation which do not assume a bounded parameter space for β. A numerical comparison based on simulation of the bias and the mean square error of the various methods studied in section 3 is given in section 4.

2. Some approaches to estimating nonestimable functionals

In this section we shall present some approaches to estimating nonestimable functionals. These approaches either assume $\Omega_\beta = R^m$ or make no specific assumption in this regard.

(i) Best Linear Minimum Bias Estimator (BLIMBE)

This approach has been studied by Chipman [5], Drygas [6], Rao and Mitra [17, p. 139], Schönfeld [19] and several other authors. We note that a nonestimable linear parametric functional $q'\beta$ has no linear unbiased estimator and that the actual bias $(X'b-q)'\beta$ of the estimator $b'Y$ is a function of unknown parameters. A sensible way of controling bias seems to be to choose b such that the coefficient vector $X'b-q$ of the expression determining bias is as close to the null vector as possible in some acceptable sense. With the norm $\|X'b-q\|$ of the vector $X'b-q$ as induced by a given positive definite matrix M, that is

$$\| X'b-q \| = \sqrt{(X'b-q)'M(X'b-q)}, \tag{2.1}$$

an optimal choice of b is given by $b = (X')^-_{2(M)}q$. The corresponding estimator may be called a minimum bias estimator. Since $Var(b'Y)=\sigma^2 b'\Lambda b$, the minimum variance minimum bias estimator uses as coefficient vector a least squares solution with the least Λ norm. The BLIMBE is accordingly given by

$$b_o'Y = q'\left[(X')^{+}_{M\Lambda}\right]'Y, \tag{2.2}$$

which simplifies to $q'X^{+}_{\Lambda^{-1}M^{-1}}Y$ when Λ is invertible.

We note that the vector $X'b = X'(X')^{-}_{\ell(M)}q = P_{X'}q$ is unique for every least squares solution b, where $P_{X'}$ denotes orthogonal projection under the inner product induced by M. This implies in particular that every minimum bias estimator has precisely the same bias. The BLIMBE is thus automatically the minimum mean square error minimum bias estimator.

(ii) Conditionally Unbiased Estimators

Consider the normal equations

$$X'\Lambda^{-1}X\beta = X'\Lambda^{-1}Y \tag{2.3}$$

associated with the Gauss Markov model $(Y, X\beta, \sigma^2\Lambda)$. It is well known that equations (2.3) have a multiplicity of solutions when the matrix X of order $n \times m$ is of rank $r < m$. One could however get a unique solution requiring for example that the solution $\hat{\beta}$ of (2.3) belongs to $\mathcal{M}(H)$ a given subspace of R^m. For uniqueness one needs additionally that the matrices H and XH be both of rank r. It is not difficult to see that this unique $\hat{\beta}$ is an unbiased estimator of β if

$$L(\Omega_\beta) = \mathcal{M}(H), \tag{2.4}$$

where $L(\Omega_\beta)$ represents the linear span. In this sense $\hat{\beta}$ may be called a conditionally unbiased estimator of β. In choosing this solution of (2.3) one is thus playing safe in that if $p'\beta$ is estimable $p'\hat{\beta}$ is still the BLUE of $p'\beta$. If $q'\beta$ is nonestimable $q'\hat{\beta}$ is conditionally the BLUE of $q'\beta$ if (2.4) is true. This observation is due to Plackett [13] (see also Scheffe [18], p. 19).

In general one may enquire if it is possible to find a matrix G such that GY is conditionally an unbiased estimator of β given that

(2.4) holds. This is equivalent to requiring that G satisfies the equation

$$GXH = H \tag{2.5}$$

which has a solution if and only if

$$\text{Rank }(XH) = \text{Rank}(H) \leq \text{Rank } X. \tag{2.6}$$

When one is sufficiently convinced about the validity of (2.4)one could actually reformulate the model as $(Y, XH\gamma, \sigma^2\Lambda)$ noting that β is of the form $H\gamma$ for some γ. If further (2.6) holds $\beta = H\gamma$ is indeed estimable and $H\hat{\gamma} = H(XH)^-_{\ell(\Lambda^{-1})}Y$ is conditionally the BLUE of β if(2.4) is true. If the equality sign holds all through in (2.6), $H\hat{\gamma}$ is in fact the unique solution $\hat{\beta}$ of (2.3) to which a reference was made earlier in this section.If on the other hand the common value of Rank XH and Rank H is strictly less than r, $\tilde{\beta} = H\hat{\gamma}$ may not even satisfy (2.3). In using $\tilde{\beta}$ as an estimator for β one may also be exposing himself to the risk that even if $p'\beta$ is otherwise estimable, $p'\tilde{\beta}$ may not even be an unbiased estimator of $p'\beta$ unless $p \in \mathcal{M}(X'\Lambda^{-1}XH)$ or (2.4) is true. If $p \in \mathcal{M}(X'\Lambda^{-1}XH)$, $p'\tilde{\beta} = p'\hat{\beta}$ which is no longer true if $p \in \mathcal{M}(X')$ but is outside $\mathcal{M}(X'\Lambda^{-1}XH)$.For such p, the use of $p'\tilde{\beta}$ which is not globally unbiased would be justified on grounds of a lower variance if the a priori evidence about the truth of (2.4) is sufficiently strong.

In other cases in addition to (2.5), one may bring in global unbiasedness of estimable functionals as an additional condition and note that the latter condition is equivalent to requiring

$$XGX = X. \tag{2.7}$$

A general solution to (2.5) and (2.7) is easily seen to be

$$G = X^+_{\Lambda^{-1}I} + V(I-XX^+_{\Lambda^{-1}I}) + (I-X^+_{\Lambda^{-1}I}X)\ H(XH)^+_{\Lambda^{-1}I} +$$
$$+ (I-X^+_{\Lambda^{-1}I}X)\ Z\,(I-XH(XH)^+_{\Lambda^{-1}I}). \tag{2.8}$$

We note further that if G satisfies (2.5) and (2.7) so does GP_X, where $P_X = X(X'\Lambda^{-1}X)^{-}X'\Lambda^{-1}$ is the orthogonal projector onto $\mathcal{M}(X)$ under the inner product induced by Λ^{-1}. Also GP_XY has the same expectation as GY and

$$D(GY) - D(GP_XY)$$

is n.n.d. This shows that in arriving at an optimal choice of G in the class (2.8) one may without any loss of generality restrict himself to the subclass

$$\begin{aligned} G = {} & X^{+}_{\Lambda^{-1}I} + (I - X^{X}_{\Lambda^{-1}I}X)H(XH)^{+}_{\Lambda^{-1}I} \\ & + (I - X^{+}_{\Lambda^{-1}I}X)Z(I - XH(XH)^{+}_{\Lambda^{-1}I})P_X \, . \end{aligned} \tag{2.9}$$

Let the sum of the first two terms in (2.9) be denoted by G_o and let $\mathcal{G}$ denote the entire class of matrices determined by (2.9) through arbitrary choice of the matrix Z. The following theorem is easily established and we omit the proof.

Theorem 2.1

$$\operatorname{tr} D(G_oY) \leqslant \operatorname{tr} D(GY) \text{ all } G \in \mathcal{G} \, . \tag{2.10}$$

Further the equality sign holds in (2.10) if and only if $G = G_o$.

Since for arbitrary g-inverses G of X, GP_X is a reflexive least squares inverse of X and every reflexive least squares inverse $X^{-}_{\ell r(\Lambda^{-1})}$ of X can be so determined, the class $\mathcal{G}$ of matrices as defined above consists precisely of such inverses $X^{-}_{\ell r(\Lambda^{-1})}$ which in addition satisfies equation (2.5) or equivalently the condition

$$\mathcal{M}(H) \subset \mathcal{M}(X^{-}_{\ell r(\Lambda^{-1})}X) = \mathcal{M}(X^{-}_{\ell r(\Lambda^{-1})}), \tag{2.11}$$

G_o is the unique member of this class for which

$$\operatorname{tr} D(GY) = \operatorname{tr} G\Lambda G' = \operatorname{tr} \Lambda G'G \tag{2.12}$$

is a minimum.

For a matrix X of order $n \times m$ and p.d. matrices N and M of order n and m, respectively, the Moore Penrose inverse X^{+}_{NM} is the unique N least squares inverse of X such that if G is any other such inverse then

$$G'MG - (X^{+}_{NM})' MX^{+}_{NM} \tag{2.13}$$

is n.n.d. The uniqueness achieved above is through a different principle that incorporates the principle of conditional unbiasedness due to Plackett and Scheffe. Accordingly we propose to call G_0Y the optimal Plackett-Scheffe estimator and G_0 the Plackett-Scheffe inverse of X. It is hoped that properties of this unique inverse will be more fully understood in future.

Let us write $\tilde{G} = H(XH)^{+}_{\Lambda^{-1}I}$ so that $\tilde{\beta} = \tilde{G}Y$. Observe that

$$G_0 = \tilde{G} + X^{+}_{\Lambda^{-1}I}\left[I-XH(XH)^{+}_{\Lambda^{-1}I}\right] \tag{2.14}$$

and

$$D(G_0Y) - D(\tilde{G}Y) = D\left\{X^{+}_{\Lambda^{-1}I}\left[I-XH(XH)^{+}_{\Lambda^{-1}I}\right]Y\right\} \tag{2.15}$$

is n.n.d. This gives us an idea about the loss of accuracy that results, when on account of lack of conviction on the truth of (2.4) one insists on global unbiasedness of estimable functional as an additional condition to be fulfilled in addition to conditional unbiasedness as enunciated in (2.5).

(iii) Hyperestimators

Consider the expectational relation

$$X\beta = \mu(=E(Y)) \tag{2.16}$$

in the Gauss Markov model $(Y, X\beta, \sigma^2\Lambda)$. When the matrix X of order $n \times m$ has rank $r < m$, β does not have a linear unbiased estimator. We may nevertheless look for an unbiased estimator of the set of points

(β) which satisfy (2.16). Thus we try to estimate the entire hyperplane (2.16) rather than the single point (the true parameter point) which belongs to this hyperplane. This approach is due to Bunke and Bunke [3], Bjerhammar [2] and Sjoberg [20]. The name hyperestimator was suggested by Bjerhammar.

The general solution to equation (2.16) is given by

$$\beta = X^{-}\mu + (I-X^{-}X)z \tag{2.17}$$

where X^{-} is an arbitrary but fixed g-inverse of X and z is an arbitrary vector in R^m. As z varies over R^m, (2.17) gives the set of points we propose to estimate. Since $(I-X^{-}X)z$ is parameter free the problem thus reduces to estimating a single point $X^{-}\mu$.

$X^{-}\mu$ is estimable and its BLUE is given by $X^{-}XX^{-}_{\ell(\Lambda^{-1})}Y$ with dispersion matrix $\sigma^2 D$. Observe that

$$X^{-}XX^{-}_{\ell(\Lambda^{-1})} \in \left\{ X^{-}_{\ell r(\Lambda^{-1})} \right\}$$

and that

$$X^{-}XX^{-}_{\ell(\Lambda^{-1})}\mu = X^{-}\mu . \tag{2.18}$$

Without any loss of generality we can assume that the single point $X^{-}\mu$ that we shall be estimating is defined in terms of a reflexive least squares inverse of X. By theorem 3.2.2 of Rao and Mitra [17] such a g-inverse has always the representation

$$(X'\Lambda^{-1}X)^{-}X'\Lambda^{-1} \tag{2.19}$$

for some g-inverse $(X'\Lambda^{-1}X)^{-}$ of $X'\Lambda^{-1}X$ and the problem reduces to choosing a g-inverse $(X'\Lambda^{-1}X)^{-}$ of $X'\Lambda^{-1}X$ suitably. Since

$$D = (X'\Lambda^{-1}X)^{-}X'\Lambda^{-1}X\left[(X'\Lambda^{-1}X)^{-}\right]' \tag{2.20}$$

an optimal choice of $(X'\Lambda^{-1}X)^{-}$ that minimizes for example

$$\operatorname{tr} \Delta D, \tag{2.21}$$

where Δ is a given p.d. matrix of order m, is given by $(X'\Lambda^{-1}X)^{-}_{m(\Delta)}$ leading to the optimal hyperestimator

$$(X'\Lambda^{-1}X)^{-}_{m(\Delta)}\ X'\Lambda^{-1}Y + \alpha = X^{+}_{\Lambda^{-1}\Delta}\ Y + \alpha, \tag{2.22}$$

where α represents an arbitrary vector in $\mathcal{N}(X)$, the null space of X. The same choice is also optimal in the sense of minimizing

$$\operatorname{tr} D'\Delta D\ . \tag{2.23}$$

3. A bounded parameter space for β

In this section we shall consider the case where the parameter space Ω_β of β is given by

$$\Omega_\beta = \left\{\beta \in R^m : \beta' H\beta \leq \delta^2\right\}. \tag{3.1}$$

The analysis we propose with trivial modifications, will extend itself to the case where the parameter space is an ellipsoid not centred at the origin. In this set up the BLIMBE we introduced in section 2 has a more natural interpretation.

(i) BLIMBE revisited

If H is positive semidefinite we note that an estimator $b'Y$ of $p'\beta$ has a finite maximum bias over Ω_β if and only if

$$X'b - p \in \mathcal{M}(H). \tag{3.2}$$

Thus the parametric functional $p'\beta$ will admit an estimator with a finite maximum bias on Ω_β if and only if

$$p \in \mathcal{M}(X' : H). \tag{3.3}$$

Further if $p = X'b + Ha$, the maximum bias of the estimator $b'Y$ on Ω_β, is given by $\delta\sqrt{a'Ha}$. Hence determining a minimax bias estimator for $p'\beta$ requires finding a minimum H_e seminorm solution of the consistent equation

$$(X':H)\binom{b}{a} = p, \tag{3.4}$$

where H_e = diag $(0,H)$. If $\begin{pmatrix} G_1 \\ G_2 \end{pmatrix}$ is $(X':H)^-_{m(H_e)}$, using known properties of such inverses it is seen that $G_1' \in \{X^-_{m(H)}\}$ and in fact the class $\{G_1'\}$ of such matrices is precisely equal to $\{X^-_{m(H)}\}$. Hence if $b'Y$ is a minimax bias estimator of $p'\beta$, $b = (X^-_{m(H)})'p = (X')^-_{\ell(H^{-1})}p$ when H is p.d. In this particular case the minimum variance minimax bias estimator is given by $b'Y$ where

$$b = (X')^+_{H^{-1}\Lambda}\, p \tag{3.5}$$

which indeed leads to the BLIMBE as introduced in section 2, if H^{-1} is used for M.

When H is singular it was shown in Rao and Mitra [16] that if M denotes a n.n.d. g-inverse of $(H + X'X)$, then

$$\{(X')^-_{\ell(M)}\} \subset \{[X^-_{m(H)}]'\} \tag{3.6}$$

Nonnegative definiteness of the chosen g-inverse of $H+X'X$ is not essential but will nevertheless be preferred to avoid conceptual difficulties that may otherwise arise in using M to define a seminorm Generally the set inclusion in (3.6) will be a 'proper' one. So the class of minimax bias estimators obtained through arbitrary choice of a M semileast squares inverse of X' cannot be expected to exhaust all minimax bias estimators. Hence the estimator $p'[(X')^+_{M\Lambda}]'Y$ will have the minimum variance property in a smaller subclass of minimax bias estimators. In the present case however using the general solution to a minimum seminorm g-inverse of X given in Rao and Mitra [16] it is seen that for $p \in \mathcal{M}(X' : H)$

$$\{[X^-_{m(H)}]'p\} = \{(X')^-_{\ell(M)}\;p\}. \tag{3.7}$$

Hence $p'[(X')^+_{M\Lambda}]'Y$ is the minimum variance minimax bias estimator of $p'\beta$.

(ii) Minimax estimator

Let us assume that the parameter space for (β, σ^2) be $\Omega = \Omega_\beta \times \Omega_\sigma$ where Ω_β is as defined earlier and Ω_σ is a subset of the positive half of the real line with a known finite maximal element σ_u^2. The estimator $b_o'Y$ of $p'\beta$ is called a minimax (linear) estimator (Rao [15] if

$$\max_{\beta,\sigma^2\in\Omega} E(b_o'Y - p'\beta)^2 = \min_{b\in R^n} \max_{\beta,\sigma^2\in\Omega} E(b'Y - p'\beta)^2. \quad (3.8)$$

As in the earlier section we observe that $p'\beta$ will admit an estimator with a finite maximum mean square error if and only if

$$p \in \mathcal{M}(X' : H)$$

and that if $p = X'b + Ha$

$$\max_{\beta,\sigma^2\in\Omega} E(b'Y - p'\beta)^2 = \sigma_u^2\, b'\Lambda b + \sigma^2\, a'Ha = \delta^2(\Theta^2 b'\Lambda b + a'Ha) \quad (3.9)$$

Hence if $H_e = \text{diag}\,(\Theta^2\Lambda, H)$ and $\begin{pmatrix} G_1 \\ G_2 \end{pmatrix}$ is a minimum H_e seminorm g-inverse of $(X':H)$, $b_o'Y = p'G_1'Y$ is a minimax estimator of $p'\beta$. We note that a minimum diag $[\Theta^2\ (\Lambda+XX'), H]$ g-inverse of $(X':H)$ is also a minimum H_e seminorm g-inverse of $(X': H)$ and that

$$\mathcal{M}\begin{pmatrix} X \\ \cdots \\ H \end{pmatrix} \subset \mathcal{M}\left(\text{diag}\left[\Theta^2\ (\Lambda+XX'), H\right]\right). \quad (3.10)$$

Hence using formula (3.1.7) of Rao and Mitra [17] it is seen that the following is one choice of G_1

$$G_1 = (\Lambda+XX')^- X\left[X'(\Lambda+XX')^- X + \Theta^2 H\right]^-, \quad (3.11)$$

When Λ is p.d. $\Lambda + XX'$ could be replaced by Λ and $(\Lambda+XX')^-$ by Λ^{-1} which leads to the expression for the minimax estimator given by Rao [15] for the case where both Λ and H are p.d. If K denotes the matrix $[X'(\Lambda+XX')^- X + \Theta^2 H]^-$ and $L = X'(\Lambda+XX')^- X$ the maximum mean square error for the minimax estimator (the L.H.S. of (3.8)) is given by

$$\sigma_u^2\left[p'Kp - p'K'L^2Kp\right]. \tag{3.12}$$

For the simultaneous estimation of several parametric functionals, say $P\beta$ (P being a matrix), one defines the minimax linear estimator (MILE) as $C_*Y + d_*$, where

$$\max_{\beta,\sigma^2\in\Omega} E(C_*Y+d_* - P\beta)' A(C_*Y + d_* - P\beta)$$
$$= \min_{C,d} \max_{\beta,\sigma^2\in\Omega} E(CY+d-P\beta)' A(CY+d-P\beta),$$

A being a given n.n.d. matrix. In this definition, if one restricts the class to estimators of the form CY then one obtains the minimax homogeneous linear estimator (MIHLE). We shall not consider this problem here, except for briefly indicating the results that are available.

Using the notation of optimal inverse (Mitra [12]),

$$\beta^{(m)} = \left[(X')^{\dagger}_{\Theta} \,{}^{-2}H^{-1}\oplus\Lambda\right]'Y \text{ is MILE of } \beta \text{ when } A = pp',$$

as was noted by Kuks [8]. Considering the matrix loss function $(CY-\beta)(CY-\beta)'$ Bunke [4] shows that $\beta^{(m)}$ is minimax. Lauter [10] derives an explicit expression for the MILE of β when $H = A = I$. But his expression for the more general situation cannot be used due to the absence of a computational algorithm. However, for certain situations there is an iterative procedure for computing MILE suggested by Kuks and Olman [9].

(iii) Restricted least squares estimator (maximum likelihood estimator)

If Y follows a multivariate normal distribution and Λ is p.d. the maximum likelihood estimate $\hat{\beta}_{ml}$ minimizes

$$G(\beta) = (Y-X\beta)' \Lambda^{-1}(Y-X\beta) \tag{3.13}$$

subject to the condition $\beta' H \beta \leq \delta^2$. Even without the distributional assumption the same method yields a least squares estimate subject to the quadratic constraint $\beta' H \beta \leq \delta^2$.

This and similar optimization problems were considered by Balakrishnan [1], Forsythe and Golub [7] and Majumdar and Mitra [11] Let us write $\hat{\beta}_0 = X^{+}_{\Lambda^{-1}H} Y$. If $\hat{\beta}_0' H \hat{\beta}_0 \leq \delta^2$ then $\hat{\beta}_0$ is one choice of $\hat{\beta}_{ml}$ and in fact any solution of the normal equation

$$X'\Lambda^{-1}X\beta = X'\Lambda^{-1}Y, \tag{2.3}$$

which belongs to Ω_β could serve as the ML estimate. To avoid confusion in such cases we shall choose and fix $\hat{\beta}_{ml} = \hat{\beta}_0$. If $\hat{\beta}_0' H \hat{\beta}_0 > \delta^2$ the ML estimate $\hat{\beta}_{ml}$ will obviously lie on the surface of the closed ellipsoid Ω_β. To compute $\hat{\beta}_{ml}$ in such cases one has to minimize $G(\beta)$ subject to the condition $\beta' H \beta = \delta^2$. The solution $\hat{\beta}_\lambda$ will satisfy the equation

$$(X'\Lambda^{-1}X + \lambda H)\beta = X'\Lambda^{-1}Y, \tag{3.14}$$

where λ is a root of the equation

$$F(\lambda) = Y'\Lambda^{-1}X(X'\Lambda^{-1}X + \lambda H)^{-}H(X'\Lambda^{-1}X + \lambda H)^{-}X'\Lambda^{-1}Y = \delta^2. \tag{3.15}$$

Let W be a non singular matrix such that

$$H = W' \begin{pmatrix} I_r & 0 \\ 0 & 0 \end{pmatrix} W \tag{3.16}$$

and

$$X'\Lambda^{-1}X = W' \begin{pmatrix} D_1 & 0 \\ 0 & D_2 \end{pmatrix} W, \tag{3.17}$$

where $D_1 = \text{diag}(d_1, d_2, \ldots, d_r)$, $D_2 = \text{diag}(d_{r+1}, d_{r+2}, \ldots, d_m)$.

It is well known that such a matrix W exists for every pair of n.n.d. matrices (see e.g. Rao and Mitra [17], p. 122]). Assume without any loss of generality that $d_1 \geq d_2 \geq \ldots \geq d_u > 0$, $d_{u+1} = \ldots = d_r = 0$ Put $\gamma = W\hat{\beta}_0$ and check that

$$F(\lambda) = \sum_{i=1}^{u} \frac{d_i^2 \ \gamma_i^2}{(d_i + \lambda)^2} \tag{3.18}$$

and $\sum_{i=1}^{u} \gamma_i^2 = \hat{\beta}_o' \ H\hat{\beta}_o > \sigma^2$. By Theorem 3.1 of Majumdar and Mitra [11], there exists a unique positive root λ_o of the equation $F(\lambda) = \sigma^2$. With this choice of λ, $\hat{\beta}_{\lambda_o}$ as defined in (3.14), gives the minimum of $G(\beta)$ on the surface of Ω_β. Similar arguments will establish the existence of a unique negative root which is $< -d_1$. This leads to the maximum of $G(\beta)$ on the surface of Ω_β. Other real roots if any belong to the open interval $(-d_1, -d_r)$ and correspond to saddle points of $G(\beta)$ on Ω_β.

Using the notation of an optimal inverse (Mitra [12]) we have

$$\hat{\beta}_{\lambda_o} = (X'\Lambda^{-1}X + \lambda_o H)^- X'\Lambda^{-1}Y = X^+_{\Lambda^{-1} \oplus \lambda_o H} Y. \tag{3.19}$$

and the required ML estimator of β is given by

$$\hat{\beta}_{ml} = \begin{cases} \hat{\beta}_o & \text{if } \hat{\beta}_o' H \hat{\beta}_o \leq \sigma^2, \\ \beta_{\lambda_o} & \text{if } \hat{\beta}_o' H \hat{\beta}_o > \sigma^2. \end{cases} \tag{3.20}$$

(iv) Bayes Homogenous Linear Estimator (BHLE, Chipman [5], Rao, [15]).

Consider a specific prior distribution for β over Ω_β and denote $E(\beta)$, $D(\beta)$, and $E(\beta\beta')$ by a, $\sigma^2\Delta$ and $\sigma^2 E$, respectively. Clearly $\sigma^2 E = \sigma^2\Delta + aa'$. The BHLE $b'Y$ of $p'\beta$ minimizes $E(b'Y - p'\beta)^2 = \sigma^2 [b'\Lambda b + (X'b-p)' E(X'b-p)]$ and is given by the choice

$$b = (X')^+_{E \oplus \Lambda} \, p \tag{3.21}$$

in terms of the $E \oplus \Lambda$ optimal inverse of X'. Using Corollary 4.1(b) of Mitra [12] the expected mean square error of the BHLE is seen to be equal to

$$\sigma^2 p' \left[E - EX'(X')^+_{E \oplus \Lambda}\right] p. \tag{3.22}$$

The Bayes Linear estimator BLE, $b'Y+c$ similarly minimizes

$$E(b'Y+c-p'\beta)^2 = \sigma^2\left[b'\Lambda b + (X'b-p)'\Delta(X'b-p)\right] +$$
$$+\left[(X'b-p)'a+c\right]'\left[(X'b-p)'a+c\right] \tag{3.23}$$

and is given by the choice

$$b = (X')^{+}_{\Delta\oplus\Lambda}\,p,\quad c = p'\left[I-X'(X')^{+}_{\Delta\oplus\Lambda}\right]a. \tag{3.24}$$

The expexted mean square error of the BLE is similarly equal to

$$\sigma^2 p'\left[\Delta - \Delta X'(X')^{+}_{\Delta\oplus\Lambda}\right]p\ . \tag{3.25}$$

Since $E-\Delta$ is n.n.d., so is $E\oplus\Lambda - \Delta\oplus\Lambda$. Corollary 4.1(b) of Mitra [12] therefore implies that

$$\left[E-EX'(X')^{+}_{E\oplus\Lambda}\right]-\left[\Delta - \Delta X'(X')^{+}_{\Delta\oplus\Lambda}\right] \tag{3.26}$$

is n.n.d., as is to be even otherwise expected.

Note: One may wonder if we would have similarly fared better with the BLIMBE and the minimax estimator by considering the wider class of nonhomogeneous linear estimators. However, if $p \in \mathcal{M}(X':H)$ and $p=X'b + Ha$, the maximum bias of the estimator $b'Y+c$ on Ω_β is seen to be equal to $\sigma\sqrt{a'Ha} + |d|$. This shows that the BLIMBE and the minimax estimator as derived earlier are indeed best and minimax in the wider class of nonhomogeneous linear estimators.

4. Comparison of rival estimators

The alternative interpretation for BLIMBE given in section 3 is based on the well known fact that an inner product defined on a real (or complex) vector space V induces a dual inner product on the vector space of linear functionals on V. For example if V is R^m and one considers the inner product defined by the p.d. matrix H, then the dual inner product on the vector space of linear functionals is induced by the p.d. matrix H^{-1} if one considers the representations of linear functionals in terms of the dual basis ($p'\beta \leftrightarrow p$). Consider now the resolution of the vector space of linear parametric functionals into the vector subspace of estimable functionals and its orthogonal complement under the dual inner product. A parametric functional in the orthogonal complement may be called totally nonestimable. We have seen that $p'\beta$ is estimable if $p \in \mathcal{M}(X')$. The above resolution corresponds to the following resolution of p

$$p = p_e + p_{ne},$$

where $p_e = P_{X'}p$, $p_{ne} = (I-P_{X'})p$, and $P_{X'}$ represents the orthogonal projector onto $\mathcal{M}(X')$ under the inner product induced by H^{-1}. Thus $p_e'\beta$ is the estimable part of $p'\beta$ and $p_{ne}'\beta$ is its totally nonestimable part. The above resolution also corresponds to the following resolution of β

$$\beta = \beta_e + \beta_{ne},$$

where $\beta_e = P_{X'}'\beta$ and $\beta_{ne} = (I-P_{X'}')\beta$ and this indeed is an orthogonal resolution under the original inner product. Clearly each coordinate of β_e is estimable. Hence β_e may be called the estimable part of β. The justification for calling β_{ne} the totally nonestimable part of β is seen from the fact that if β is totally nonestimable, i.e

$$\beta = (I-P_{X'}')\beta$$

every linear functional in Y has identically a zero expectation. The name totally nonestimable given to $p_{ne}'\beta = p'\beta_{ne}$ is therefore like a transferred epithet which seems to be quite appropriate.

The interesting fact to note is that a totally nonestimable parametric functional is estimated by zero by the BLIMBE, MLE or the minimax estimator given in section 3. Equivalently the result could be stated thus

<u>Theorem 4.1.</u> Write $\hat{\beta}_0$ for $X^+_{\Lambda^{-1}H}Y$ which provides the BLIMBE for

$p'\beta$, $\hat{\beta}_1$ for $X^{\dagger}_{\Lambda^{-1} \oplus \theta^2 H} Y$ which provides the minimax estimator for $p'\beta$ and let $\hat{\beta}_2$ denote the MLE as defined in (3.20). Then

$$(I-P'_{X'})\,\hat{\beta}_i = 0, \quad i = 0,1,2, \tag{4.1}$$

P r o o f: Theorem 4.1 follows from the representation of the minimum norm least square inverse given in (3.3.11)of Rao,Mitra [17] and from the property of optimal inverse described in Theorem 3.2(v) of Mitra [12].

We now proceed to compare $\hat{\beta}_0$, $\hat{\beta}_1$ and $\hat{\beta}_2$ on the basis of their biases and mean square errors at $\beta \in \Omega_{\beta}$. Our criteria will be norm (induced by H) of the bias vector and trace $H\Delta$ where Δ is the m.s. p.e. matrix. Put

$$B_i = \left[E(\hat{\beta}_i) - \beta\right]' H \left[E(\hat{\beta}_i) - \beta\right], \tag{4.2}$$

$$M_i = E(\hat{\beta}_i - \beta)\; H(\hat{\beta}_i - \beta). \tag{4.3}$$

Theorem 4.2 gives certain inequalities among the B_i's and M_i's that holds when σ^2 is known.

<u>Theorem 4.2.</u> (a) $B_o \leq \min(B_1, B_2)$; (4.4)

(b) $M_1 \leq M_0$. (4.5)

P r o o f: Let $\hat{\beta}$ denote an estimator which is such that

$$P'_{X'}\,\hat{\beta} = \hat{\beta} \tag{4.6}$$

then

$$\begin{aligned}\left[E(\hat{\beta})-\beta\right] H\left[E(\hat{\beta})-\beta\right] &= \left[E(\hat{\beta}) - \beta\right] H\left[P'_{X'} + (I-P'_{X'})\right]\left[E(\hat{\beta}) - \beta\right] \\ &= \left[E(P'_{X'}\hat{\beta})-\beta_e\right]' H\left[E(P'_{X'}\hat{\beta})-\beta_e\right] + \beta'_{ne}\, H\, \beta_{ne} \\ &= \left[E(\hat{\beta})- \beta_e\right]' H\left[E(\hat{\beta})-\beta_e\right] + \beta'_{ne} H\beta_{ne} \geq \beta'_{ne}\, H\beta_{ne}\end{aligned} \tag{4.7}$$

using known properties of such projections (see e.g. Theorem 5.2.1 of [17] and the fact that

$$P'_{X'}E(\hat{\beta}) = E(P'_{X'}\hat{\beta}) = E(\hat{\beta}),$$

$$(I-P'_{X'})E(\hat{\beta}) = E\left[(I-P'_{X'})\hat{\beta}\right] = 0.$$

Since $\hat{\beta}_i$ satisfies (4.6) we have from (4.7)

$$B_i \geqslant \beta'_{ne}\ H\beta_{ne}.$$

However for i = 0, the equality holds since $E(\hat{\beta}_o) = \beta_e$. This establishes (4.4).

To prove the (b) part check similarly that

$$M_i = E(\hat{\beta}_i - \beta_e)'H(\hat{\beta}_i - \beta_e) + \beta'_{ne}\ H\beta_{ne}. \tag{4.8}$$

Since $E(\hat{\beta}_o) = \beta_e$, we have

$$M_o = \mathrm{tr}\ H\ D(\hat{\beta}_o) + \beta'_{ne}\ H\beta_{ne}.$$

Let T be a nonsingular matrix such that

$$X'\Lambda^{-1}X = T'DT,\ H = T'T,$$

where $D = \mathrm{diag}\ (d_1, d_2, \ldots, d_r, 0, \ldots, 0)$. Let T be partitioned as

$$T = \begin{pmatrix} T_1 \\ \cdots \\ T_2 \end{pmatrix},$$

where T_1 has r rows and T_2 has (m-r). Let T^{-1} be partitioned as $T^{-1} = (S'_1 : S'_2)$ where S'_1 and S'_2 have respectively r and(m-r)columns. Then it is easily checked that

$$H = T'_1T_1 + T'_2T_2,$$

$$\mathcal{M}(T'_1) = \mathcal{M}(X'\Lambda^{-1}X) = \mathcal{M}(X'),$$

$$P_{X'} = T'_1S_1, \quad I-P_{X'} = T'_2S_2,$$

$$\beta'_{ne}\, H\, \beta_{ne} = \beta' T'_2\, S_2 (T'_1 T_1 + T'_2 T_2) S'_2 T_2 \beta = \beta' T'_2 T_2\, \beta = \eta'_2 \eta_2 .$$

Writing η_1 for $T_1\beta$ and η_2 for $T_2\beta$, $\eta = T\beta$.

Further $D(\hat{\beta}_0) = D(P'_{X'}\ \hat{\beta}_0) = \sigma^2 P'_{X'} (X' \Lambda^{-1} X)^{-} P_{X'}$

$$= \sigma^2 S'_1 T_1 (S'_1 : S'_2)\ D^{-} \begin{pmatrix} S_1 \\ .. \\ S_2 \end{pmatrix}\ T'_1\ S_1$$

$$= \sigma^2\ (S'_1\ \ 0)\ D^{-} \begin{pmatrix} S_1 \\ 0 \end{pmatrix} = \sigma^2 T^{-1} D^{+} (T^{-1})'$$

where $D^{+} = \text{diag}\ (d_1^{-1},\ d_2^{-1}, \ldots,\ d_r^{-1},\ 0, \ldots, 0)$.

Hence

$$M_0 = \sigma^2 \sum_{i=1}^{r} \frac{1}{d_i} + \sum_{i=r+1}^{m} \eta_i^2 .$$

Similarly

$$M_1 = \text{tr}\ H\ D(\hat{\beta}_1) + \left[E(\hat{\beta}_1) - \beta\right]' P_{X'} H P'_{X'} \left[E(\hat{\beta}_1) - \beta\right] + \beta'_{ne}\ H \beta_{\ ne}$$

$$= \text{tr}\ \sigma^2 (D+\theta^2 I)^{-1} D (D+\theta^2 I)^{-1} + \theta^4 \sum_{i=1}^{r} \frac{\eta_i^2}{(d_i+\theta^2)^2} + \sum_{i=r+1}^{m} \eta_i^2$$

$$= \sigma^2 \sum_{i=1}^{r} \frac{1}{(d_i+\theta^2)^2} \left[d_i + \frac{\sigma^2}{\delta^4}\ \ \eta_i^2 \right] + \sum_{i=r+1}^{m} \eta_i^2 .$$

Note that $\beta' H \beta \leqslant \delta^2 \Longleftrightarrow \eta'\eta \leqslant \delta^2 \Longleftrightarrow \eta_i^2 \leqslant \delta^2$ all i.

Hence

$$M_1 \leqslant \sigma^2 \sum_{i=1}^{r} \frac{1}{(d_i+\ \theta^2)^2} \left[d_i + \ \theta^2 \right] + \sum_{i=r+1}^{m} \eta_i^2$$

$$= \sigma^2 \sum_{i=1}^{r} \frac{1}{d_i + \ \theta^2} + \sum_{i=r+1}^{m} \eta_i^2$$

$$\leqslant \sigma^2 \sum_{i=1}^{r} \frac{1}{d_i} + \sum_{i=(r+1)}^{m} \eta_i^2 = M_0 \ \text{ since }\ \theta^2 \geqslant 0.$$

Since the MLE is nonlinear it is difficult to carry out a similar exercise on $\hat{\beta}_2$. Simulation was therefore resorted to for a comparison of these estimates. For simplicity Ω_β was taken to be a sphere rather than an ellipsoid, and the dispersion matrix of Y was taken to be $\sigma^2 I$. Also since the nonestimable part affects the B_i and M_i values of all the three estimators in the same way, we shall compare them only on a full rank model (i.e. Rank X = m). Further keeping in view the possibility of a singular value decomposition of X

$$X = U'\begin{pmatrix} A \\ 0 \end{pmatrix} V,$$

where $A = \text{diag}\ (\alpha_1, \alpha_2, \ldots, \alpha_m)$, U and V are orthogonal matrices and noting that a parameter transformation $\beta \to \tau = V\beta$ replaces a sphere in β by a sphere in τ of equal radius and an orthogonal transformation of the observations $Y \to Z = UY$ keeps the dispersion matrix $(\sigma^2 I)$ invariant, one could without any loss of generality restrict himself to a simpler model as the one that holds for Z, and even omit the last n-m observations as they contribute only to the estimation of σ^2. For the simulation the following canonical model was considered

$$E(y_i) = \alpha_i \beta_i, \quad \text{Cov}(y_i, y_j) = \sigma^2 \delta_{ij} \tag{4.9}$$

α_i given positive numbers, $\sum_{i=1}^{m} \beta_i^2 \leq \delta^2$, δ_{ij} is the Kronecker symbol, $i, j = 1, 2, \ldots, m$.

In the computations m was chosen to be 4. Four different design matrices were chosen. One had large α_i values, one moderate but $\alpha_i \geq 1$ the third had $\alpha_i \leq 1$ and the last one had α_i on either side of 1. For each design matrix, three points in Ω_β, one near the circumference, on well inside and the third near the origin were chosen. The entire experiment was done with 3 values of σ, $\sigma = 1, 2$ and 3. Since σ was varied, δ^2 was fixed at 17, their ratio only matters. For each of the 36 sets thus obtained N = 1000 samples were chosen. Random normal deviates were obtained from Herman Wold's Random Normal Deviates: Tracts for Computers No. XXV, Cambridge University Press, 1954.

Besides B_i and M_i, the distribution of λ the MLE shrinkage factor, was also obtained. For lack of space only an abridged version of this distribution could be included in tables published here. Note that when $\lambda = 0$, MLE is same as BLIMBE.

Tables 1,2 and 3 give the B_i and M_i values for i = 0,1,2 and $\sigma = 1,2$ and 3. Each table corresponds to one particular value of σ. Besides these the frequency of cases where $\lambda = 0$ and $\lambda > 0$ are also tabulated. For BLIMBE and the minimax, theoretical expressions for B_i and M_i are available. Theoretical values computed from these expressions are shown in parenthesis in addition to values estimated by simulation. A comparison of these two entries will indicate the accuracy attained by simulation.

Table 4 gives a ranking of these three procedures from the point of view of bias and also from the point of view of mean square error. Biasewise, BLIMBE appears to be distinctly superior to the other two methods. The second place is shared by Minimax and MLE about equally frequently. However, interestingly enough the one that does better on bias fares poorly on m.s.e. In the whole table there are only four exceptions to this rule. We have already noted that when $\lambda = 0$, MLE is same as BLIMBE. Hence in those cases where $\lambda = 0$ has a large frequency it was to be expected that minimax would have a lower m.s.e. compared to MLE. However there are cases where minimax has a lower m.s.e. compared to MLE even when the frequency at $\lambda = 0$ is small. These happen to be cases where the model itself does not allow for a precise estimation of parameters either on account of small singular values of the design matrix or large values of σ or both. One could, for example, take M_0 as a measure of this characteristic of the model. Subject to this reservation for parameter values near the surface of the ellipsoid the MLE seems to have a lower mean square

Table 1. Showing the bias and mean square error of BLIMBE, MINIMAX and MLE

$(n = m = 4,\ \delta^2 = 17,\ \sigma = 1)$

Sl. No.	β/DR				BLIMBE		MINIMAX		MLE		Frequency of λ at	
					BIAS	MSE	BIAS	MSE	BIAS	MSE	0	>0
1	2 4	2 5	2 7	2 10	$.0^{3}15$ (0)	.1363 (.1329)	$.0^{3}40$ ($.0^{4}82$)	.1359 (.1323)	.0020	.1185	738	262
2	1 4	2 5	1 7	2 10	$.0^{4}91$ (0)	.1352 (.1329)	$.0^{3}16$ ($.0^{4}38$)	.1345 (.1322)	$.0^{4}91$	.1352	1000	0
3	1 4	-1 5	-1 7	1 10	$.0^{4}49$ (0)	.1337 (.1329)	$.0^{4}76$ ($.0^{4}21$)	.1331 (.1322)	$.0^{4}49$	.1337	1000	0
4	2 1.0	2 3.0	2 3.5	2 5.0	.0017 (0)	1.2522 (1.2327)	.0023 (.0126)	1.1614 (1.1349)	.1114	.8499	599	441
5	1 1.0	2 3.0	1 3.5	2 5.0	$.0^{3}25$ (0)	1.2662 (1.2327)	.0031 (.0033)	1.1559 (1.1256)	$.0^{3}43$	1.1867	950	50
6	1 1.0	-1 3.0	-1 3.5	1 5.0	.0011 (0)	1.2757 (1.2327)	.0076 (.0032)	1.1671 (1.1255)	.0012	1.2724	995	5
7	2 0.25	2 0.50	2 0.75	2 1.00	.0082 (0)	23.4844 (22.7777)	1.1405 (1.1336)	10.6609 (10.3501)	1.9740	8.8110	207	793
8	1 0.25	2 0.50	1 0.75	2 1.00	.0391 (0)	22.7410 (22.7777)	.2686 (.4015)	9.5188 (9.6180)	.4959	8.2916	282	718
9	1 0.25	-1 0.50	-1 0.75	1 1.00	.0040 (0)	23.1457 (22.7777)	.2908 (.2834)	9.5572 (9.4999)	.2618	10.7166	445	555
10	2 0.25	2 0.75	2 1.00	2 5.00	.0258 (0)	19.1109 (18.8177)	1.1617 (.9885)	9.2224 (7.6235)	1.8065	6.7288	250	750
11	1 0.25	2 0.75	1 1.00	2 5.00	.0047 (0)	19.3796 (18.8177)	.2719 (.2740)	7.1004 (6.9090)	.3333	6.6350	389	611
12	1 0.25	-1 0.75	-1 1.00	1 5.00	.0177 (0)	19.3637 (18.8177)	.3211 (.2471)	7.0908 (6.8821)	.2695	9.0579	547	453

Note 1. Theoretical values are shown in parenthesis. Other values are based on simulation using a sample of 1000 from $N(X\beta, \sigma^2 I)$. The first column with the heading β/DR has two entries in each cell. The upper entries are the values of β_1, β_2, β_3 and β_4. The lower entries are the singular values of the X matrix.

Table 2. Showing the bias and mean square error of BLIMBE, MINIMAX and MLE

$(n = m = 4, \delta^2 = 17, \sigma = 2)$

Sl. No.	β/DR				BLIMBE		MINIMAX		MLE		Frequency of λ at	
					BIAS	MSE	BIAS	MSE	BIAS	MSE	0	> 0
1	2 4	2 5	2 7	2 10	$.0^341$ (0)	.4971 (.5316)	$.0^371$ (.0013)	.4872 (.5218)	.0142	.3877	572	428
2	1 4	2 5	1 7	2 10	$.0^336$ (0)	.5405 (.5316)	.0012 ($.0^360$)	.5299 (.5211)	$.0^338$	.5379	994	6
3	1 4	-1 5	-1 7	1 10	$.0^319$ (0)	.5348 (.5316)	$.0^356$ ($.0^333$)	.5240 (.5208)	$.0^319$	.5348	1000	0
4	2 1.0	2 3.0	2 3.5	2 5.0	.0069 (0)	5.0090 (4.9310)	.2035 (.1495)	3.7745 (3.6643)	.5756	3.0010	432	568
5	1 1.0	2 3.0	1 3.5	2 5.0	.0010 (0)	5.0644 (4.9310)	.0387 (.0396)	3.6519 (3.5543)	.0367	3.4938	720	280
6	1 1.0	-1 3.0	-1 3.5	1 5.0	.0047 (0)	5.0646 (4.9310)	.0616 (.0374)	3.6608 (3.5521)	.0216	4.3777	894	106
7	2 0.25	2 0.50	2 0.75	2 1.00	.0327 (0)	93.9208 (92.1109)	4.0281 (3.9305)	17.7834 (17.1521)	4.7617	16.4960	48	952
8	1 0.25	2 0.50	1 0.75	2 1.00	.1514 (0)	90.9601 (91.1109)	1.4377 (1.7967)	14.8443 (15.0183)	1.9268	14.2518	73	927
9	1 0.25	-1 0.50	-1 0.75	1 1.00	.0161 (0)	92.5692 (91.1109)	1.0389 (.9826)	14.2610 (14.2042)	.9946	15.5000	112	888
10	2 0.25	2 0.75	2 1.00	2 5.00	.1031 (0)	76.4459 (75.2709)	3.2608 (2.9906)	12.6214 (12.1231)	3.7257	12.2599	104	896
11	1 0.25	2 0.75	1 1.00	2 5.00	.0189 (0)	77.5149 (75.2709)	1.0684 (1.0089)	10.4787 (10.1414)	1.1729	10.9519	136	864
12	1 0.25	-1 0.75	-1 1.00	1 5.00	.0709 (0)	77.4561 (75.2709)	.8565 (.7476)	10.0552 (9.8802)	.7832	13.7544	185	815

Note 1. Theoretical values are shown in parenthesis. Other values are based on simulation using a sample of 1000 from $N(X\beta, \sigma^2 I)$. The first column with heading β/DR has two entries in each cell. The upper entries are the values of $\beta_1, \beta_2, \beta_3$ and β_4. The lower entries are the singular values of the X matrix.

T a b l e 3. Showing the bias and mean square error of BLIMBE, MINIMAX and MLE

($n = m = 4$, $\delta^2 = 17$, $\sigma = 3$)

Sl. No.	β/DR				BLIMBE		MINIMAX		MLE		Frequency of λ at	
					BIAS	MSE	BIAS	MSE	BIAS	MSE	0	>0
1	2 4	2 5	2 7	2 10	$.0^389$ (0)	1.1185 1.1962	.0038 .0064	1.0706 (1.1475)	.0485	.8297	508	492
2	1 4	2 5	1 7	2 10	$.0^381$ (0)	1.2162 (1.1962)	.0045 (.0030)	1.1635 (1.1441)	.0013	1.1723	941	59
3	1 4	-1 5	-1 7	1 10	$.0^340$ (0)	1.2023 (1.1962)	.0022 (.0016)	1.1485 (1.1427)	$.0^340$	1.2023	1000	0
4	2 1.0	2 3.0	2 3.5	2 5.0	.0155 (0)	11.2687 (11.0947)	.6173 (.5002)	6.4763 (6.2601)	1.1946	5.4038	333	667
5	1 1.0	2 3.0	1 3.5	2 5.0	.0052 (0)	11.5097 (11.0947)	.1193 (.1356)	6.0942 (5.8955)	.1463	5.7310	494	506
6	1 1.0	-1 3.0	-1 3.5	1 5.0	.0102 (0)	11.3974 (11.0947)	.1761 (.1251)	6.0578 (5.8850)	.1139	7.5504	702	298
7	2 0.25	2 0.50	2 0.75	2 1.00	.0736 (0)	211.3267 (204.9996)	6.6153 (6.4650)	20.6562 (19.8680)	6.9253	20.8978	15	985
8	1 0.25	2 0.50	1 0.75	2 1.00	.3404 (0)	204.6011 (204.9996)	2.9140 (3.3598)	16.5696 (16.7629)	3.2267	17.5061	24	976
9	1 0,25	-1 0.50	-1 0.75	1 1.00	.0361 (0)	208.2756 (204.9996)	1.7295 (1.6162)	15.1542 (15.0193)	1.6268	17.3724	43	957
10	2 0.25	2 0.75	2 1.00	2 5.00	.2318 (0)	171.9734 (169.3596)	4.9235 (4.6212)	15.2234 (14.6657)	5.2546	15.7386	41	959
11	1 0.25	2 0.75	1 1.00	2 5.00	.0426 (0)	174.3936 (169.3596)	1.9919 (1.8618)	12.3709 (11.9063)	2.1188	13.5888	54	943
12	1 0.25	-1 0.75	-1 1.00	1 5.00	.1597 (0)	173.6389 (169.3596)	1.2727 (1.1553)	11.2667 (11.1998)	1.1798	15.6814	82	918

Note 1. Theoretical values are shown in parenthesis. Other values are based on simulation using a sample of 1000 from $N(X\beta, \sigma^2 I)$. The first column with the heading β/DR has two entries in each cell. The upper entries are the values of β_1, β_2, β_3 and β_4. The lower entries are the singular values of the X matrix.

T a b l e 4. Showing ranking by order of magnitude of bias and mean square error of BLIMBE (1), MINIMAX (2) and MLE (3), based on simulation results

S		DR		c	4	5	7	10	1.0	3.0	3.5	5.0	.25	.50	.75	1.00	.25	.75	1.00	5.00
												BIAS								
2	2	2	2	1	1	2	3	(738)	1	2	3	(559)	1	2	3	(207)	1	2	3	(250)
				2	1	2	3	(572)	1	2	3	(432)	1	2	3	(48)	1	2	3	(104)
				3	1	2	3	(508)	1	2	3	(333)	1	2	3	(15)	1	2	3	(41)
1	2	1	2	1	1	3	2	(1000)	1	3	2	(950)	1	2	3	(282)	1	2	3	(389)
				2	1	3	2	(994)	1	3	2	(720)	1	2	3	(73)	1	2	3	(136)
				3	1	3	2	(941)	1	2	3	(494)	1	2	3	(24)	1	2	3	(54)
1	-1	-1	1	1	1	3	2	(1000)	1	3	2	(995)	1	3	2	(445)	1	3	2	(547)
				2	1	3	2	(1000)	1	3	2	(894)	1	3	2	(112)	1	3	2	(185)
				3	1	3	2	(1000)	1	3	2	(702)	1	3	2	(43)	1	3	2	(82)
				σ								MSE								
2	2	2	2	1	3	2	1	(738)	3	2	1	(559)	3	2	1	(207)	3	2	1	(250)
				2	3	2	1	(572)	3	2	1	(432)	3	2	1	(48)	3	2	1	(104)
				3	3	2	1	(508)	3	2	1	(333)	2	3	1	(15)	2	3	1	(41)
1	2	1	2	1	2	3	1	(1000)	2	3	1	(950)	3	2	1	(282)	3	2	1	(389)
				2	2	3	1	(994)	3	2	1	(720)	3	2	1	(73)	2	3	1	(136)
				3	2	3	1	(941)	3	2	1	(494)	2	3	1	(24)	2	3	1	(54)
1	-1	-1	1	1	2	3	1	(1000)	2	3	1	(995)	2	3	1	(445)	2	3	1	(547)
				2	2	3	1	(1000)	2	3	1	(894)	2	3	1	(112)	2	3	1	(185)
				3	2	3	1	(1000)	2	3	1	(702)	2	3	1	(43)	2	3	1	(82)

Note 1. Figures in parenthesis indicate the frequency of λ at 0, out of a total frequency of 1000.

T a b l e 5. Showing a possible analysis of a lower mean square error
(2 indicates MINIMAX and 3 MLE)

M_o \ f_o	.133	.532	1.196	1.233	4.931	11.095	18.818	22.778	75.271	91.111	169.360	205.000
.015												2
.024												2
.041											2	
.043												2
.048										3		
.054											2	
.073										3		
.082											2	
.104									3			
.112										2		
.136									2			
.185									2			
.207								3				
.250							3					
.282								3				
.333						3						
.389							3					
.432					3							
.445								2				
.494						3						
.508			3									
.547							2					
.559				3								
.572		3										
.702						2						
.720					3							
.738	3											
.894					2							
.941			2									
.950				2								
.994		2										
.995				2								
1.000	2,2	2	2									

Note 1. f_o denotes the relative frequency of λ at zero.

error compared to minimax while for parameter values closer to the centre of the ellipsoid minimax is better on this score. These facts are further brought out more clearly in Table 5. Another fact worth recording is that BLIMBE has the largest m.s.e. of all three methods. That $M_1 \leq M_0$ was proved in Theorem 4.2. It would be nice if one could similarly show that $M_2 \leq M_0$. This the authors have failed so far.

References

[1] Balakrishnan, A.V., An operator theoretic formulation of a class of control problems and a steepest descent method of solution. J. Soc. Indust. Appl. Math. Ser. A: Control, 1, 109-127 (1963).

[2] Bjerhammar, A., Hyper-estimators. Paper presented at the 41st session of the International Statistical Institute, New Delhi 1977.

[3] Bunke, H. and Bunke, O., Identifiability and estimability. Math. Operationsforsch. Statist., 5, 223-233 (1974).

[4] Bunke, O., Minimax linear, ridge and shrunken estimators for linear parameters. Math. Operationsforsch. Statist., 6, 817--829 (1975).

[5] Chipman, J. S., On least squares with insufficient observations. J. Amer. Statist. Assoc., 59, 1078-1111 (1964).

[6] Drygas, H., Gauss-Markov estimation and best linear minimum bias estimation. Unpublished Technical Report, University of Heidelberg, May 1969.

[7] Forsythe, G.E. and Golub, G. H., On the stationary values of a second degree polynomial on the unit sphere. SIAM J. Appl. Math., 13, 1050-1068 (1965).

[8] Kuks, J., A minimax estimator of regression coefficients (in Russian). Izv. Akad. Nauk Eston. SSR, 21, 73-78 (1972).

[9] K u k s, J. and O l m a n, W., Minimax linear estimation of regression coefficients (in Russian). Izv. Akad.Nauk Eston.SSR, 21, 66-72 (1972).

[10] L a u t e r, H., A minimax linear estimator for linear parameters under restrictions in form of inequalities. Math.Operationsforsch. Statist., 5, 689-696 (1975).

[11] M a j u m d a r, D. and M i t r a, S. K., Least squares under quadratic constraints. Optimizing methods in statistics (ed. J. S. Rustagi) Academic Press (1979).

[12] M i t r a, S. K., Optimal inverse of a matrix, Sankhya. A, 37, 550-563 (1975).

[13] P l a c k e t t, R. L., Some theorems in least squares. Biometrika, 37, 149-157 (1950).

[14] R a o, C. R., Unified theory of linear estimation. Sankhya A, 33, 371-394 (1971).

[15] R a o, C. R., Estimation of parameters in a linear model, Ann. Statist., 4, 1023-1037 (1976).

[16] R a o, C. R. and M i t r a, S. K., Further contributions to the theory of generalized inverse of matrices and its applications. Sankhya A, 33, 289-300 (1971).

[17] R a o, C. R. and M i t r a, S. K., Generalized inverse of matrices and its applications. Wiley, New York 1971.

[18] S c h e f f e, H., Analysis of variance. Wiley, New York 1959.

[19] S c h o n f e l d, P., Best linear minimum bias estimation in linear regressions. Econometrika, 39, 531-544 (1971).

[20] S j o b e r g, L., Are least squares estimators biased? Technical Report. Division of Geodesy, Royal Institute of Technology, Stockholm 1976.

A CORRECTING NOTE TO "STATISTICAL ANALYSIS OF NONESTIMABLE FUNCTIONALS PART I: ESTIMATION"

by

Dibyen Majumdar and Sujit Kumar Mitra

(i) Read equation (3.11) as

$$G_1 = (F_{11}X + F_{12}H)[X'F_{11}X + X'F_{12}H + HF_{21}X + HF_{22}H]^- , \qquad (3.11)$$

where

$$\begin{bmatrix} F_{11} & F_{12} \\ F_{21} & F_{22} \end{bmatrix} = \begin{bmatrix} + XX' & XH \\ HX' & H + H^2 \end{bmatrix}$$

(ii) Read expression (3.12) as

$$\delta^2 p'[K-I]p , \qquad (3.12)$$

where K is a g-inverse of the matrix

$$X'F_{11}X + X'F_{12}H + HF_{21}X + HF_{22}H .$$

ESTIMATION FOR SOME CLASSES OF GAUSSIAN MARKOV PROCESSES

by

Marek Musiela and Roman Zmyślony

Polish Academy of Sciences, Wrocław

1. Introduction

In the paper we consider a class of Gaussian Markov processes $y(t)$. For a fixed $t \in [0,T]$, the random vector $y(t)$ may be interpreted as a vector of observations in a classical linear model. We give a minimal sufficient statistic and obtain a simple characterization of its completeness. Using the terminology of stochastic integrals we give explicit formulas for estimators of regression coefficients and variance components. Moreover, we prove that they are the best unbiased estimators. Such problems are considered in detection, modulation, communication and control. We use here the same notation and terminology as in [7].

2. A minimal sufficient statistic

Let $(\Omega, \mathcal{F}, \mathcal{P})$ be a complete probability space and let $\{\mathcal{F}_t\}$, $t \geq 0$, be a nondecreasing right continuous family of σ- fields contained in $\mathcal{F}$. Moreover, let $y = (y(t), \mathcal{F}_t)$ be an n-dimensional continuous stochastic process and let $W = (W(t), \mathcal{F}_t)$ be an n-dimensional Wiener process. We assume that the processes y and W are connected by the equation

$$y(t) = X(t)\beta + t\, e + \omega(t). \tag{1}$$

Here $X(t)$ is an $n \times p$ matrix of known twice - continuously differentiable functions, β is a $p \times 1$ vector of parameters running over $\Theta \subset R^p$ and e is an $n \times 1$ normal random vector. We assume that e is $\mathcal{F}_0$ measurable with expectation zero and covariance $\Gamma = \sum_{i=1}^{k} \sigma_i V_i$, where $\sigma = (\sigma_1,\dots,\sigma_k)'$ is a vector of parameters running over $\Omega \subset R^k$, while $V_1,\dots, V_k$ are known symmetric matrices. Without loss of generality we may assume that $V_1,\dots, V_k$ are linearly independent. Moreover, we assume that Θ and Ω have non-empty interiors in R^p and R^k, respectively, and e and W are independent. Throughout the paper we assume also that

$$X(0) = X(T) = 0. \tag{2}$$

First we shall derive a minimal sufficient statistic for parameters β and σ. It is convenient for us to use for this problem the factorization theorem [5]. So we have to find a density function of the measure generated by the process y. Note that $z(t) = t\, e + W(t)$ is a Gaussian Markov process. Moreover, the covariance operator of $z(t)$ has the following form

$$R(t,s) = t \cdot s\Gamma + t \wedge s\, I, \tag{4}$$

where I stands for the identity matrix. In view of Theorem 1 in [7] (see also [6]) there exists a Wiener process $v = (v(t), \mathcal{F}_t)$ such that

$$d\, y(t) = (A(t)\beta + \Gamma(I + t\Gamma)^{-1} y(t))dt + dv(t), \tag{5}$$

where

$$A(t) = X_1(t) - \Gamma(I + t\Gamma)^{-1} X(t), \tag{6}$$

while the elements of matrix $X_1(t)$ are the first derivatives of elements of $X(t)$, i.e. $X_1(t)$ is the derivative of $X(t)$.

Now let $\mathcal{C}$ denote the space of all continuous functions from $[0,T]$ to R^n and let $\mathcal{B}$ denote the σ-field of Borel subsets of $\mathcal{C}$. Moreover, let $\mu^{\beta,\Gamma}$ and μ stand for the measures induced by processes $y = (y(t), \mathcal{F}_t)$ and $v = (v(t), \mathcal{F}_t)$, respectively; that is $\mu^{\beta,\Gamma}(B)=P(y\in B)$ and $\mu(B) = P(v\in B)$ for every $B\in\mathcal{B}$. Finally, let $\frac{d\mu^{\beta,\Gamma}}{d\mu}$ denote the Radon-Nikodym derivative.

Theorem 1. The measure $\mu^{\beta,\Gamma}$ is absolutely continuous with respect to the measure μ for every β and Γ. Moreover, the Radon-Nikodym derivative has the form

$$\frac{d\mu^{\beta,\Gamma}}{d\mu}(y) = c(\beta,\Gamma)\, h(T, y(T))\times$$
$$\times \exp\left\{-\frac{1}{2T}\, y'(T)(I+T\Gamma)^{-1}y(T) + \beta'\int_0^T X_1'(t)\, dy(t)\right\}, \tag{7}$$

where

$$\log c(\beta,\Gamma) = -\frac{1}{2}\int_0^T \operatorname{tr}\Gamma(I+t\Gamma)^{-1}dt - \frac{1}{2}\beta'\int_0^T A'(t)\, A(t)\, dt\,\beta$$

and

$$\log h(T, y(T)) = -\frac{1}{2T}\, y'(T)\, y(T).$$

P r o o f: Since y has the differential representation given by (5) $\mu^{\beta,\Gamma}$ is absolutely continuous with respect to μ for every β and Γ and

$$\frac{d\mu^{\beta,\Gamma}}{d\mu}(y) = \exp\left\{\int_0^T \left(A(t)\beta + \Gamma(I+t\Gamma)^{-1}y(t)\right) dy(t) + \right.$$
$$\left. -\frac{1}{2}\int_0^T (A(t)\beta + \Gamma(I+t\Gamma)^{-1}y(t))'(A(t)\beta + \Gamma(I+t\Gamma)^{-1}\, y(t))\, dt\right\}, \tag{8}$$

(see [6], [7]).

Now, let $f(t, x)$, $t\in[0,T]$, $x\in R^n$ stand for the real function defined by formula

$$f(t, x) = \frac{1}{2} x'\Gamma(I+t\Gamma)^{-1} x + \beta' A'(t) x +$$
$$- \frac{1}{2} \int_0^t \operatorname{tr}\Gamma(I+s\Gamma)^{-1} ds - \frac{1}{2}\beta' \int_0^t A'(s) A(s) ds\beta . \qquad (9)$$

Let $A_1(t)$ and $X_2(t)$ stand for the derivatives of $A(t)$ and $X_1(t)$, respectively. From (6) we have

$$A_1'(t) = X_2(t) - A'(t)\Gamma(I + t\Gamma)^{-1} \qquad (10)$$

The Ito formula and formulas (5) and (8) yield

$$df(t, y(t)) = - \frac{1}{2} y'(t)\Gamma(I+t\Gamma)^{-1}\Gamma(I+t\Gamma)^{-1} y(t) +$$
$$+ \beta' A_1'(t)y(t) - \frac{1}{2} \beta' A'(t) A(t)\beta +$$
$$+ (A(t)\beta + \Gamma(I+t\Gamma)^{-1}y(t))' (A(t)\beta + \Gamma(I+t\Gamma)^{-1}y(t))dt +$$
$$+ (A(t)\beta + \Gamma(I+t\Gamma)^{-1}y(t))' d v(t). \qquad (11)$$

In view of (9), (10) and (11) we have

$$f(T, y(T)) - \beta'\int_0^T X_2'(t) d v(t) +$$
$$= \int_0^T (A(t)\beta + \Gamma(I+t\Gamma)^{-1} y(t))' d v(t) + \qquad (12)$$
$$+ \frac{1}{2}\int_0^T (A(t)\beta + \Gamma(I+t\Gamma)^{-1}y(t))' (A(t)\beta + \Gamma(I+t\Gamma)^{-1}y(t)) dt.$$

From the Ito formula it follows that

$$\beta' X_1'(T)y(T) - \beta'\int_0^T X_2'(t) y(t) dt = \beta'\int_0^T X_1'(t) dy(t). \qquad (13)$$

Moreover, it is clear that

$$(I+t\Gamma)^{-1} = \frac{1}{t}(I - (I+t\Gamma)^{-1}). \qquad (14)$$

Finally, using (2), (5), (8), (12), (13) and (14) we obtain the required result.

Note that in view of (7) and the factorization theorem [5] it follows that $((\int_0^T X_1'(t) dy(t))', y'(T))$ is a sufficient statistic for the family $\{\mu^{\beta,\Gamma}\}$. However, in general, this statistic is not minimal sufficient. Now, we state a lemma which will be used to the minimal sufficiency and completeness of this statistic.

Lemma 1. Vectors $\int_0^T X_1'(t)\, dy(t)$ and $y(T)$ are independent.

P r o o f: Note that $\int_0^T X_1'(t)\, dy(t)$ and $y(T)$ are Gaussian random vectors and that $E\, y(T) = 0$. Then it is sufficient to prove that

$$E \int_0^T X_1'(t)\, dy(t)\, y'(T) = 0. \tag{15}$$

The Ito formula yields

$$E \int_0^T X_1'(t) dy(t) y'(T) = X_1'(T) Ey(T) y'(T) - \int_0^T X_2'(t) Ey(t) y'(T) dt. \tag{16}$$

But, in view of (2), it follows that for every $t \in [0,T]$ we have

$$E\, y(t)\, y'(T) = tI + tT\Gamma. \tag{17}$$

So, using (16) and (17) we obtain

$$E \int_0^T X_1'(t)\, dy(t)\, y'(T) = X_1'(T)(I + T\Gamma)T - \int_0^T X_2'(t)(tI + tT\Gamma)\, dt. \tag{18}$$

Since the expression on the right hand side of (18) is equal to $\int_0^T X_1'(t)\, dt\, (I + T\Gamma)$ the lemma follows from (2) and (15).

Now, we shall find a minimal sufficient statistic for $\{\mu^{\beta,\Gamma}\}$. Note that set $\left\{ I + T\Gamma \,\middle|\, \Gamma = \sum_{i=1}^{k} \sigma_1 V_i, \sigma \in \Omega \right\}$ contains an open subset of a linear manifold $\vartheta = I + \vartheta_1$, where ϑ_1 is a subspace generated by $V_1, \ldots, V_k$. Let ω be the smallest manifold such that

$$\left\{ (I + T\Gamma)^{-1} \,\middle|\, \sigma \in \Omega \right\} \subset \omega .$$

Then there exists a matrix W_0 such that $\omega = W_0 + \omega_1$, where ω_1 is a subspace parallel to ω.

Let $W_1, \ldots, W_r$ be a basis for ω_1. Then $(I + T\Gamma)^{-1}$ can be uniquely represented in the following way

$$(I + T\Gamma)^{-1} = \sum_{i=1}^{r} c_i(\sigma)\, W_i + W_o. \tag{19}$$

<u>Theorem 2.</u> Let $\int_0^T X_1'(t)\, X_1(t)\, dt$ be positive definite. Then the vector

$$Z(T) = \left(\left(\int_0^T X_1'(t)\, dy(t)\right)', y'(T)\, W_i y(T),\ i = 1, \ldots, r\right) \tag{20}$$

is a minimal sufficient statistic for $\{\mu^{\beta,\Gamma}\}$.

P r o o f: In view of (19) the density function given by (7) can be represented in the following form

$$\frac{d\mu^{\beta,\Gamma}}{d\mu}(y) = c(\beta,\Gamma) h_o(T, y(T)) \exp\left\{-\frac{1}{2T} \sum_{i=1}^{r} c_i(\sigma) y'(T) W_i\, y(T) + \right.$$

$$\left. + \beta' \int_0^T X_1'(t)\, dy(t)\right\}, \tag{21}$$

where $c(\beta,\Gamma)$ is defined by (7) and

$$\log h_o(T, y(T)) = -\frac{1}{2T} y'(T)\, (I + W_o)\, y(T).$$

Note that $c_1(\sigma), \ldots, c_r(\sigma), \beta_1, \ldots, \beta_p$ are linearly independent. Moreover, from Lemma 1 it follows that the distribution of $Z(T)$ defined by (20) is not degenerated and the coordinates of this vector are linearly independent. Thus, in view of Theorems 5.2 and 5.3 in [4], we obtain the required result.

<u>3. Completeness and estimation</u>

In this section we shall consider the problem of estimation of parameters β and σ. First we shall derive a necessary and sufficient conditions for the completeness of the minimal sufficient statistic $Z(T)$ defined in (20).

Recall that a subspace $\mathcal{D}$ of all symmetric matrices is called a quadratic subspace if $V \in \mathcal{D}$ implies $V^2 \in \mathcal{D}$ (see [8]).

Theorem 3. Let $\int_0^T X_1'(t)\, X_1(t)\, dt$ be positive definite. Then, the following conditions are equivalent:

(a) statistic $Z(T)$ is complete,

(b) ϑ_1 is a quadratic subspace,

(c) $\vartheta_1 = \omega_1$.

P r o o f: The equivalence of (b) and (c) follows from Theorem 5 given in the Appendix provided we put $V_0 = I$. To prove that (b) implies (c) we note that from Corollary 1 in the Appendix it follows that $r = k$ and that the set $\Omega_0 = \{(c_1(\sigma), \ldots, c_k(\sigma))' \mid \sigma \in \Omega\}$ contains an open subset of R^k, where $c_i(\sigma)$ are defined by (19). In view of Lemma 1 the density function of $Z(T)$ is given by (21). Then, $Z(T)$ is exponentially distributed and the natural parameter space $\{(\beta', c'(\sigma)) \mid \beta \in \Theta, \sigma \in \Omega\}$ contains an open subset of R^{p+k}. Hence the statistic $Z(T)$ is complete (see [5]).

Now, suppose that ϑ_1 is not equal to ω_1. Since $\vartheta_1 \subset \omega_1$, there exists a vector $W \in \omega_1$, $W \neq 0$, such that $tr(W(I + T\Gamma)) = 0$ for all $\Gamma = \sum_{i=1}^{k} \sigma_i V_i$. Moreover, $E\, y'(T)\, W\, y(T) = tr(W\, E\, y(T)\, y'(T)) =$ $= T\, tr\,(W(I + T\Gamma))$. Hence $E\, y'(T)\, W\, y(T) = 0$. Because $y'(T)\, W\, y(T)$ is a function of the minimal sufficient statistic $Z(T)$ and because $P(y'(T)\, W\, y(T) \neq 0) = 1$, statistic $Z(T)$ is not complete.

Remark 1. If ϑ_1 is a quadratic subspace, then

$$Z_1(T) = \left(\left(\int_0^T X_1'(t)\, dy(t)\right)',\ y'(T)\, V_i\, y(T),\ i = 1, \ldots, k\right)$$

is a minimal sufficient and complete statistic for $\{\mu^{\beta,\Gamma}\}$.

Now, we are able to describe the best unbiased estimators (B U E for short) of β and σ. An estimator $\hat{\beta}$ of β is called unbiased if $E\hat{\beta} = \beta$ for every β and σ. Moreover, an unbiased estimator $\hat{\beta}$ of β is called B U E if the matrix $E(\bar{\beta} - \beta)(\bar{\beta} - \beta)' - E(\hat{\beta} - \beta)(\hat{\beta} - \beta)'$ is nonnegative definite for every unbiased estimator $\bar{\beta}$ of β and for every β and σ.

Let $A = (a_{ij})$ stand for a symmetric matrix, where $a_{ij} = tr(V_i V_j)$ for $i, j = 1, \ldots, k$. Note that A^{-1} exists because $V_1, \ldots, V_k$ are linearly independent. Moreover, let $u(T) = (u_1(T), \ldots, u_k(T))'$ stand for a random vector, where $u_i(T) = \frac{1}{T^2} y'(T) V_i y(T) - \frac{1}{T} tr V_i$ for $i = 1, \ldots, k$. We prove the following

Theorem 4. Let $\int_0^T X_1'(t) X_1(t) dt$ be positive definite. Then

(i) $\hat{\beta}_T = (\int_0^T X_1'(u) X_1(u) du)^{-1} \int_0^T X_1'(t) dy(t)$

and

(ii) $\hat{\sigma}_T = A^{-1} u(T)$

are the B U E for β and σ, respectively, if and only if ϑ_1 is a quadratic subspace.

P r o o f: Notė that

$$E \int_0^T X_1'(t) dy(t) = \int_0^T X_1'(t) X_1(t) dt \beta.$$

Hence in view of (i) we have $E\hat{\beta}_T = \beta$. Moreover, after simple calculations we obtain

$$E u(T) = \frac{1}{T}(TA\sigma + v - v) = A\sigma,$$

where $v = (tr V_1, \ldots, tr V_k)'$. This implies that $E \hat{\sigma}_T = \sigma$. Since ϑ_1 is a quadratic subspace $Z(T)$ is complete. Thus, $\hat{\beta}_T$ and $\hat{\sigma}_T$ are the B U E because they are functions of a complete sufficient statistic (see Remark 1).

Now let $\hat{\beta}_T$ and $\hat{\sigma}_T$ be the B U E for β and σ, respectively. Then, in particular, $\hat{\sigma}_T$ is the best unbiased estimator in the class of quadratic estimators of the form $y'(T) A y(T)$, where A is a symmetric matrix. But from Theorem 2 in [10] it follows that the subspace generated by

$$\left\{ I + T\Gamma \mid \Gamma = \sum_{i=1}^{k} \sigma_i V_i, \sigma \in \Omega \right\}$$

is a quadratic subspace of all symmetric matrices. This implies that ϑ_1 is also a quadratic subspace.

Remark 2. In view of Theorem 3 in [9] there exists B U E for σ if and only if there exists a best quadratic unbiased estimator for σ.

4. Appendix

Let V_0 be an $n \times n$ regular and symmetric matrix and let ϑ_1 be a linear subspace of all $n \times n$ symmetric matrices. Then

$$\vartheta = V_0 + \vartheta_1 = \{V_0 + V \mid V \in \vartheta_1\}$$

is a linear manifold. If $V_0 \in \vartheta_1$, then $\vartheta = \vartheta_1$. Let $\mathfrak{M}$ be the set of all $M \in \vartheta_1$ such that $V_0 + M$ is regular and let ω be the smallest linear manifold containing the set $\{V^{-1} \mid V \in V_0 + \mathfrak{M}\}$. Since $V_0^{-1} \in \omega$, the manifold ω can be represented in the form $\omega = V^{-1} + \omega_1$, where ω_1 is a subspace parallel to ω. Let $\mathcal{L}$ be the set of all $L \in \omega_1$ such that the matrix $V_0^{-1} + L$ is regular. Moreover, let $(V_0^{-1} + \mathcal{L})^{-1}$ stand for the set of all V^{-1} such that $V \in V_0^{-1} + \mathcal{L}$. We shall describe the subspaces $\mathfrak{M}$ for which the mapping $\pi(V) = V^{-1}$ from $V_0 + \mathfrak{M}$ into $V_0^{-1} + \mathcal{L}$ is open and one to one. For any subspace $\mathcal{D}$ of all symmetric matrices and for any symmetric matrix V let $V\mathcal{D}V = \{VBV \mid B \in \mathcal{D}\}$. Let Σ stand for a symmetric matrix.

Definition 1. A subspace $\mathcal{D}$ of all symmetric matrices is called a Σ-quadratic subspace if $A \in \mathcal{D}$ implies $A\Sigma A \in \mathcal{D}$. If Σ is the identity matrix then $\mathcal{D}$ is called a quadratic subspace (see [8]). Each Σ - quadratic subspace is a Jordan Algebra, where

$$A \circ B = (A\Sigma B + B\Sigma A)/2.$$

Theorem 5. The following conditions are equivalent

(i) $\vartheta_1 = V_0 \omega_1 V_0$,

(ii) $\vartheta = V_0 \omega V_0$,

(iii) $V_o + \mathfrak{M} = (V_o^{-1} + \mathcal{L})^{-1}$,

(iv) ϑ_1 is V_o^{-1} - quadratic subspace.

P r o o f: The equivalence of (i) and (ii) is obvious. We prove that (i) implies (iii). It is sufficient to show that $(V_o^{-1}+\mathcal{L})^{-1} \subset V_o + \mathfrak{M}$. If $L \subset \mathcal{L}$, then $V_o + L$ is regular and in view of (i) we have that $(V_o^{-1} + L)^{-1} = V_o(V_o + V)^{-1} V_o$ for some $V \in \vartheta_1$. So, from (ii) it follows $V_o(V_o + V)^{-1} V_o \subset \vartheta$. Since the above matrix is regular we conclude that it belongs to $V_o + \mathfrak{M}$. Now we prove that (iii) implies (i). If $\Gamma \in \vartheta_1$, then we have that $V_o - \lambda\Gamma$ is regular for sufficiently small λ. Moreover, we have that

$$(V_o - \lambda\Gamma)^{-1} = \sum_{n=0}^{\infty} (V_o^{-1} \lambda \Gamma)^n V_o^{-1} \tag{22}$$

belongs to $V_o^{-1} + \mathcal{L}$, where A^o stands for the identity matrix. Hence we have

$$\sum_{n=1}^{\infty} (V_o^{-1} \lambda\Gamma)^n V_o^{-1} \in \omega_1.$$

So, it is clear that

$$V_o^{-1} \Gamma V_o^{-1} + \sum_{n=2}^{\infty} \lambda^{n-1} (V_o^{-1}\Gamma)^n V_o^{-1} \in \omega_1.$$

It means that $\vartheta_1 \subset V_o \omega_1 V_o$. Similary, one can prove $\vartheta_1 \supset V_o \omega_1 V_o$ putting in (22) V_o^{-1} instead of V_o and taking $\Gamma \in \omega_1$.

We prove that (iv) implies (i). Let ϑ_1 be a V_o^{-1} quadratic subspace. It is clear that for every $n \geqslant 1$ we have $(V V_o^{-1} V)^n \in \vartheta_1$ provided $V \in \vartheta_1$. Since $(V_o + V)^{-1}$ can be represented in the form $V_o^{-1} + \sum_{n=1}^{k} a_n (V_o^{-1} V V_o^{-1})^n$, for some a_n and k, it is clear that $\omega_1 \subset V_o^{-1} \vartheta_1 V_o^{-1}$. Thus, because $\omega_1 \supset V_o^{-1} \vartheta_1 V_o^{-1}$ holds we have $\omega_1 = V_o^{-1} \vartheta_1 V_o^{-1}$.

Now, we prove that (i) implies (iv). Let $V \in \vartheta_1$. Then $\Gamma = V_o^{-1} V V_o^{-1}$ is an element of ω_1 and for sufficiently small λ we have that $V_o^{-1} - \lambda V_o^{-1} V V_o^{-1}$ is regular. Since conditions (i) and (iii) are equivalent we have $(V_o^{-1} - \lambda V_o V V_o)^{-1} \in V_o + \mathfrak{M}$. This implies that $V V_o^{-1} V + \sum_{i=2}^{\infty} \lambda^{i-1} (V V_o)^i V \in \vartheta_1$. Thus letting λ tend to zero we obtain $V V_o^{-1} V \in \vartheta_1$. This terminates the proof of the theorem.

We have the following consequence of Theorem 5.

Corollary 1. If ϑ_1 is a V_o^{-1} - quadratic subspace, then function $\pi: V_o + \mathfrak{M} \longrightarrow V_o^{-1} + \mathcal{L}$ defined by $\pi(V) = V^{-1}$ is an open mapping.

References

[1] A r a t o , M., On the statistical examination of continuous state Markov processes I, II, III,IV, Selected Transl.in Math. Statist. and Probability, Vol, 14 (1978).

[2] B a r n d o r f f - N i e l s e n, O., Exponential families; exact theory, Aarhus University Mathematics Institute, Various Publication Series, No. 19 (1970).

[3] L e B r e t o n, A.,Parameter estimation in a linear stochastic differential equation, Trans. of the Seventh Prague Conference on Information Theory (1974).

[4] J e n s e n, S. T., Covariance hypotheses which are linear in both the covariance and the inverse covariance, Institute of Mathematical Statistics, Univ. of Copenhagen, Preprint,1(1975).

[5] L e h m a n n, E. L., Testing statistical hypotheses,New York 1959.

[6] L i p c e r, R. S. and S h i r y a y e v, A. N. Statistics of random processes I/II, Springer-Verlag Berlin 1977 (1978).

[7] M u s i e l a, M. and Z m y ś l o n y, R., Estimation of regression parameters of Gaussian Markov processes, in this volume, Preprint No 155, Institute of Polish Academy of Sciences(1978).

[8] S e e l y, J. Quadratic subspaces and completeness. Ann. Math. Statist., 42, 710-721 (1971).

[9] S e l l y, J. Minimal sufficient statistics and completeness for multivariate normal families, Sankhya, Ser. A, 39, 170-185 (1977).

[10] Z m y ś l o n y, R., On estimation of parameters in linear models. Applicationes Mathematicae, 22, 271-276 (1976).

ESTIMATION OF REGRESSION PARAMETERS OF GAUSSIAN MARKOV PROCESSES

by

Marek Musiela and Roman Zmyślony
Polish Academy of Sciences, Wrocław

1. Introduction

Let $\{\Omega, \mathcal{F}, \mathcal{P}\}$ be a complete probability space and let for $t \in [0,\infty)$

$$y(t) = X(t)\beta + z(t), \quad y(0) = 0 \tag{1}$$

be an n-dimensional continuous stochastic process. $X(t)$ stands for a matrix of known functions, β for a p-dimensional vector of unknown parameters and $z(t)$ for an n-dimensional Gaussian Markov process with expectation zero and covariance matrix $R(t,s)$. Let $R_o(t)=R(t,t)$. The following limits

$$X_1(t) = \lim_{h \to 0} \frac{1}{h} (X(t+h) - X(t)), \tag{2}$$

$$R_1(t) = \lim_{h \to 0} \frac{1}{h} (R_o(t+h) - R_o(t)) \tag{3}$$

and

$$R_2(t) = \lim_{h \to 0+} \frac{1}{h} (R(t+h,t) - R_o(t)) \tag{4}$$

are assumed to exist for $t \in [0,\infty)$ and to be continuous. In this paper we study the problem of estimation of β when one sample path $y(t)$ on $[0, T]$ is available. Two cases are considered, one deals with a known covariance matrix $R(t,s)$ and the other one with an unknown covariance matrix $R(t,s)$.

This model is somewhat more general than the model considered by

Hajek [7, 8]. He assumed that y is a one dimensional process and that β is an unknown number. In order to find a sufficient statistic Hajek used methods of Hilbert spaces. In 1969 Holevo considered the estimation of regression parameters of process (1) in the one dimensional case. Moreover, he assumed that the process z(t)is stationary. At the begining of the seventies modern theory of martingales and stochastic integrals has been used in many statistical problems arising in stochastic processes, (see [1, 2, 3, 5, 10, 11, 13]). Using the new techniques we obtain best unbiased estimators of regression coefficients.

2. Two auxiliary lemmas

First we prove two lemmas concerning the covariance operator which will be used repeatedly. Let A^+ stand for the Moore-Penrose general inverse of matrix A.

Lemma 1. Whatever be u, $s \leq u \leq t$, we have

$$R(t,s) = R(t,u)\, R_0^+(u)\, R(u,s). \tag{5}$$

P r o o f: Since z(t) belongs to the image of $R_0(t)$ and since $R_0(t)\, R_0^+(t)$ is the projection on $R_0(t)$, it is clear that $z(t) = R_0(t)\, R_0^+(t)\, z(t)$. From the above and a theorem on normal correlation [13] it follows that for $u \leq t$

$$E(z(t)\,|\,z(u)) = R(t,u)\, R_0^+(u)\, z(u). \tag{6}$$

Because z(t) is a Markov process it is clear that $R(t,s) = E(E(z(t)\,|\,z(u))z'(s))$ for $s \leq u$. Thus the lemma follows from (6).

Now let

$$B(t) = R_1(t) - R_2(t) - R_2'(t), \tag{7}$$

where $R_1(t)$ and $R_2(t)$ are defined by (3) and (4), respectively. Symbol A' stands for the transpose of matrix A.

Lemma 2. Matrix B(t) is non-negative definite.

P r o o f: Simple calculations show that

$$B(t) = \lim_{h \to 0^+} \frac{1}{h} E(z(t+h) - z(t))(z(t+h) - z(t))'.$$

Because $\frac{1}{h} E(z(t+h) - z(t))(z(t+h) - z(t))'$ is non-negative definite for every $h > 0$ the lemma follows.

3. Differential representation of the process

In this section we show that the process defined in (1) is a solution of a stochastic differential equation. Let $\bar{y}(t)$ be a solution of the following equation

$$d\,\bar{y}(t) = (A(t)\beta + C(t)\,\bar{y}(t))\,dt + B^{\frac{1}{2}}(t)\,d\,\vec{W}(t), \qquad (8)$$

$$\bar{y}(0) = 0,$$

where

$$A(t) = X_1(t) - R_2(t)\,R_o^+(t)\,X(t), \qquad (9)$$

$$C(t) = R_2(t)\,R_o^+(t), \qquad (10)$$

while $\bar{w}(t)$ is an n-dimensional Wiener process. Moreover, let $\bar{m}(t)$, $\bar{R}(t,s)$ and $\bar{R}_o(t)$ be the expectation, the covariance and the variance matrices of $y(t)$, respectively.

Lemma 3. The distributions of the processes $y(t)$ and $\bar{y}(t)$ coincide.

P r o o f: Because $\bar{y}(t)$ is a Gaussian process it is sufficient to prove that the first two moments of $\bar{y}(t)$ and $y(t)$ coincide. One can prove that $\bar{m}(t)$ and $\bar{R}(t,s)$ are solution of the following differential equations (compare [13], Theorem 15.2).

$$\frac{d\,\bar{m}(t)}{dt} = A(t)\beta + C(t)\,\bar{m}(t), \quad \bar{m}(0) = 0, \qquad (11)$$

$$\frac{d\,\bar{R}_o(t)}{dt} = C(t)\,\bar{R}_o(t) + \bar{R}_o(t)\,C'(t) + B(t), \quad \bar{R}_o(0) = 0, \tag{12}$$

and

$$\bar{R}(t,s) = \begin{cases} D(t)\,D^{-1}(s)\,\bar{R}_o(s) & \text{if } t \geqslant s, \\ \bar{R}_o(t)\,D^{-1}(t)\,D(s) & \text{if } t < s, \end{cases} \tag{13}$$

where

$$\frac{d\,D(t)}{dt} = C(t)\,D(t), \qquad D(0) = I_n. \tag{14}$$

Here I_n stands for the identity matrix. Letting $f(t) = \bar{m}(t) - X(t)\beta$ we obtain from (9) and (11) that

$$\frac{d\,f(t)}{dt} = C(t)\,f(t),\ f(0) = 0\,.$$

Thus we have $\bar{m}(t) = X(t)\beta$. Moreover, letting $G(t) = \bar{R}_o(t) - R_o(t)$, and using (4), (10) and (12) we obtain

$$\frac{d\,G(t)}{dt} = R_2(t)(R_o^+(t)\,\bar{R}_o(t) - I_n) + (\bar{R}_o(t)\,R_o^+(t) - I_n)R_2'(t),$$
$$G(0) = 0. \tag{15}$$

By Lemma 1 we have $R_2(t) = R_2(t)\,R_o^+(t)\,R_o(t)$. Since $R_o^+(t) = R_o^+(t)\,R_o(t)\,R_o^+(t)$ we can rewrite (15) in the following form

$$\frac{d\,G(t)}{dt} = C(t)\,G(t) + G(t)\,C'(t), \quad G(0) = 0.$$

This implies that $\bar{R}_o(t) = R_o(t)$.

Finally, we prove that $\bar{R}(t,s) = R(t,s)$. Without loss of generality we may assume that $s < t$. From Lemma 1 it follows that $\partial R(t,s)/\partial t = C(t)\,R(t,s)$. On the other hand in view of (13) and (14) we have

$$\frac{\partial \bar{R}(t,s)}{\partial(t)} = \frac{d\,D(t)}{dt}\,D^{-1}(s)\,\bar{R}_o(s) = C(t)\,\bar{R}(t,s).$$

This terminates the proof of the lemma.

Now we shall prove that the process defined in (1) is the solution of a stochastic differential equation.

Theorem 1. There exists a standard Wiener process w(t) such that for $t \in [0,T]$

$$d\,y(t) = (A(t)\beta + C(t)\,y(t))\,dt + B^{\frac{1}{2}}(t)\,d\,w(t), \quad y(0) = 0. \tag{16}$$

P r o o f: From Lemma 3 it follows that the distributions of the processes

$$u(t) = y(t) - \int_0^t (A(u)\beta + C(u)\,y(u))\,du \tag{17}$$

and

$$\bar{u}(t) = \bar{y}(t) - \int_0^t (A(u)\beta + C(u)\,y(u))\,du$$

coincide. Using (8) we have

$$\bar{u}(t) = \int_0^t B^{\frac{1}{2}}(u)\,d\,\bar{w}(u).$$

This implies that u(t) has independent increaments. For $t \in [0,T]$ let

$$w(t) = \int_0^t (B^{\frac{1}{2}}(s))^+\,d\,u(s) + \int_0^t (I_n - B^{\frac{1}{2}}(s))^+\,B^{\frac{1}{2}}(s)\,d\,v(s),$$

where v(s) is a standard Wiener process which is independent of y(t). From Lemma 10.4 in [13] it follows that w(t) is a Wiener process and

$$u(t) = \int_0^t B^{\frac{1}{2}}(u)\,d\,w(u).$$

This together with (17) completes the proof of the theorem.

4. Sufficient statistics and estimation

In this section we assume that the vector β runs over a subset $\Theta \subset R^p$ such that the interior of Θ is non-empty. Moreover, we assume that the covariance matrix is known. In this case we shall prove that there exists a complete sufficient statistic for the vector of parameters β.

Let $\mathcal{C}$ stand for the space of continuous functions $f:[0,T] \rightarrow R^n$ endowed with the topology of uniform convergence and let $\mathcal{B}$ denote the σ-field of Borel subsets of $\mathcal{C}$. Moreover, let

$$\mu^{\beta}(B) = P_{\beta}(y \in B) \quad \text{and} \quad \mu(B) = P(u \in B),$$

where $B \in \mathcal{B}$, while y and u are defined by (1) and (17), respectively.

Theorem 2. Suppose that

$$\Sigma(T) = \int_0^T A'(t)\, B^+(t)\, A(t)\, dt$$

is a positive definite matrix. If int $\Theta \neq \emptyset$, then

$$S(T) = \int_0^T A'(t)\, B^+(t)\, [d\, y(t) - C(t)\, y(t)\, dt] \tag{18}$$

is a complete and sufficient statistic for $\{\mu^{\beta} \mid \beta \in \Theta\}$.

P r o o f: Since $y(t)$ has the representation given by (16) it follows from Theorem 7.20 in [13] that $\mu^{\beta} \ll \mu$ for every $\beta \in \Theta$ and that

$$\frac{d\,\mu^{\beta}(y)}{d\mu} = \exp\left\{ \int_0^T (A(t)\beta + C(t)\, y(t))'\, B^+(t)\, d\, y(t) + \right.$$

$$\left. - \frac{1}{2} \int_0^T (A(t)\beta + C(t)\, y(t))' B^+(t) (A(t)\beta + C(t)\, y(t))\, dt \right\}. \tag{19}$$

After a simple calculation the factorization theorem [12] implies that $S(T)$ defined by (18) is sufficient. Moreover, formula (18) implies that $S(T)$ is normally distributed $N(\Sigma(T)\beta,\ \Sigma(T))$. So, $S(T)$ is complete because $\Sigma(T)$ is positive definite and int Θ is non-empty.

Note that Theorem 1 is valid provided $\Sigma(T)$ is positive definite. In the following lemma we give a simple characterization of this assumption.

Lemma 4. Let $M(t)$ be an $n \times n$ non-negative definite symmetric matrix for all $t \in [0,T]$. Let the mapping $t \to M(t)$ be continuous. The integral $\int_0^T M(t)\, dt$ is positive definite if and only if there exist $t_1, \ldots, t_k \in (0,T)$, where $k \leq n(n+1)/2 + 1$, such that $\sum_{i=1}^{k} M(t_i)$ is positive definite.

P r o o f: Suppose that $\int_0^T M(t)\, dt$ is positive definite. Let $\mathcal{M}^*$ stand for the convex hull of $\mathcal{M} = \{M(t) \mid t \in (0,T)\}$. Since $\frac{1}{T}\int_0^T M(t)\, dt \in \mathcal{M}^*$

it follows from the Caratheodory theorem that there exist numbers $t_1, \ldots, t_k \in [0,T]$, where $k \leqslant n(n+1)/2 + 1$, and $p_i > 0$, $\sum_{i=1}^{k} p_i = 1$, such that

$$\int_0^T M(t)\, dt = T \sum_{i=1}^{k} p_i\, M(t_i).$$

Since $M(t_i)$ is non-negative definite and since $T \sum_{i=1}^{k} p_i\, M(t_i)$ is positive definite $\sum_{i=1}^{k} M(t_i)$ is positive definite, too.

Now, let $\sum_{i=1}^{k} M(t_i)$ be positive definite. Then the mapping $(s_1, \ldots, s_k) \longrightarrow \sum_{i=1}^{k} M(s_i)$ is continuous. Hence, there exists $\varepsilon > 0$ such that $\sum_{i=1}^{k} M(s_i)$ is positive definite for every $s_i \in (t_i - \varepsilon,\ t_i + \varepsilon)$, $i = 1, \ldots, k$. It is clear that

$$\int_0^T (M(t)x, x)\, dt \geqslant \int_0^{\varepsilon} \left(\sum_{i=1}^{k} M(t_i + t)x, x \right) dt > 0.$$

This terminates the proof.

Now we consider the problem of estimation of β and $Ey(t) = X(t)\beta$.

Theorem 3. Suppose that $\Sigma(T)$ is positive definite. If $\operatorname{int} \Theta$ is non-empty, then

$$\hat{\beta}_T = \Sigma^{-1}(T) S(T) \tag{20}$$

is the best unbiased estimator of β.

P r o o f: Since $\Sigma(T)$ is positive definite, $\hat{\beta}_T$ is an unbiased estimator of β [14]. Moreover, in view of Theorem 2, $S(T)$ is a complete and sufficient statistic. Hence the Rao-Blackwell Theorem implies that $\hat{\beta}_T$ is the best unbiased estimator of β.

Remark 1. Note that $X(t)\sum^{-1}(T)\,S(T)$ is the best unbiased estimator of $Ey(t)$.

Remark 2. If $\lim_{T\to\infty}\sum^{-1}(T) = 0$, then $\hat{\beta}_T$ is a consistent estimator of β, i.e $\lim_{T\to\infty}\hat{\beta}_T = \beta$ a.s. for each $\beta\in\Theta$.

When $R(t,s)$ is unknown μ depends on $R(t,s)$, thus μ does not dominate μ^{β}. In order to find out whether or not μ dominates μ^{β} we need to have an explicit form of $B(\cdot)$. We shall prove that there exists $\mathcal{F}_t^y = \sigma(y(s),\ s\le t)$ - measurable modification of B, say $\hat{B}$, such that B and $\hat{B}$ are indistinguishable (see [4] for definition)

Let $t_k^{(n)} = \frac{kT}{2^n}$, where $k = 0, 1, \dots, 2^n$.

Define

$$Q_n(t) = \sum_{k\in I_n(t)} \left(y(t_{k+1}^n) - y(t_k^n)\right)\left(y(t_{k+1}^n) - y(t_k^n)\right)',$$

where $I_n(t) = \left\{k:\ t_k^n \le t\right\}$. Clearly, $Q_n(t)$ converges to $Q(t)$ with probability one, where $Q(t) = \int_0^t B(u)du$ (see [6], [13]). Since $B(t)$ is continuous it follows that

$$\hat{B}(t) = \lim_{h\to 0}\frac{1}{h}\left(Q(t+h) - Q(t)\right)$$

is a $\mathcal{F}_t^y$ measurable modification of $B(t)$. So we have proved the following lemma.

Lemma 5. Processes $\hat{B}(\cdot)$ and $B(\cdot)$ are indistinguishable.

Remark 3. By Lemma 5 process $B(\cdot)$ can be determined from an observation of the process in the interval $[0,T]$. So, we can assume for futher considerations that $B(t)$ is known. If this is the case we have, in view of (16), that $\mu^{\beta} \ll \mu$ for every β and $R(t,s)$. Moreover, if $C(t) = R_2(t)\,R_0(t)$ is known, it is easy to see that $S(T)$ is still a sufficient statistic for μ^{β}, and that an estimator of β can be obtained in the same way as in (20) by putting $\hat{B}(t)$ instead of $B(t)$.

5. Examples

Three examples of applications of our results will be given. When $z(t)$ is a Wiener process the best estimators have an analogous form as those obtained in the case of the classical linear models. We deal with this situation in Example 1. Example 2 illustrates how one may apply Lemma 5 for certain Gaussian processes. In these two examples $C(t) = 0$. Example 3 is free of this assumption, however it deals with the case when a best estimator exists and does not depend on $C(\cdot)$.

Example 1. Let $y(t) = X(t)\beta + w(t)$ be a continuous stochastic process, where $w(t)$ is an n-dimensional Wiener process. Then $A(t) = X_1(t)$, $B(t) = I$ and $C(t) = 0$. If $\int_0^T X_1'(t)\, X_1(t)\, dt$ is positive definite, then, in view of (20),

$$\hat{\beta}_T = \left(\int_0^T X_1'(t)\, X_1(t)\, dt\right)^{-1} \int_0^T X_1'(t)\, d\, y(t).$$

In the particular case when $X(t) = t\,X$ this formula reduces to $\hat{\beta}_T = \frac{1}{T}\,(X'X)^{-1}\, X'y(T)$.

Example 2. A covariance operator $R(\cdot,\cdot)$ of a Gaussian Markov process can be represented as $R(t,s) = U(t)\,V(s)$ for $t \geqslant s$, where $U(t)$ and $V(s)$ are $n{\times}n$ matrices. This result was obtained by Timoszyk [15] in the one dimensional case. For a generalization to the n-dimensional case see [13].

Let us now consider a process for which $R(t,s) = V(s)$. Assume that $V(\cdot)$ is unknown, and that $V_1(t) = \lim_{h \to 0} \frac{1}{h}\,(V(t+h) - V(t))$ is continuous. From Lemma 5 it follows that there exists $\mathcal{F}_t^y$-measurable process $\hat{V}(t)$ such that V and $\hat{V}$ are indistinguishable. If

$$\int_0^T X_1'(t)\, \hat{V}_1^{+}(t)\, X_1(t)\, dt$$

is positive definite then the best unbiased estimator of β is of the following form

$$\hat{\beta}_T = \left(\int_0^T X_1(t)\, V_1^+(t)\, X_1(t)dt\right)^{-1} \int_0^T X_1(t)\, V_1^+(t)\, d\, y(t).$$

E x a m p l e 3. Let us consider the process $y(t) = t\,x + w(t)$ where $w(t)$ is a Wiener process, while x is normally distributed with an unknown expectation β and a known covariance matrix Γ. Finally, we assume that x and $w(\cdot)$ are independent. Then $y(t) = t\beta + z(t)$, where $z(t) = t(x-\beta) + w(t)$ is a Gaussian Markov process. For this process we have

$$R(t,s) = t\,s\,\Gamma + t \wedge s\, I.$$

Easy calculations show that $A(t) = (I + t\,\Gamma)^{-1}$, $B(t) = I$ and that $C(t) = \Gamma(I + t\Gamma)^{-1}$. Moreover, using

$$\frac{d\, A^{-1}(t)}{dt} = -\, A^{-1}(t)\, \frac{d\, A(t)}{dt}\, A^{-1}(t),$$

we obtain

$$\textstyle\sum(T) = T(I + T\,\Gamma)^{-1}.$$

In view of (21) and (20) we have

$$\hat{\beta}_T = \frac{1}{T}(I + T\Gamma) \int_0^T (I+t\Gamma)^{-1}\left[d\, y(t) - \Gamma(I+t\Gamma)^{-1} y(t)\, dt\right]. \tag{22}$$

From the Ito formula it follows that

$$d(I + t\Gamma)^{-1}(y(t) - t\beta) = (I + t\Gamma)^{-1} d\, w(t). \tag{23}$$

Combining (21), (22), (23) and using Theorem 1 we find

$$\hat{\beta}_T = \frac{1}{T}\, y(T).$$

Because $\hat{\beta}$ is independent of Γ, these results hold also when Γ is unknown.

Acknowledgement. We wish to thank T. Bednarski and A. Kozek for useful comments and discussions.

References

[1] A a l e n, O., Statistical inference for a family of counting processes, Ph. D. dissertation, Univ. of California, Berkeley. Reprinted by the Copenhagen University Institute of Mathematical Statistics (1975).

[2] A r a t o, M., Exact formulas for densities of measures of elementary Gaussian processes, Studia Scientiarum Mathematicarum Hungarica, 5, 17-27 (1970).

[3] - On estimation of parameters of stochastic linear differential equations, ibidem, 5, 11-15 (1970).

[4] D e l l a c h e r i e, C., Capacites et processus stochastiques, Springer - Verlag Berlin 1972.

[5] D i o n, J. P. and K e i d i n g, N., Statistical inference in branching processes, Preprint No 10, Institute of Mathematical Statistics University of Copenhagen (1977).

[6] F i s k, D. L., Sample quadratic variation of simple second order martingales. Z. Wahrscheinlichkeitstheorie verw. Geb., 6, 273-278 (1966).

[7] H a j e k, J., On a simple linear model in Gaussian processes, Proc. of the Second Prague Conf. on Prob. Th. (1962).

[8] - On linear statistical problems in stochastic processes, Czechoslovak Math. J., 12, 404-444 (1962).

[9] H o l e v o, A. S., On estimates of regression coefficients, Theory of Probability and its Applications 14, 79-104 (1969).

[10] L e B r e t o n, A. On continuous and discrete sampling for parameter estimation in diffusion type processes, Mathematical Programming Study 5, 124-144 (1976).

[11] L e B r e t o n, A., Parameter estimation in a linear stochastic differential equation, Trans. of the Seventh Prague Conference on Information Theory (1974).

[12] L e h m a n n, E. L., Testing Statistical Hypotheses, New York 1959.

[13] L i p c e r, R. S. a n d S h i r y a y e v, A. N., Statistics of random processes I/II, Springer - Verlag Berlin 1977, 1978.

[14] R a o, C. R., Linear statistical inference and its applications, New York 1973.

[15] T i m o s z y k, W., A characterization of Gaussian processes that are Markovian, Colloquium Mathematicum, 30, 157-167(1974).

SOME REMARKS ON THE CENTRAL LIMIT THEOREM
IN BANACH SPACES

by

Jan Rosiński

Wrocław University and Polish Academy of Sciences, Wrocław

1. Introduction

Let E be a separable Banach space. A symmetric E-valued random vector (r.v. for short) X (or the distribution $\mathcal{L}(X)$ of X) is said to be stable of order p, $p \in (0,2]$, if $\mathcal{L}(aX_1+bX_2) = \mathcal{L}((a^p+b^p)^{1/p}X)$, for all $a,b > 0$, where X_1, X_2 are independent copies of X.

Let $(\Omega, \mathcal{F}, P)$ be a fixed probability space. For each $p \in (0,2]$ let DNA_p denote the set of all symmetric E-valued r.v. X such that $n^{-1/p}(X_1+\ldots+X_n)$ converges in law whenever $X_1, X_2, \ldots$ are independent copies of X. As known the limit law of $n^{-1/p}(X_1+\ldots+X_n)$ is a symmetric stable measure of order p.

The space DNA_2 was investigated by G. Pisier in [10] (without the assumption of symmetry of X). In Section 2 we establish some properties of DNA_p spaces. In Section 3 we give a new characterization of Banach spaces of stable type $p<2$, also a full characterization of DNA_p spaces for Banach spaces of stable type p for $p<2$. Necessary and sufficient conditions for r.v. to be in the domain of attraction of a stable law were obtained by A. Araujo and E. Gine [2] which used methods based on the relations between tightness of row sums in infinitesimal arrays and tightness of the accompanying Poisson laws. We prove this result under the assumption of symmetry of X using the characterization of stable type Banach spaces given in Theorem 2. It seems that our proof is simpler than the one given in [2].

2. Domain of attraction of a stable law

For every r.v. X in E and every $p \in (0,\infty)$ define

$$\Lambda_p(X) = \sup_{t>0} t^p P\{\|X\|>t\}.$$

Λ_p is a p-homogeneous metrizable modular on $L^0(\Omega,\mathcal{F},P;E)$ so that $\Lambda_p(\Omega,\mathcal{F}, P; E) \stackrel{df}{=} \{X : \Lambda_p(X)<\infty\}$ forms the Frechet space with the F-norm

$$\|X\|_{\Lambda_p} \stackrel{df}{=} [\Lambda_p(X)]^{\frac{1}{p+1}}$$

(see for example [11] p.17). Moreover, for every $0 \leq q < p$ we have

$$L^p(\Omega, \mathcal{F}, P; E) \subset \Lambda_p(\Omega,\mathcal{F}, P; E) \subset L^q(\Omega,\mathcal{F}, P; E)$$

and the natural imbeddings are continuous.

Proposition 1. If $X \in DNA_p$, then

$$\sup_n \Lambda_p (n^{-1/p} \sum_{i=1}^n X_i) < \infty .$$

Moreover for every $\varepsilon > 0$ there exist a finite dimensional subspace F_ε of E such that

$$\sup_n \Lambda_p(\text{dist}\,(n^{-1/p} \sum_{i=1}^n X_i, F_\varepsilon)) < \varepsilon ,$$

where $X_1, X_2, \ldots$ are independent copies of X.

Proposition 2. If $E \neq \{0\}$ then DNA_p is a linear subspace of $L^0(\Omega, \mathcal{F}, P; E)$ for p = 2 only.

Now, we will define a metric in DNA_p. For each $X \in L^0(\Omega,\mathcal{F}, P; E)$ denote by X_i a r.v. defined on the product space $(\times_{n\in\underline{N}}\Omega, \times_{n\in\underline{N}}\mathcal{F}, \times_{n\in\underline{N}} P)$ by $X_i((\omega_1,\ldots)) = X(\omega_i)$. Then

$$\Lambda_p(X,Y) = \sup_n \Lambda_p(n^{-1/p} \sum_{i=1}^{n} (X_i - Y_i)),$$

$X, Y \in DNA_p$ is a metric in DNA_p.

Now let DNA_p^f be the set of all r.v. $X \in DNA_p$ which take a.e. values in a finite dimensional subspace of E.

Theorem 1. For every $p \in (0,2]$ the space (DNA_p, ρ_p) is a complete metric space. Moreover, DNA_2 is a linear space and the set of simple r.v. is dense in DNA_2. If $p>2$ and if E has the bounded approximation property, then DNA_p^f is dense in DNA_p.

Finally we mention the following property (called D_p) of probability measures on E.

(D_p) For every symmetric stable measure γ_p of order p on E holds true: if X is symmetric and if $\lim \mathcal{L}(n^{-1/p} \sum_{i=1}^{n} x^* X_i) = x^* \gamma_p$ for every $x^* \in E^*$, then $\lim_{n \to \infty} \mathcal{L}(n^{-1/p} \sum_{i=1}^{n} X_i) = \gamma_p$ (here $X_1, X_2, \ldots$ are independent copies of X).

Proposition 4. If E has property D_p for some $p<2$, then E is a finite dimensional space. A Banach space statisfies D_2 if and only if it is of Rademacher cotype 2.

3. Banach spaces of stable type $p<2$

A Banach space E is said to be of stable-type p if for every sequence $\{x_n\} \subset E$ such that $\sum \|x_n\|^p < \infty$ the series $\sum g_n x_n$ converges a.e. in E, where g_n are independent stable real r.v.'s with characteristic functions $E \exp(itg_n) = \exp(-|t|^p)$, $n \geq 1$. It is known that every Banach space is of stable type $p<1$ and that Hilbert space is of stable-type p for every $p \in (0,2]$. L^q spaces are of stable-type p for every $p<q$.

Theorem 2. A Banach space E is of stable-type p, for $p<2$, if and only if there exists a constant $C = C(E)$ such that

$$\Lambda_p\left(\sum X_i\right) \leqslant C \sum \Lambda_p(X_i) \tag{1}$$

for every finite sequence $\{X_i\}$ of independent symmetric r.v. in E such that $\Lambda_p(X_i) < \infty$.

Proposition 5. If E is of stable-type p for $p<2$ and if F is a closed subspace of E, then inequality (1) holds with the same constant $C = C(E)$ for every finite sequence $\{Y_i\}$ of independent symmetric r.v. in the quotient space E/F.

Theorem 3. Let $p \in (0,2)$. The following proprties of Banach space E are equivalent:

(i) For every r.v. X in E, X belongs to DNA_p if and only if

$$\lim_{t\to\infty} t^p\, P\left\{|x^*X| > t\right\} \text{ exists for every } x^* \in E^* \tag{2}$$

and for every $\varepsilon > 0$ there exists a finite dimensional subspace F_ε of E such that

$$\Lambda_p(\mathrm{dist}(X, F_\varepsilon)) < \varepsilon \tag{3}$$

(ii) E is stable-type p.

P r o o f: First we notice that the condition (2) is equivalent to the existence of the limit $\lim_{n\to\infty} \mathcal{L}(n^{-1/p} \sum_{i=1}^{n} x^*(X_i))$ for every $x^* \in E$ ([3], chapter 7, § 35). By Proposition 1 conditions (2) and (3) are necessary for X to be in DNA_p for every Banach space.

We show that (ii) implies (i). Suppose that (2) and (3) are satisfied. In view of (2) it is sufficient to show that for every $\delta > 0$

there exists a finite dimensional subspace F of E such that

$$\sup_n P\left\{\operatorname{dist}(n^{-1/p}\sum_{i=1}^{n} X_i, F) > \delta\right\} \leq \delta.$$

Let F be a closed subspace of E and $\pi_{E/F}$ denote the canonical mapping of E onto E/F. Using Proposition 5 we obtain

$$\Lambda_p(\operatorname{dist}(n^{-1/p}\sum_{i=1}^{n} X_i, F)) = \Lambda_p(\pi_{E/F}(n^{-1/p}\sum_{i=1}^{n} X_i)) \leq$$

$$C\Lambda_p(\pi_{E/F}(X)) = C\Lambda_p(\operatorname{dist}(X,F)),$$

where C does not depend on F.

Let $\delta > 0$. Take $\varepsilon = \delta^{p+1} C^{-1}$ and put $F = F_\varepsilon$, where F_ε appears in (3). We have

$$P\left\{\operatorname{dist}(n^{-1/p}\sum_{i=1}^{n} X_i, F) > \delta\right\} \leq \delta^{-p}\Lambda_p(\operatorname{dist}(n^{-1/p}\sum_{i=1}^{n} X_i, F)) \leq \delta$$

which ends the proof of (ii) $\Rightarrow$ (i).

The inverse implication may be proved as follows: If X is a symmetric r.v. such that $\lim_{t\to\infty} t^p P\{\|X\| > t\} = 0$, then X satisfies conditions (2) and (3). Now X belongs to DNA_p by (i), and obviously in this case $n^{-1/p}\sum_{i=1}^{n} X_i \xrightarrow{P} 0$. Thus we have that every symmetric r.v. X such that $\lim_{t\to\infty} t^p P\{\|X\| > t\} = 0$ satisfies the weak law of large numbers. In view of a result of B. Marcus and W. Woyczyński on the weak law of large numbers [7] E is of stable type p.

References

[1] de Acosta, A. Asymptotic behaviour of stable measures, Ann. Prob., 5, 494-499 (1977).

[2] A r a u j o, A. and G i n é, E., On tails and domains of attraction of stable measures in Banach spaces (to appear).

[3] G n e d e n k o, B. V. and K o l m o g o r o w, A. N. Limit Distributions for Sums of Independent Random Variables, Moscow (1949), Eng. transl., Addison-Wesely Publishing Co.Inc., Reading, Mass., (1954).

[4] H o f f m a n n-J ø r g e n s e n, J., Sums of independent Banach space valued random variables, Aarhus U. Preprint Series 15, (1972/1973).

[5] H o f f m a n n-J ø r g e n s e n, J. and P i s i e r, G., The law of large numbers and the central limit theorem in Banach spaces, Ann. Prob., 4, 587-599 (1976).

[6] K u e l b s, J. and M a n d r e k a r, V., Domains of attraction of stable measures on a Hilber space, Studia Math. 50, 149--162 (1974).

[7] M a r c u s, M. B. and W o y c z y ń s k i, W. A., Stable measures and central limit theorems in space of stable type, Trans. Amer. Math. Soc. (to appear).

[8] M a u r e y, B. and P i s i e r, G., Séries de variables aléatoires vectorielles indépendantes et propriétés géométriques des espaces de Banach, Studia Math., 58, 45-90 (1976).

[9] P i s i e r, G., Sur les espaces qui ne cotiennent pas de l_n^1 uniformement, Comptes Rendus Acad. Sci., Paris 277, 991-994(1973)

[10] P i s i e r, G., Sur le theoreme limite central et la loi du logarithme itere dans les espaces de Banach, Sem. Maurey-Schwartz, 1975/1976, Exp. III-IV.

[11] R o l e w i c z, S., Metric Linear Spaces, Polish Sci. Publ. Warszawa 1972.

CHARACTERIZATION OF COVARIANCE OPERATORS WHICH GUARANTEE THE CLT

by

V.I. Tarieladze

Academy of Sciences of the Georgian SSR, Tbilisi

1. Introduction

In the paper we consider weak second order mean zero random elements. The problem considered is under what conditions imposed on covariance operators these random elements satisfy the central limit theorem. A characterization of covariance operators having these property will be given and used to establish corresponding results concerning continuous processes.

Let X be a real Banach space. Let $(\Omega, \mathcal{A}, P)$ be a (sufficiently reach) probability space. A Bochner measurable map $\xi : \Omega \to X$ is called a random element (r.e., for short). The distribution μ_ξ of a r.e. ξ, which is defined by

$$\mu_\xi(B) = P(\xi^{-1}(B))$$

is a Radon probability measure on X. If Radon probability measures μ_n converge weakly to μ as $n \to \infty$, we write $\mu_n \xrightarrow{w} \mu$.

Throughout the paper X^* stands for the dual space of X and $\langle x, x^* \rangle$ for the value of the linear functional $x^* \in X^*$ at $x \in X$. If $E \langle \xi, x^* \rangle^2 < \infty$, then ξ is said to be a weak second order r.e. A second order r.e. has a mean value $x_\xi \in X$ (that is $E\langle \xi, x^* \rangle = \langle x_\xi, x^* \rangle$ for all $x^* \in X^*$) and it has a covariance operator $R_\xi : X^* \to X$ (that is $\langle R_\xi x^*, y^* \rangle = E\langle \xi, x^* \rangle \langle \xi, y^* \rangle - \langle x_\xi, x^* \rangle \langle x_\xi, y^* \rangle$ for all x^*, $y^* \in X^*$. Every covariance operator R is a symmetric and a positive linear operator.

It may be easily established that every covariance operator has a separable range. Conversely, any symmetric and positive linear operator $R: X^* \longrightarrow X$ with a separable range is a covariance operator of a weak second order r.e. [1]. A r.e. Γ is said to be Gaussian, if $<\Gamma, x^*>$ is a Gaussian random variable for every $x^* \in X^*$.

Let ξ be a r.e. in X. Let $\{\xi_n\}$ be a sequence of independent copies of ξ and let $Z_n = n^{-1/2} \sum_{k=1}^{n} \xi_k$, where n = 1,2,... . We say r.e. ξ satisfies the central limit theorem (CLT, for short), if there exists a Gaussian r.e. Γ in X such that $\mu_{Z_n} \xrightarrow{w} \mu_\Gamma$ as $n \to \infty$.

Proposition 1. Suppose that r.e. ξ in X satisfies the CLT. Then ξ has the following properties:

(i) $E<\xi, x^*>^2 < \infty$ and $E<\xi, x^*> = 0$ for all $x^* \in X^*$,

(ii) there exists a Gaussian r.e. Γ such that $R_\xi = R_\Gamma$,

(iii) $r^2 P[\|\xi\| > r] \longrightarrow 0$ as $r \longrightarrow \infty$.

Properties (i) and (ii) follow immediately from the one dimensional CLT. Property (iii) has been proved in [2].

From part (iii) of Proposition 1 if follows that $E\|\xi\|^p < \infty$ for all $0 < p < 2$. This particular result was already proved in [3].

For convenience of the reader we recall the notions of type 2 and cotype 2 Banach spaces. They are needed to make some historical comments.

Let $\{\gamma_n\}$ be a sequence of independent standard Gaussian random variables. A Banach space X is said to be of type 2 if $\sum x_n \gamma_n$ converges a.s. in X for any sequence $\{x_n\} \subset X$ such that $\sum \|x_n\|^2 < \infty$ and is said to be cotype 2 if $\sum \|x_n\|^2 < \infty$ for any sequence $\{x_n\} \subset X$ such that $\sum x_n \gamma_n$ converges a.s. in X.

A central limit theorem for the infinite dimensional case was first established in [6]. It asserts that if X is a separable Hilbert space, then condition $E\,\|\xi\|^2 < \infty$ is necessary and sufficient for ξ to satisfy the CLT. Later it has been established that $E\|\xi\| < \infty$ is a sufficient condition in reflexive type 2 spaces provided they have a basis [7]. In [8] it has been shown that condition (ii) of Proposition 1 is sufficient in case $X = l_p$ for $1 \leqslant p \leqslant 2$. A counter-example given in [9] has illustrated the difficulties on the way to formulate necessary and sufficient conditions for the CLT. It has been shown that there exists a r.e. ξ in C(0,1) meeting condition (ii) of Proposition 1 and the condition $\|\xi\| \leqslant 1$ a.s., but not satisfying the CLT. In [10] it was shown that condition $E\,\|\xi\|^2 < \infty$ is sufficient for the CLT if and only if X is of type 2, and in [11-12] that condition (ii) of Proposition 1 implies the CLT if and only if X is of cotype 2. Recently [2] conditions (ii) and (iii) have been shown to be sufficient for the CLT in l_p for $2 < p < \infty$.

Finally we would like to recall that a characterization of the Banach spaces in which it is possible to construct a r.e. ξ fulfilling (ii) and $\|\xi\| \leqslant 1$ a.s., but not satisfying the CLT, has been given in [13].

In Section 2 we will prove a theorem which describes covariances that have desirable properties from the point of view of the CLT. Using these results we will investigate in Section 3 the CLT for continuous processes.

2. Structure of covariance operators which quarantees the CLT

The main results of this section is the following.

Theorem 1. Let X be a real Banach space let $R\colon\ X^* \longrightarrow X$ be a symmetric and positive linear operator with a separable range. Then the following statements on R are equivalent:

(1) each r.e. ξ in X with $R_\xi = R$ satisfies the CLT;

(2) each r.e. ξ with $R_\xi = R$ has a finite positive strong moment, i.e. there exists $p = p_\xi > 0$ such that $E\,\|\xi\|^p < \infty$;

(3) R can be written as $R = usu^*$, where $s\colon l_2 \to l_2$ is a positive and symmetric nuclear linear operator and $u\colon l_2 \to X$ is a continuous linear operator;

(4) If ξ is a weak second order mean zero r.e. in X with $R_\xi = R$, then $\xi = u\eta$ a.s., where η is a r.e. in l_2, $E\,\|\eta\|^2 < \infty$, $E\eta = 0$ and $u\colon l_2 \to X$ is a continuous linear operator which depends on R only;

(5) Let X be a vector subspace of X^* which separates points of X and let $T\colon Y \to L_2(\Omega, \mathcal{A}, P)$ be a linear operator. If $\|Tx^*\|^2 = \langle R x^*, x^* \rangle$ for all $x^* \in Y$, then there exists a r.e. ξ in X such that $T x^* = \langle \xi, x^* \rangle$ holds a.s. for all $x^* \subset Y$.

The proof of this theorem is based on the following result.

Proposition 2 [14]. Let H be a Hilbert space, X a Banach space and let $v\colon H \to X$ be a continuous linear operator such that: $\sum \|ve_n\|^2 < \infty$ for each orthonormal sequence $\{e_n\}$ in H. Then v admits the factorization $v = uA$, where $u\colon l_2 \to X$ is a continuous linear operator and $A\colon H \to l_2$ is a Hilbert-Schmidt operator.

P r o o f of Theorem 1: Obviously (1) implies (2) by Proposition 1. Next we show that (2) implies (3). Operator R can be written as $R = vv^*$, where $v\colon H \to X$ is a continuous linear operator, while H is a separable Hilbert space ([1], p. 135). It is enough to prove that v satisfies the assumption of Proposition 2. Because then $R = usu^*$, where $s = AA^*$ is a symmetric positive nuclear operator. Now let $\{e_n\}$ be an orthonormal basis in H, and let $x_n = ve_n$, $n = 1, 2, \ldots$. We may assume, that $0 < \|x_n\| < 1$. Now we need that to show that $\sum \|x_n\|^2 < \infty$. Suppose to the contrary that $\sum \|x_n\|^2 = \infty$. Then there would exist a sequence $\{\beta_n\}$ such that $\beta_n > 0, \beta_n \downarrow 0$ as $n \to \infty$ and such that $\sum \|x_n\|^2 \beta_n = 1$ and $\sum \|x_n\|^2 \beta_n^\delta = \infty$ for every $0 < \delta < 1$. Let ξ be a r.e. in X such that for $n = 1, 2, \ldots$

$$P\left[\xi = x_n\|x_n\|^{-1}\beta_n^{-\frac{1}{2}}\right] = P\left[\xi = -x_n\|x_n\|^{-1}\beta_n^{-\frac{1}{2}}\right] = \frac{1}{2}\|x_n\|^2\beta_n .$$

Then $R_\xi = R$ and $E\|\xi\|^p = \infty$ for all $p>0$, which contradicts (2). Thus (3) is established.

To show that (3) implies (4) we proceed as follows. Let ξ be a r.e. in X with $R_\xi = R$. Consider the operator $T: X \to L_2$ defined for each $x^* \in X^*$ by $Tx^* = \langle \xi, x^* \rangle$. Then $\|Tx^*\|^2 = \langle Rx^*, x^* \rangle$ for each $x^* \in X^*$. We may assume that the range of $s^{\frac{1}{2}}u^*$ is dense in l_2. Consider the isometry $v: l_2 \to L_2$ defined by $vs^{\frac{1}{2}}u^* = T$. Clearly $vs^{1/2}: l_2 \to L_2$ is a Hilbert-Schmidt operator and thus it has the representation $vs^{1/2}h = (\eta, h)$ a.s. for all $h \in l_2$, here η is a r.e. in l_2 with $E\|\eta\|^2 < \infty$. Now we observe easily that $E\eta = 0$ and $u\eta = \xi$ a.s. And ξ satisfies the CLT, because η satisfies it in l_2. Thus (4) implies (1).

Now we show that (3) implies (5). Since $\|Tx^*\|^2 = \langle Rx^*, x^* \rangle$, $x^* \in Y$ it follows that T admits a unique continuous extension $\bar{T}: X^* \to L_2$. As above we may construct an isometry $v: l_2 \to L_2$ such that $vs^{1/2}u^* = \bar{T}$ and next find a r.e. η in l_2 such that $vs^{1/2}h = (\eta, h)$ a.s. for all $h \in l_2$. Clearly $u\eta = \xi$ has all the required properties.

To end the proof we need to show that (5) implies (3). Consider again $R = vv^*$. We prove that the linear continuous operator $v: H \to X$ satisfies the assumption of Proposition 1. Let $\{e_n\}$ be an orthonormal basis of H and let $x_n = v e_n$ for $n=1,2,\ldots$. As above we may assume that $0 < \|x_n\| < 1$.

Let $f_1\ f_2, \ldots,$ be independent random variables such that for all n

$$P\left[f_n = \frac{1}{\|x_n\|}\right] = P\left[f_n = -\frac{1}{\|x_n\|}\right] = \frac{1}{2}\|x_n\| \ , \ P\left[f_n = 0\right] = 1 - \|x_n\|^2.$$

We define a linear operator $T: X^* \to L_2$ by

$$T x^* = \sum < x_n, x^* > f_n, \quad x^* \in X^*.$$

Note that $\|T x^*\|^2 = < R x^*, x^* >$, for all $x^* \in X^*$. Under assumption (5) there exists a ξ in X such that $T x^* = < \xi , x^* >$ a.s.for all $x^* \in X^*$. As known [15] it follows that $\sum x_n f_n$ is a.s. convergent in X. Hence $\|x_n\| |f_n| \to 0$ a.s. as $n \to \infty$. By the Borel-Cantelli lemma $\sum P [\|x_n\| |f_n| > \frac{1}{2}] < \infty$, which yields $\sum \|x_n\|^2 < \infty$. The proof of the theorem is terminated.

Remark 1. The proof of the fact that (2) implies (3) is based on a technical device used in an example given in [16].

Remark 2. It has been shown in [12] that only in cotype 2 spaces all Gaussian covariances have the property (3) of Theorem 1. Thus for cotype 2 spaces Theorem 1 yields results on the CLT already stablished in [11-12].

3. CLT for sample continuous random processes

Let S be a compact metric space and let C(S) be the space of all continuous real functions defined on S with the maximum norm. If $\xi(t)$, $t \in S$, is a sample continuous random process (r.p. for short), then it induces a r.e. in C(S) to be denoted by ξ. If $\xi(t)$, $t \in S$, is a Gaussian, then the corresponding is Gaussian too. A sample continuous r.p. is said to satisfy the CLT if the corresponding r.e. in C(S) satisfies the CLT. If $\xi(t)$, $t \in S$, satisfies the CLT, then by virtue of Proposition 1

(i) $E \xi^2(t) < \infty$, $E\xi(t) = 0$ for all $t \in S$,

(ii) there exists a sample continuous Gaussian r.p. $\Gamma(t)$, $t \in S$, such that $r_\xi(t, s) = E \xi(t) \xi(s) = E\Gamma(t) \Gamma(s) = r_\Gamma(t, s)$ for all t, $s \in S$.

A central limit theorem for sample continuous r.p. was proved first in [9]. At presend there are two surveys papers [4, 17] on this subject. After publishing these surveya, interesting results have appeared in [2, 18, 19].

Now let us consider the sample continuous r.p. with property (i). It may be shown that if $E|\xi(t)-\xi(s)|^2 \leqslant |t-s|^{\alpha}$ for some $\alpha > 1$ and for all $t, s \in S = [0, 1]$, then ξ satisfies the CLT. For $0<\alpha<1$ this is not true in general as shown for $0 \leqslant \alpha < \frac{1}{2}$ in [20] and for $\frac{1}{2} \leqslant \alpha < 1$ in [21].

We will give a description of covariances which guarantee the CLT without further restrictions on the process.

Theorem 2. Let S be a compact metric space and let $r: S \times S \to \underline{R}$ be a symmetric positive definite continuous function.

The following statements on r are equivalent:

(i) every sample continuous second order mean zero r.p. $\xi(t), t \in S$, with $r_{\xi} = r$ satisfies the CLT,

(ii) every second order r,p. $\xi(t)$, $t \in S$, with $r_{\xi} = r$ has a sample continuous modification,

(iii) if $r(t, s) = \sum \varphi_k(t)\varphi_k(s)$, $\varphi_k \in C(S)$, for all $t, s \in S$, then $\sum\limits_{k} \max\limits_{t \in S} \varphi_k^2(t) < \infty$.

Proof: First we show that (i) implies (ii). Let $R: C(S) \to C(S)$ be defined by

$$(R\,\nu)(t) = \int_S r(t,s)\, d\,\nu(s), \quad \nu \in C(S)^*, \quad t \in S. \tag{1}$$

Obviously R is a symmetric and a positive linear operator with a separable range ($C(S)$ is separable). From part (i) of Theorem 1 it follows that $R = usu^*$. Now let $Y \subset C(S)^*$ be the linear span of the family of Dirac measures δ_t, $t \in S$. Let $\xi(t)$, $t \in S$, be a second order r.p. with $r_{\xi} = r$. We construct a linear operator $T: Y \longrightarrow L_2$ putting $T\,\delta_t = \xi(t)$, $t \in S$. Note that $\| T\nu \|^2 = \langle R\nu, \nu \rangle$ for all $\nu \in Y$.

Now Theorem 1 implies that there exists a r.e. $\bar{\xi}$ in C(S) such that $T\nu = \langle \bar{\xi}, \nu \rangle$ a.s. for all $\nu \in Y$. Hence $\bar{\xi}(s)$, $s \in S$, is a sample continuous r.p. with $\bar{\xi}(s) = \xi(s)$ a.s. for all $s \in S$.

In order to show that (ii) implies (i) let R be defined by (1). It might be easily seen that condition (ii) Theorem 2 implies condition (4) of Theorem 1.

Now we show that (i) implies (iii). Let $r(t,s) = \sum' \varphi_n(t)\varphi_n(s)$ for all $t,s \in S$, where $\varphi_n \in C(S)$. It is sufficient to prove that $\sum \|\varphi_n\|^2_{C(S)} < \infty$. Consider an operator $v: l_2 \longrightarrow C(S)$ such that $vh = \sum h_n \varphi_n$, where $h = \{h_n\} \in l_2$. Note that $vv^* = R$, where R is defined by (1). Now (i) implies that $\sum \|\varphi_n\|^2_{C(S)} < \infty$ as shown in the proof of implication (1) $\Longrightarrow$ (3) in Theorem 1.

Finally to show that (iii) implies (ii) we consider again R defined by (1). Let H be a separable Hilbert space. It is enough to show that an arbitrary linear continuous operator $v: H \longrightarrow C(S)$ which meets the condition $vv^* = R$ satisfies the assumption of Proposition 2. Let $\{e_n\}$ be an orthonormal basis in H. Put $ve_n = \varphi_n$, for $n = 1,2,\ldots$. Then $r(t,s) = \sum \varphi_n(t)\varphi_n(s)$ for all $t, s \in S$. Now from (iii) it now follows that $\sum \|\varphi_n\|^2_{C(S)} < \infty$. The proof is terminated.

The following corollary gives and answer to a question posed in [21]:

<u>Corollary 1.</u> There exists a sample continuous second order mean zero $\xi(t)$, $t \in [0, 1]$, such that $E\xi(t)\xi(s) = \min(t,s)$, $t,s \in [0,1]$ and which does not satisfy the CLT.

P r o o f: If $\{\varphi_n\}$ are the Schauder's functions, then $\min(t,s) = \sum \varphi_n(t)\varphi_n(s)$, $t,s \in [0,1]$. Because $\sum \|\varphi_n\|^2_{C[0,1]} = \infty$, Theorem 2 guarantees the existence of ξ which does not satisfy the CLT. Also one can give easily an explicite construction of such ξ. Let $\{\beta_n\}$ be a sequence of positive numbers with $\sum \beta_n^2 = 1$ and

$\sum \beta_n \|\varphi_n\|_{C[0,1]} = \infty$. Now consider r.p. $\xi(t)$, $t \in [0,1]$, with possible values $\pm \beta_n^{-1} \varphi_n(t)$ with probabilities $\frac{1}{2}\beta_n^2$. Then $E\,\xi(t)\,\xi(s) = \min(t.s)$. Because $E \|\xi\|_{C[0,1]} = \sum \beta_n \|\varphi_n\|_{C[0,1]} = \infty$. Proposition 1 implies that ξ does not satisfy the CLT.

Let $r : [0, 1] \times [0, 1] \longrightarrow \underline{R}$ be a positive definite symmetric continuous function. Moreover let $d(s, t) = r(t, t) - 2\,r(s, t) + r(s, s)$, for $t,s \in [0,1]$ and let for $u \in [0,1]$

$$g_r(u) = \left(\iint_{|t-s| \leqslant u} d(t,s)\;dt\;ds \right)^{1/2}.$$

Using a result which may be found in [22] and the Theorem 2 the following result may be established.

Corollary 2. Let $r(s,t)$ where $t,s \in [0,1]$, be a positive definite symmetric continuous function and let

$$\int_0^1 \frac{g_r(u)}{u^2}\,du < \infty . \tag{2}$$

Then every sample continuous second order mean zero r.p. $\xi(t)$, $t \in [0,1]$, with $r_\xi = r$ satisfies the CLT.

P r o o f: From the assumption it follows that each second order r.p. $\xi(t)$, $t \in [0,1]$ has a sample continuous modification provided $r_\xi = r$. Thus one may apply Theorem 2.

Remark 1. If $d(t, s) \leqslant f(|t - s|)$, where f is nondecreasing, then $g_r(u) \leqslant (2\,u\,f(u))^{1/2}$. Moreover, if

$$\int_0^1 \frac{f(u)}{u^{3/2}}\,du < \infty ,$$

then r satisfies the assumption of Corollary 2. This shows that Corollary 2 extends Theorem 1.7 given in [23].

Remark 2. By virtue of Theorem 2 and Corollary 2 condition (2) imposed on r ensures (iii) of Theorem 2. It would be interesing to give a direct proof of this fact.

Finally we state two open problems.

Problem 1. Does there exist a sample continuous mean zero r.p. $\xi(t)$, $t \in [0,1]$, $|\xi(t)| \leq 1$ a.s. for $t \in [0,1]$, with $E\xi(t)\xi(s) = \min(t, s)$, which does not satisfy the CLT?

Problem 2. Let X be a separable Banach space. Give a characterization of symmetric and positive linear operators $R: X^* \longrightarrow X$ which have the following property: if ξ is a r.e. in X with $\|\xi\| \leq 1$ and $E\xi = 0$, then $R_\xi = R$ implies that ξ satisfies the CLT. Note that for type 2 and cotype 2 spaces all Gaussian convariances have this property.

References

[1] V a k h a n i a, N. N., Probability distributions in linear spaces (Russian), Tbilisi 1971.

[2] P i s i e r, G. and Z i n n, J., On the limit theorems for random variables with values in the spaces L_p $(2 \leq p < \infty)$, Z. Wahrscheinlichkeitstheorie Verw. Gebiete, 41, 4, 289-304(1978).

[3] J a i n, N. C., Central limit theorem in a Banach space, Lecture Notes in Mathematics, 526, 113-130 (1975).

[4] J a i n, N. C., Central limit theorem and related questions in Banach spaces, Proceedings of the AMS Probab. Sympos. Urbana, Illinois 1976.

[5] A l d o u s, D. J., A characterization of Hilbert space using the central limit theorem, J. London Math. Soc., 14, 2, 376-380 (1976).

[6] M o u r i e r, E., Élément aléatoires dans un espace de Banach, Ann. Inst. H. Poincaré 13, 161-244 (1953).

[7] F o r t e t, R. et M o u r i e r, E., Les fonctions aleatoires comme èlements aleatoires dans les espaces de Banach, Studia Math., 15, 1, 62-79, (1955).

[8] V a k h a n i a, N. N., Sur les répartitions de probabilités dans les espaces des suites numeriques, C. R. Acad. Sci. Paris 260, 6, 1560-1562 (1965).

[9] S t r a s s e n, V. and D u d l e y R.M., The central limit theorem and ε-entropy, Lecture Notes in Mathematics 89, 224-231 (1969).

[10] H o f f m a n n-Jørgensen, J. and P i s i e r, G., The strong law of large numbers and the central limit theorem in Banach spaces, Ann. Probability, 4, 588-599 (1976).

[11] P i s i e r, G., Le theoreme de la limite centrale et la loi du logarithme itere dans les espaces de Banach, Seminaire Maurey-Schwartz, 1975-1976, Exp. IV.

[12] C h o b a n j a n, S. A. and T a r i e l a d z e, V. I., Gaussian characterization of certain Banach spaces, J. Multivar. Anal. 7, 1, 183-203 (1977).

[13] C h o b a n j a n, S. A. and T a r i e l a d z e, V. I., A counter-example concerning CLT in Banach space, Lecture Notes in Mathematics 656, 25-30 (1978).

[14] S ł o w i k o w s k i, W., Absolutely 2-summing mappings from and to Hilbert space and Sudakov theorem, Bul. Acad. Pol. Sci., 17, 381-386 (1969).

[15] I t o, K. and N i s i o, M., On the convergence of sums of independent Banach space valued random variables, Osaka Math. J. 5, 35-48 (1968).

[16] R o s i ń s k i, J. and S u c h a n e c k i, Z., On the space of vector valued functions which are integrable with respect to the white noise, Preprint.

[17] M a r c u s, M. B., Some new results on central limit theorems for C(S)-valued random variables, Lecture Notes in Mathematics 526, 167-186 (1975).

[18] M a r c u s, M. B., Continuity and the central limit theorem for random trigonometric series, Z. Wahrscheinlichkeitstheorie Verw. Gebiete 42, 35-56 (1978).

[19] F e r n i q u e, X., Continuite et théoreme central limite pour les transformées de Fourier des mesures aléatoires du second ordre, ib. 42, 57-66 (1978).

[20] D u d l e y, R. M., Metric entropy and the central limit theorem in C(S), Ann. Inst. Fourier, Grenoble, 24, 49-60 (1974).

[21] H a h n, M., What second order Lipschitz conditions imply the CLT?, Lecture Notes in Math. 526, 107-112 (1975).

[22] G a r s i a, A. M., A remarkable inequality and the uniform convergence of Fourier series, Indiana Univ. Math. J. 25, 85--102 (1975).

[23] H a h n, M., Conditions for sample-continuity and the central limit theorem, Ann. Probability 5, 3, 351-360 (1977).

FIXED PRECISION ESTIMATE OF MEAN OF A GAUSSIAN SEQUENCE WITH UNKNOWN COVARIANCE STRUCTURE

by

Ryszard Zieliński

Polish Academy of Sciences, Warsaw

Summary

Let $C(\delta)$, $\delta > 0$, be the class of sequences of covariance matrices $K = (K_t(u,v),\ u,\ v=1,\dots,t)_{t=1,2,\dots}$ such that

$$\sum_{u=1}^{t}\sum_{v=1}^{t} K_t(u,v) = O(t^{2-\delta}) \text{ as } t \to \infty$$

Consider discrete-time stochastic process $X_t = \mu + \xi_t$, $t=1,2,\dots$, where μ is an (unknown) constant and (ξ_t) is a Gaussian sequence such that $E\xi_t = 0$ and $K = (E\xi_u\xi_v,\ u,v=1,\dots,t)_{t=1,2,\dots}$ is an (unknown) covariance structure of (ξ_t). Let $P_{(\mu,K)}$ be the probability measure induced by (X_t). Suppose that for every $k=1,2,\dots$ we can observe simultaneously k independent copies of (X_t). The result is: for every $\delta>0$, $\varepsilon>0$ and $\gamma \in (0,1)$ there exist a sequential estimate $(\hat{\mu}_t)_{t=1,2,\dots}$ and a finite stopping rule τ such that

$$P_{(\mu,K)}\left\{\left|\hat{\mu}_\tau - \mu\right| > \varepsilon\right\} < \gamma \text{ for all } \mu \in R^1 \text{ and } K \in C(\delta).$$

Consider a discrete-time stochastic process $(X_t,\ t=1,2,\dots)$ of the form $X_t = \mu + \xi_t$, where μ is an (unknown) constant and (ξ_t) is a sequence of random variables distributed normally with zero mean. The covariance structure of (ξ_t) is described by the sequence $K = K_t(u,v)_{t=1,2,\dots}$ where $K_t(u,v)$ denotes the covariance matrix $[E\xi_u\xi_v]_{u,v=1,\dots,t}$ of the t-dimensional random variable $(\xi_1,\dots,\xi_t)$. The probability measure induced by (X_t) will be denoted by $P_{(\mu,K)}$.

Let $C(\delta)$, $\delta>0$, be the class of all sequences K such that

$$\sum_{u=1}^{t}\sum_{v=1}^{t} K(u,v) = O(t^{2-\delta}).$$

Following an idea presented in [5] suppose that for every $k = 1,2,\dots$ we can observe simultaneously k independent copies $(X_t^{(i)})$ of (X_t). The theorem below states that a suitable estimate $\hat{\mu}_t = \hat{\mu}_t(X_1^{(1)},\dots,X_1^{(k)},X_2^{(1)},\dots,X_t^{(k)})$ and a stopping variable τ give us fixed precision estimate $\hat{\mu}_\tau$ of the parameter μ provided K is known to belong to $C(\delta)$ for some (known) positive δ. It is well-known (see [1] and [4]) that for k=1 no solution exists even in a smaller class of Gaussian sequences.

For any given k denote

$$\bar{X}_t^{(i)} = \sum_{n=1}^{t} X_n^{(i)}/t, \quad \hat{\mu}_t = \sum_{i=1}^{k} \bar{X}_t^{(i)}/k, \quad S_t^2 = \sum_{i=1}^{k} (\bar{X}_t^{(i)} - \hat{\mu}_t)^2/(k-1)$$

and for any given $\varepsilon>0$, $a>0$ and any integer T denote

$$\tau(a,T) = \begin{cases} \text{first } t \geq T \text{ such that } t^a S_t \leq \varepsilon, \\ \infty \text{ if no such } t \text{ exists.} \end{cases}$$

<u>Theorem.</u> For every $\delta>0$, $\varepsilon>0$ and $\gamma\in(0,1)$ there exist a and T such that

$$P_{(\mu,K)}\left\{\left|\hat{\mu}_{\tau(a,T)} - \mu\right| > \varepsilon\right\} < \gamma$$

for all $\mu\in R^1$ and all $K\in C(\delta)$.

P r o o f: It is sufficient to construct an appropriate stopping rule $\tau(a,T)$ for the sequential estimate $\hat{\mu}_t$, $t=1,2,\ldots$

For any given a and T we have

$$P_{(\mu,K)}\{|\hat{\mu}_{\tau(a,T)}-\mu|>\varepsilon\} =$$

$$= \sum_{t=T}^{\infty} P_{(\mu,K)}\{|\hat{\mu}_t-\mu|>\varepsilon, \tau(a,T)=t\} + P_{(\mu,K)}\{|\hat{\mu}_\infty-\mu|>\varepsilon, \tau(a,T)=\infty\}$$

$$\leqslant \sum_{t=1}^{\infty} P_{(\mu,K)}\{|\hat{\mu}_t-\mu|>\varepsilon, T^a S_t\leqslant\varepsilon\} + P_{(\mu,K)}\{|\hat{\mu}_\infty-\mu|>\varepsilon, \tau(a,T)=\infty\}$$

$$\leqslant \sum_{t=T} \mathrm{Prob}\{|t_{k-1}|\geqslant t^a\sqrt{k}\} + P_{(\mu,K)}\{|\hat{\mu}_\infty-\mu|>\varepsilon, \tau(a,T)=\infty\},$$

where t_{k-1} is the standard t-Student random variable with k-1 degrees of freedom.

Using the approximation for tails areas given in [2] we get

$$\mathrm{Prob}\{|t_{k-1}|\geqslant t^a\sqrt{k}\} \leqslant c(k)\cdot t^{-(k-1)a}$$

so that the series $\sum \mathrm{Prob}\{|t_{k-1}|\geqslant t^a\sqrt{k}\}$ converges whenever $(k-1)a>1$.

For the random variable S_t^2 we have

$$ES_t^2 = D^2\bar{X}_t^{(i)} = t^{-2}\sum_{u=1}^{t}\sum_{v=1}^{t} K(u,v)$$

and by the Tchebycheff inequality, for every $\eta>0$

$$P_{(\mu,K)}\{t^{2a}S_t^2>\eta\} \leqslant \eta^{-1}t^{2(a-1)}\sum_{u=1}^{t}\sum_{v=1}^{t} K(u,v)$$

so that for every $\mu\in R^1$ and every $K\in C(\delta)$ the sequence $t^{2a}S_t^2$ tends to zero in probability $P_{(\mu,K)}$ whenever $2(a-1)<\delta$. Hence for every T the stopping rule $\tau(a,T)$ is finite a.s. whenever $a<\delta/2$; this implies that

$$P_{(\mu,K)}\{|\hat{\mu}_\infty - \mu| > \varepsilon, \tau(a,T) = \infty\} = 0.$$

Now, to construct the stopping rule it is enough to take any $a < \delta/2$ and next an integer k such that $(k-1)a > 1$ and, eventually, T such that

$$\sum_{t=T}^{\infty} \text{Prob}\left\{|t_{k-1}| \geqslant t^a\sqrt{k}\right\} \leqslant \gamma.$$

Remark 1. A special case of time series (X_t) has been considered in [1] where the proof of nonexistence of fixed precision estimate when k=1 was given. Existence of the estimates for $k \geqslant 4$ has been proved in [5].

Remark 2. The proof is based on an approximation for tails areas in the Student distribution. A recent paper [3] shows that the t statistic is robust over a quite large family of distribution so that the assumption that ξ_t are normally distributed could be relaxed.

References

[1] Blum, J. R. and Rosenblatt, J., On fixed precision estimation in time series. Ann. Math. Statist., 40, 1021–1032 (1969).

[2] Gross, A. J. and Hosmer, D. W., Jr, Approximating tail areas of probability distributions. Ann. Statist., 6, 1352–1359 (1978).

[3] Posten, H. O., The robustness of the two-sample t-test over the Pearson system. J. Statist. Comput. Simul. 6, 295–311 (1978).

[4] Zacks, S., The theory of statistical inference. Wiley (1971).

[5] Z i e l i ń s k i, R., A class of stopping rules for fixed precision sequential estimates. Inst. Math. Polish Acad. Sci. Preprint 147 (1978).

A CHARACTERIZATION OF BEST LINEAR UNBIASED ESTIMATORS IN THE GENERAL LINEAR MODEL

by

Roman Zmyślony

Polish Academy of Sciences, Wrocław

Summary

A characterization of best linear unbiased estimators is given in the case of the general linear model. In addition necessary and sufficient conditions are derived for a given estimable function to have a best linear unbiased estimator. In particular models for which each estimable function has a best linear unbiased estimator are characterized. The conditions stated are given in a computational atractive form. The problems are discussed from the coordinate-free point of view. This is important for the results can be easily adopted in the case of estimation of variance components. The problem of estimation of either treatment or block effects in a mixed model serves as an example which illustrates the applicability of the results.

1. Introduction

One of the classical results in the theory of linear models is a theorem of Lehmann and Scheffé [3] which states, that for a linear model $M(\mathcal{E},\mathcal{V})$, where $\mathcal{E}$ = span $\{E\ y\}$ and $\mathcal{V}$ = span $\{\mathrm{cov}\ y\}$, a function (a,y) is a best linear unbiased estimator of $E(a,y)$ if and only if $Va \in \mathcal{E}$ for all $V \in \mathcal{V}$, or equivalently, iff $a \in \mathcal{A} = \bigcap_{V \in \mathcal{V}} V^{-1}(\mathcal{E})$. Here $(.,.)$

stands for an inner product and $V^{-1}(\mathcal{E})$ stands for the set of all elements a, such that $V a \in \mathcal{E}$.

The aim of this paper is to give a computational atractive form for $\mathcal{A}$. Moreover, if $\mathcal{E} = \{X\beta \mid \beta \in \mathcal{L}\}$, where $\mathcal{L}$ is a finite-dimensional space with an inner product $[\cdot,\cdot]$, a necessary and sufficient condition for the existence of a best linear unbiased estimator for $[c,\beta]$ is given.

2. An essentially complete class of estimators

Let $\mathcal{X}$ be a finite-dimensional vector space with an inner product $(\cdot,\cdot)$. Let $\{\mathcal{F}, S, P_\Theta\}$, $\Theta \in \Omega$ be a probability space and let y be a mapping from $\mathcal{F}$ to $\mathcal{X}$ such that expected vector $E_\Theta\, y$ and covariance operator $\mathrm{cov}_\Theta\, y$ exist for each $\Theta \in \Omega$. In the sequel let $\mathcal{E} = \mathrm{span}\{E_\Theta\, y \mid \Theta \in \Omega\}$ and let $\mathcal{V} = \mathrm{span}\{\mathrm{cov}_\Theta\, y \mid \Theta \in \Omega\}$.

Definition 1. A subset $\mathcal{B}$ of $\mathcal{X}$ is called an essentially complete class if for every $a \notin \mathcal{B}$ there exists a vector $b \in \mathcal{B}$ such that $E_\Theta(a,y) = E_\Theta(b,y)$ and $\mathrm{var}_\Theta(b,y) \leq \mathrm{var}_\Theta(a,y)$ for every $\Theta \in \Omega$.

Let V_0 be a maximal element in $\mathcal{V}$, i.e. $V_0 \in \mathcal{V}$ and $R(V) \subset R(V_0)$ for all $V \in \mathcal{V}$. Here $R(V)$ stands for the range of V. Such an element exists (see [2]). Moreover let X be a linear operator from a finite dimensional inner product space $\mathcal{L}[\cdot,\cdot]$ into $\mathcal{X}$ such that $R(X) = \mathcal{E}$. If $\mathcal{E} \not\subset R(V_0)$, define $W = V_0 + XX'$, where X' stands for the adjoint operator of X, and if $\mathcal{E} \subset R(V_0)$, define $W = V_0$. We shall prove the following result.

Lemma 1. R(W) is an essentially complete class.

P r o o f: Let $a_0 \notin R(W)$. Then a_0 can be uniquely decomposed as $a_0 = a_1 + a_2$, where $a_1 \in R(W)$ and $a_2 \in N(W)$, while $N(W)$ denotes the kernel of W. Since $\mathcal{E} \subset R(W)$ we have that $N(W) \subset \mathcal{E}^\perp$, where $\mathcal{E}^\perp$ denotes the orthogonal complement of $\mathcal{E}$. This in turn implies that $E_\Theta(a_0,y) =$

$= E_\Theta(a_1,y)$. Moreover, since $R(V) \subset R(V_0)$ for all $V \in \vartheta$ and since XX' is non-negative definite we have that $N(W) \subset N(V)$ for all $V \in \vartheta$. This implies that $\mathrm{var}_\Theta(a_2,y) = 0$ for all $\Theta \in \Omega$ and so $\mathrm{var}(a_0,y) = \mathrm{var}(a_1,y)$. This terminates the proof of the lemma.

Remark 1. In the definition of W the operator XX' may be replaced by any symmetric and non-negative definite operator B such that $R(B) \subset \mathcal{E}$ and $\mathcal{E} \subset R(W)$.

Lemma 2. If (a,y) and (b,y), where $a,b \in R(W)$, are best linear unbiased estimators of a given function $\varphi(\Theta)$, then $a = b$.

P r o o f: The first assumption yields

$$a - b \in R(W). \tag{1}$$

Since (a,y) and (b,y) are unbiased for $\varphi(\Theta)$, it follows that

$$E_\Theta(a,y) = E_\Theta(b,y) \quad \text{for all } \Theta \in \Omega. \tag{2}$$

The above mentioned Lehmann Scheffé Lemma implies that

$$(c,Va) = (c,Vb) = 0 \text{ for all } c \in \mathcal{E}^\perp \text{ and } V \in \vartheta. \tag{3}$$

Note that (2) and (3) are equivalent to

$$a - b \in \mathcal{E}^\perp \tag{4}$$

and

$$V(a - b) \in \mathcal{E} \quad \text{for all } V \in \vartheta, \tag{5}$$

respectively. Using the fact that $XX'c = 0$ for all $c \in \mathcal{E}^\perp$, it is easy to find that (4) and (5) imply

$$W(a - b) \in \mathcal{E}. \tag{6}$$

Since W is non-negative definite, (4) and (6) imply $W(a - b) = 0$. Since $a - b \in R(W)$ by (1) we have $a - b = 0$.

3. Best linear unbiased estimators

Now our considerations will be limited to an essentially complete class of estimators $\{(a,y) \mid a \in R(W)\}$. In this class the set of all best linear unbiased estimators will be characterized. First we prove the following result.

Lemma 3. Let $a \in R(W)$. If (a,y) is the best linear unbiased estimator of $E(a,y)$, then $a \in W^+(\mathcal{E})$.

P r o o f: From the Lehmann Scheffé Lemma it follows that (a,y) is a best linear unbiased estimator of $E(a,y)$ if and only if $Va \in \mathcal{E}$ for all $V \in \mathcal{V}$. Since $XX'a \in \mathcal{E}$ it follows from the above that $Wa \in \mathcal{E}$ and this implies that $W^+Wa \in W^+(\mathcal{E})$. Note that W^+W is the projection on $R(W^+) = R(W)$. Hence for $a \in R(W)$ we have $W^+WA = a$. This terminates the proof of the lemma.

In the sequel we will need the following notation. Let $P = XX^+$ and let $M = I - P$, where I stands for the identity operator. Moreover, let $\{V_o, V_1, \ldots, V_k\}$ be a spanning set for $\mathcal{V}$. Define

$$U = X'W^- \sum_{i=1}^{k} V_i M V_i W^- X, \tag{7}$$

where W^- stands for a general inverse operator of W.

Theorem 1. Let $a \in R(W)$. The function (a,y) is a best linear unbiased estimator of $E(a,y)$ if and only if $a \in W^+X(\mathcal{U})$, where $\mathcal{U} = N(U)$.

P r o o f. From Lemma 3 and the Lehmann Scheffé Lemma [3] if follows that (a,y) is a best linear unbiased estimator if and only if for some $b \in \mathcal{L}$

$$a = W^+Xb \tag{8}$$

and, for $i = 1, \ldots, k$, $V_i W^+ Xb \in \mathcal{E}$.

The last condition is equivalent to $MV_iW^+Xb = 0$ for $i = 1, \ldots, k$, and this one in turn is equivalent to

$$X'W^+V_iMV_iW^+Xb = 0, \ i = 1,\ldots,k. \tag{9}$$

Since $N(A) \cap N(B) = N(A + B)$, for A and B are symmetric and non-negative definite, it is easy to see that (9) is equivalent to

$$X'W^+ \sum_{i=1}^{k} V_iMV_iW^+Xb = 0. \tag{10}$$

Since $R(X) \subset R(W) = R(W^+)$ and $R(V_i) \subset R(W)$ we have $X = WW^+X$ and $V_i = WW^+V_i$ for $i = 1,\ldots,k$. Now from Lemma 2.3 in [1] it follows that (10) is equivalent to $Ub = 0$, where U is defined by (7). Thus the theorem follows from (8).

Remark 2. Since $N(U) = R(I - U^-U)$ the condition $a \in W^+X(\mathcal{U})$ is met iff $a \in R(W^+XZ)$, where $Z = I-U^-U$.

Let $\mathcal{C}$ stand for the set of all $a \in R(W)$ such that (a,y) is a best linear unbiased estimator of $E(a,y)$. Clearly $\mathcal{C}$ is a linear subspace of $\mathcal{X}$ and in view of Remark 2 we have $\mathcal{C} = R(W^+XZ)$. Denoting by c and e the dimensions of $\mathcal{C}$ and $\mathcal{E}$, respectively, and using Theorem 1 we easily arrive at the following conclusions.

Corollary 1. If the rank of U is equal u, then $c = e - u$.

Corollary 2. $\mathcal{E} = \{0\}$ if and only if $u = e$.

Let tr U stand for the trace of U.

Corollary 3. The following statements are equivalent:

(i) for every estimable function there exists a best linear unbiased estimator,

(ii) $U = 0$,

(iii) $\mathrm{tr}\, U = 0$.

4. Best linear unbiased estimable functions

Consider a fixed linear model $y = X\beta + e$, where y is an $n\times 1$ random vector such that $Ey = X\beta$, $\operatorname{var} y = \sigma^2 V$, $V > 0$. It is known that for a parametric function $c'\beta$ there exists a best linear unbiased estimator if and only if $c \in R(X'V^{-1}X)$. Moreover, $c'\hat{\beta}$, where $\hat{\beta}$ is a solution of $X'V^{-1}X\beta = X'V^{-1}y$, is a best linear unbiased estimator for $c'\beta$. In this section similar results for the general linear model are given.

Theorem 2. There exists a best linear unbiased estimator for $[c,\beta]$ if and only if $c \in R(X'W^{-}XZ)$.

P r o o f: In view of Remark 2 there exists a best linear unbiased estimator for $[c,\beta]$ if and only if there exists a vector $b\in\mathcal{L}$ such that $a=W^{+}XZb$ and $E(a,y) = [c,\beta]$ for all $\beta\in\mathcal{L}$. These two conditions are met iff $(W^{+}XZb,\ X\beta) = [c,\beta]$ for all $\beta\in\mathcal{L}$. This in turn is met iff $[X'W^{+}XZb-c,\beta] = 0$ for all $\beta\in\mathcal{L}$, and, consequently, iff $c\in R(X'W^{+}XZ)$. Since $X'W^{+}X = X'W^{-}X$ for every W^{-} the theorem follows.

Theorem 3. Suppose that there exists a best linear unbiased estimator for $[c,\beta]$. Then $[c,\hat{\beta}]$, where $\hat{\beta}$ is a solution of the equation

$$Z'X'W^{-}X\beta = Z'X'W^{-}y, \tag{11}$$

is a best linear unbiased estimator of $[c,\beta]$.

P r o o f: Since equation (11) is consistent, $\hat{\beta}=(Z'X'W^{-}X)^{-}Z'X'W^{-}y$ is a solution of (11). Moreover, $X'W^{-}y$ is equal to $X'W^{+}y$ with probability one, and so $[c,\hat{\beta}] = (a,y)$ for some $a\in R(W^{+}XZ)$. In view of Remark 2 we conclude that $[c,\hat{\beta}]$ is a best linear unbiased estimator of $E[c,\hat{\beta}]$. Now we shall prove that $E[c,\hat{\beta}] = [c,\beta]$. From Theorem 2 it follows that $c = X'W^{-}XZb$ for some $b\in\mathcal{L}$. Hence

$$E\,[c,\hat{\beta}] = E\,[X'W^{-}XZb,\ (Z'X'W^{-}X)^{-}Z'X'W^{-}y\,]$$
$$= [b,\ Z'X'W^{-}\,X(Z'X'W^{-}X)^{-}Z'X'W^{-}X\beta]$$
$$= [b,\ Z'X'W^{-}X\beta\,] = [X'W^{-}XZb,\beta] = [c,\beta]$$

what completes the proof of the theorem.

5. Examples

Let us consider an experiment with t treatments arranged in b blocks according to the incidence matrix N. Let y be an $n\times 1$ vector. Assume that

$$y = (D'\Delta')(\beta'\ \gamma')' + e, \tag{12}$$

where D' is the $n\times b$ matrix, whose elements d_{ij} are 1 or 0, according as y_i is in j^{th} block or not; Δ' is the $n\times t$ matrix, whose elements δ_{il} are 1 or 0, according as y_i receives the lth treatment or not; γ and β are vectors of treatment effects and blocks effects, respectively. The $n\times 1$ random vector e satisfies the following conditions: $Ee = 0$, $Eee' = \sigma_e^2 I$, where σ_e^2 is unknown. Note that $N = \Delta D'$. We shall consider the following two cases.

(a) γ is a vector of fixed parameters and β is a random vector such that: $E\beta = 0$, $E\beta\beta' = \sigma_\beta^2 I$, where σ_β^2 is unknown, and $E\beta e' = 0$.

(b) β is a vector of fixed parameters and γ is a random vector such that: $E\gamma = 0$, $E\gamma\gamma' = \sigma_\gamma^2 I$, where σ_γ^2 is unknown, and $E\gamma e' = 0$.

Let $\underline{r} = N1_b$ and $\underline{k} = N'1_t$, where 1_b and 1_t are $b\times 1$ and $t\times 1$ vectors of 1's, respectively. It is clear that $\underline{r}$ is the vector of treatment replications and $\underline{k}$ is the vector of block sizes. For any vector $m = (m_1,\dots,x_p)'$ let m^δ stand for the diagonal matrix with elements $m_1,\dots,m_p$, and let $m^{-\delta}$ stand for the inverse of m^δ. Putting $\mathcal{X} = R^n$ with the usual inner product, it is easy to find that in case (a) we have $U = Nk^\delta N' - NN'\underline{r}^{-\delta}NN'$, and in case (b) we obtain $U = N'\underline{r}^\delta N - N'N\underline{k}^{-\delta}N'N$. Suppose that $\underline{k}^\delta = kI$ and $\underline{r}^\delta = rI$ for some $k\geq 1$ and $r\geq 1$.

Case (a). For each treatment parameter there exists a best linear unbiased estimator if and only if $\frac{1}{rk}NN'$ is idempotent.

Case (b). For each block parameter there exists a best linear unbiased estimator if and only if $\frac{1}{rk}N'N$ is idempotent.

Now let us consider a special case of model (12) with 2 treatments and j+k blocks. We assume that the incidence matrix N has the following form

$$N = \begin{bmatrix} 1'_j & 0'_k \\ 1'_j & 1'_k \end{bmatrix}.$$

Here $1'_i$ stands for an $i \times 1$ vector of one's and $0'_i$ stands for an $i \times 1$ zero vector. In case (a)

$$U = \begin{bmatrix} \frac{jk}{j+k} & 0 \\ 0 & 0 \end{bmatrix}.$$

From Theorem 1 it follows that if $k \geqslant 1$, then there exists a best linear unbiased estimator for the second treatment parameter, and there does not exist a best linear unbiased estimator for the first treatment parameter.

In case (b) we have

$$U = \frac{j}{2}\begin{bmatrix} \underset{j\times j}{0} & \underset{j\times k}{0} \\ \underset{k\times j}{0} & 1_k 1'_k \end{bmatrix}.$$

If $k \geqslant 1$, then there exist best linear unbiased estimators for $\beta_1, \ldots, \beta_j$ only. However if k=0, then there exist best linear unbiased estimators for all estimable functions.

References

[1] G n o t, S., K l o n e c k i, W., Z m y ś l o n y, R., Best unbiased estimation, a coordinate free approach, Preprint 124, Institute of Mathematics, Polish Academy of Sciences (1977).

[2] L a M o t t e, L. R., A canonical form for the general linear model, Ann. Statist., 4, 787-789 (1977).

[3] L e h m a n n, E., S c h e f f é, H., Completeness, similar regions and unbiased estimation, Part 1, Sankhya 10, 305-340 (1950).